APPLIED INTRODUCTION TO DIGITAL SIGNAL PROCESSING

J. Philippe Déziel

Algonquin College

Upper Saddle River, New Jersey
Columbus, Ohio

Library of Congress Cataloging-in-Publication Data

Déziel, J. Philippe.
    Applied introduction to digital signal processing / J. Philippe Déziel—1st ed.
        p.   cm.
    ISBN 0-13-775768-9
    1. Signal processing—Digital techniques.   I. Title.

TK5102.9.D49   2001
621.382´2--dc21   99-058865
                                                                                        CIP

Vice President and Publisher: Dave Garza
Editor in Chief: Stephen Helba
Assistant Vice President and Publisher: Charles E. Stewart, Jr.
Production Editor: Alexandrina Benedicto Wolf
Production Supervision: Custom Editorial Productions, Inc.
Design Coordinator: Robin G. Chukes
Cover Designer: Jeff Vanik
Cover Image: Jeremy Stanton
Production Manager: Matthew Ottenweller
Marketing Manager: Barbara Rose

This book was set in Times Roman by Custom Editorial Productions, Inc. It was printed and
bound by R.R. Donnelley & Sons Company. The cover was printed by Phoenix Color Corp.

10   9   8   7   6   5   4   3   2   1
ISBN: 0-13-775768-9

PREFACE

The digital signal processing (DSP) revolution has drastically changed the way electronic circuits are designed. It has brought new possibilities that were once deemed impossible using conventional analog circuitry. Digital signal processors are used in CD players, cellular telephones, music synthesizers, and high-speed modems, to name just a few of the items that are now considered among the necessities of life. What magic makes these marvelous devices work? What do readers need to know to start using this technology? The goal of this book is to introduce readers to DSP so that they can incorporate some of this technology into their designs.

This book is based on seven years of experience teaching this subject at the college level. The goal is for readers to understand the fundamentals of DSP so that they may design and implement signal synthesis, signal analysis, filters, and modulators on any digital signal processor. The book provides readers the background necessary to attend seminars and to further their studies.

It is widely recognized that the mathematics supporting DSP prevents many good technically-oriented people from studying the subject. The challenge was to find an approach that would be both intuitive and familiar to undergraduate students. Since vector arithmetic is used very early in most technology programs to explain the basics of most scientific topics, this book relies on these simple arrows along with some of the basic rules of mathematics to introduce the subject. This allows the use of an intuitive graphical approach. Readers will be surprised at how fast they become familiar with the material. They will quickly acquire enough DSP literacy to understand the basics that drive this technology and the major issues that drive practical designs. In the process, they will acquire the basic knowledge to implement simple DSP applications.

This book delivers enough of the basics so readers can use many features of the readily available filter design software packages that automatically perform most of the math. For example, readers will be able to use filter design software packages to design almost any of the standard filters, including Butterworth and Chebyshev filters. Readers will then be able to program these filters onto a digital signal processor of their choice. Additionally, the equations that are developed in this book may be exercised using popular number-calculating software such as spreadsheets.

TARGET AUDIENCE

It is assumed that readers are either engineers who have forgotten a good deal of math or undergraduate students who understand the basics of passive circuits such as RC networks. Readers should also have some elementary programming background and an understanding of simple binary systems. Two appendices thoroughly review vector arithmetic and binary manipulation concepts to help readers comprehend the material.

Most of the complicated math and calculus is used only on rare occasions to provide a reference for the curious. In cases of mathematical derivations, readers can skip to the bottom line where the result is explained in simple, practical terms.

This book is well suited to cover an undergraduate course in the second year of a technology or engineering program. It is particularly well adapted to computer engineering or electronics technology programs. The material covered will provide graduates with a fundamental understanding of one of the essential high-tech tools to face today's technological challenges.

ORGANIZATION OF THE TEXT

Chapter 1 answers the most common question newcomers ask about DSP: "How do digital signal processors differ from other types of processors?" The chapter links readers' elementary programming knowledge to the DSP environment. It also provides readers the terminology required to comprehend most seminars that cover new processors. The content of this chapter is not essential to cover Chapters 2 through 8, but it does provide some of the background necessary to cover Chapter 9.

Chapter 2 provides a painless introduction to the things that can be done with strings of numbers. Strictly speaking, this chapter does not cover digital signal *processing;* however, it does provide some of the basics about signals, which readers need to understand the following chapters. It introduces continuous-time and discrete-time concepts and explains periodic signals, as well as harmonic components, by using the practical application of signal synthesis. The material covered in this chapter also provides opportunities to explore many applications in a laboratory environment.

Chapter 3 focuses on some of the most important properties that relate to digital signals. The material covered here is essential to understanding all of the other chapters. Subjects such as the sampling theorem, antialiasing input filters, and reconstructing output filters are concepts that must be learned in order to deal with any DSP system.

Chapter 4 deals with the basic hardware aspects of DSP systems. It examines the system philosophically as a black box and develops theories that explain what happens inside the box. The chapter presents the difference equation, which is used to define the operation of filter systems. It also introduces the impulse response technique that may be used to test systems and develops a simple criterion to determine system stability. The important convolution operation is also developed in this chapter.

Chapter 5 is devoted entirely to spectral analysis. The approach is completely unconventional because it relies on some of the synthesis rules to develop the analysis technique. It is also unconventional in its uses of vectors to illustrate the operation of the discrete Fourier transform.

Chapter 6 is concerned with the frequency response of systems. It develops a technique that uses the value of the difference equation coefficients to compute the gain and phase response of systems. The discrete-time Fourier transform is an invaluable tool to determine the frequency domain behavior of systems. The chapter also examines the response of DAC systems, which are part of almost all DSP systems. The concept of an equalizer is introduced and used to compensate for undesired frequency characteristics.

Chapter 7 develops the Z-transform, which is used to design the response of systems. It uses the concept of poles and zeros to build the system response. A graphical method using vectors provides readers an intuitive approach in positioning poles and zeros. Important concepts such as stability are explored, and practical aspects such as coefficient quantizing are examined. Oscillator systems are designed based on the positioning of poles and on trigonometric identities.

Chapter 8 is a filter primer that describes the various popular types of filters. It covers filter concepts such as filter bands, linear-phase filters, phase equalizers (all pass), and describes the characteristics of standard filters such as the Bessel, Butterworth, and Chebyshev. Design techniques, such as the bilinear transform, windowing, and the positioning of poles and zeros to implement narrowband notch filters, are also described.

Chapter 9 covers the implementation of DSP systems using some of the most popular programming structures. It develops the various structures and identifies the strengths and weaknesses of each. FIR filters are implemented using both the direct form and the cascaded form. IIR filters are implemented using a cascade of both direct form II and transposed structures. Basic concepts of noise control are introduced and applied to programming techniques. Issues of scaling are addressed using practical examples.

Appendix A reviews complex arithmetic as applied to vectors. It covers complex numbers, complex exponentials, and Euler's identity, and develops the algebra necessary to manipulate complex numbers and vectors. It also defines magnitude, angle, and argument and shows how to plot complex numbers and vectors on an Argand diagram.

Appendix B reviews the binary systems, formats, and manipulations necessary for programming DSP applications. It covers fixed-point and floating-point numbers, addition, multiplication, quantization, and tolerance.

ACKNOWLEDGMENTS

Writing a book verifies one of Murphy's fundamental laws: "Completing a project will take at least twice as much time as your worst-case estimate." One cannot bring such an endeavor to successful completion and keep one's sanity without the precious support of colleagues, friends, and family.

The material in this book could not have been assembled without the comments and thoughtful reviews of numerous people that provided input. I would like to extend my gratitude to the many students who gave me support and feedback. I am especially indebted to my friends Paul Arseneault and Bob Southern, who have been most generous with their time, spending many long hours reading the draft material.

I also thank the following reviewers for their valuable suggestions: Ray Bashnick, Texas A&M University; David Birkett, Wentworth Institute of Technology; James LeBlanc, New Mexico State University; Mike Tsalsanis, Stevens Institute of Technology; Guoliang Zeng, Arizona State University; and Omar Zia, Southern Polytechnic Institute.

A special note of appreciation goes to my good friend Gerd Schneider with whom I have been playing squash and having breakfast twice a week for more than ten years. His encouragement, support, and ever-present optimism have been important factors in maintaining my enthusiasm for this project.

I also extend my gratitude and special thanks to Brigitte and Catherine, my two teenage daughters, mostly for their patience and understanding during the long hours I spent in front of that computer.

Last but certainly not least, to my beloved Mariette, whose support and presence allowed me to bring this project to fruition. Merçi de ta compréhension, de ta patience, et de ton amour.

Philippe Déziel
dezielp@algonquincollege.com

CONTENTS

1

THE DIGITAL PROCESSING ENVIRONMENT

The first questions that technical people ask about digital signal processing relate to the hardware. Engineers are usually very curious about the toys with which they will get to play! What makes digital signal processors so incredibly fast compared with other types of processors? What is the purpose of a barrel shifter? Why do signal processors need so little memory?

Although digital signal processing theory must be understood to use the special processor features, an important first step is satisfying some of the curiosity about the hardware. This chapter describes the special features that make digital signal processors different.

Digital signal processing (DSP) relies on high-speed digital systems that specialize in the manipulation of numbers. Two types of approaches may be used to implement DSP systems. One uses pure hardware circuitry and the other, which is more popular, relies on software to program specialized processors. Both approaches achieve a level of performance that is inimitable in the analog world.

Signal processors have evolved from the microprocessor/microcontroller family of devices. They borrow characteristics found in both ancestors but append special features of their own. The overall signal processor structure is designed to achieve an incredible number-crunching performance. This performance is attained by implementing a Harvard bus architecture, by including specialized hardware, and by implementing an uncommon set of instructions.

This chapter examines the special addressing modes, the typical hardware enhancements, and some of the particular instructions that make signal processors different from conventional microprocessors and microcontrollers. The chapter concludes by examining some of the pure-hardware circuits that may be used to implement DSP.

1.1 THE DIGITAL WORLD

In the latter part of the twentieth century, digital circuits have changed the way we live. Digital devices now allow us to do things with a precision and a speed that are borderline magic. The strength of devices using digital technology resides in the fact that numbers can be stored with perfect reliably in digital memory. Once a number is properly stored, its value cannot drift; therefore, the digital memory maintains the exact value of the number as long as the storage system is functional.

1.1.1 DIGITAL SIGNAL PROCESSING SYSTEMS

A DSP system may be divided into elementary sections that take analog signals in and out of the digital environment and a central digital section that processes numerical samples. Figure 1–1 illustrates these elementary sections.

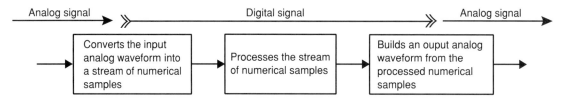

FIGURE 1–1
Elementary sections of a DSP system.

The central block of Figure 1–1 actually does all the signal processing. The processing, which defines the functionality of the system, is based on a sequence of mathematical operations referred to as the *processing algorithm*. A processing algorithm typically involves addition and multiplication operations and because of this, the complete functionality of digital processing systems relies on the value of stored numbers. Consequently, digital systems inherit the precision, stability, and reliability of the numbers that define their operation. Since the operation of a DSP system is purely mathematical, it is not plagued with the following analog circuit problems:

- Component drift, aging, and tolerance, which require calibration and adjustments.
- Size, power, and temperature limitations of components.
- Nonideal device characteristics that vary between manufacturing batches.

The precision with which we can define the operation of digital systems is such that tasks that were deemed impossible using analog circuits become possible using DSP. By changing the value of the numbers that control the processing, the signal processor may assume different roles. For example, the processor on a modem card begins its operation by assuming the role of a synthesizer, which dials the tones that establish the telephone connection. Once this is completed, the processor changes its operation to become a modulator/demodulator to generate and interpret the data carrier signal. Alternatively, the same processor could change its processing algorithm to become a facsimile machine. Since the DSP operation depends only on numbers that are loaded in memory, a system may reload and repeat different processing algorithms with perfect reliability.

What makes DSP attractive is that the technology allows the processing of signals in a *real-time* environment. This means that the processing circuitry can output the processed signal as fast as the signal is being acquired.

The mathematical complexity of the processing algorithm is the factor that limits the system's performance. Processing a signal's frequency content becomes a question of speed. The faster the digital circuitry, the higher the signal frequencies that may be processed. As technology progresses, the processing performance increases and the price drops. DSP is therefore continuously reaching into applications of higher frequency. As this happens, the use of analog technology is pushed to the high frequencies that are still beyond the reach of digital processing systems.

1.1.2 IMPLEMENTATION OF DIGITAL PROCESSING SYSTEMS

Two different approaches are used to design digital processing systems.

- *The programmed approach.* Digital signal processors execute programmed instructions that define the signal-processing application. The processing speed of these systems depends on the structure of the processor and the speed at which it can fetch and execute instructions. Programmed systems are very flexible since we can change the operation of the system anytime by loading a new program. Maintaining, upgrading, or even completely changing the operation of a programmed system takes only a few seconds.
- *The pure-hardware approach.* Specialized integrated circuits, programmable logic devices (PLDs), or application-specific integrated circuits (ASICs) perform the DSP application. The processing speed of these systems depends on the propagation delay of the logic devices and the clock rate allowed by the flip-flops. It is common to see pure-hardware systems that run a hundred times faster than their programmed counterparts. However, changing the operation of such systems usually requires that complete parts of the circuit be redesigned and replaced. Consequently, pure-hardware systems are not as flexible as processor-based systems.

Both programmed and pure-hardware processing systems bring precision, stability, and reliability at a price that outclasses the equivalent analog circuits.

The digital processor approach provides an additional bonus in that it can perform just about any control function. For example, the processor can

- perform a self-test.
- control the lights and switches of the user interface.
- manage the timing of the system peripherals.

Having a single part that does it all is every designer's dream. The signal processor is such a device at a price that is extremely affordable.

The DSP revolution brings only advantages, and a person does not need to be a math wizard to take advantage of it. Specialized software tools are available to perform the math, and new, sophisticated packages are being developed all the time. A basic understanding of DSP principles and some programming skills are all that are needed to start using this technology. Although certain processors specialize at running particular applications more efficiently, most DSP algorithms may be implemented on just about any signal processor.

1.2 DIGITAL SIGNAL PROCESSOR ARCHITECTURE

We can categorize most microprocessors as belonging to one of three groups: general-purpose microprocessors, microcontrollers, and digital signal processors.

- *General-purpose microprocessors.* Their structure satisfies the needs of general software applications such as word processing and spreadsheets. The addressing modes and register structures meet most general-purpose needs. The architecture of general-purpose microprocessors concentrates on memory management features. These processors are not very good at digital signal processing.
- *Microcontrollers.* These devices implement a computer within a single part. Typically, they include ROM, RAM, and input/output devices. The architecture of general-purpose microcontrollers concentrates on input/output features. Each microcontroller, in the large range of commercially available devices, incorporates a structure that addresses the needs of a particular environment. Some microcontrollers specifically target DSP applications.
- *Digital signal processors.* These implement specialized structures that optimize the execution of certain repetitive mathematical operations. These operations (the subject of most of this book) define how the signal is processed to implement applications such as signal synthesizers and filters. Signal processors combine features borrowed from the general-purpose processor/microcontroller structures and append innovative addressing modes and registers that specifically address the needs of the signal processing environment. The architecture of the signal processor concentrates on enhancing the arithmetic logic unit (ALU) performance.

Digital signal processors are therefore specialized devices that address the specific needs of the DSP environment. Although any processor may be used to process a signal, the specialized signal processors bring more efficiency and ease of programming.

Computing digital processing algorithms is mostly a matter of repeating short programs that use numerous multiplication and addition operations. Today's signal processor can literally perform billions of multiplication and addition operations every second. Think of it: if a person were to perform a multiplication and an addition every second of his or her life, day and night, allowing no break, it would take that person more than 30 years to perform a billion operations! Such staggering performance is at our fingertips today and amazingly keeps getting better and cheaper every year.

To achieve such performance, signal processors must include special structures, hardware components, and instructions, which make them different from conventional general-purpose processors/microcontrollers.

1.2.1 VON NEUMANN VERSUS HARVARD

The operation of a microprocessor requires the use of two memory spaces. One space is required to store the program instructions, and the other is used to store the application data. Most general-purpose microprocessors follow the Von Neumann architecture. This architecture uses a common *unified-memory space* to provide storage for both the program instructions and the application data. Figure 1–2 illustrates the Von Neumann architecture.

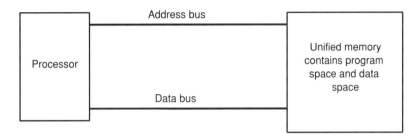

FIGURE 1–2
Von Neumann architecture.

As indicated in Figure 1–2, the processor uses the same address and data buses to access both *program instructions* and *data*. This architecture has the advantage of being simple and flexible since the system allocates parts of a *common* memory to create both the program and data spaces. If the application changes, the system may reallocate the memory resources to meet the space requirements of the new application.

The shared buses of a Von Neumann architecture imply that the system may fetch data or program instructions but not both at the same time. Unfortunately, this does not suit applications such as DSP that require great processing speeds. Such high speeds are possible only if the architecture allows the *simultaneous* access to both data and program instructions at the highest possible rate.

To achieve the high processing speeds that DSP applications require, signal processors use the Harvard *multiaddress* space architecture. This architecture calls for separate program and data spaces, a fact that implies separate address and data buses to access each space. Figure 1–3 illustrates the Harvard architecture.

Processors that implement the Harvard architecture can simultaneously fetch and execute program instructions. This architecture certainly achieves much higher processing speeds, but the following costs are associated with this higher performance:

- A Harvard processor requires more real estate to accommodate the extra buses.
- The program and data spaces cannot be reallocated with the flexibility provided by a Von Neumann architecture.
- To achieve the best possible performance, the processor must fetch and execute instructions in a single machine cycle. This puts constraints on the coding of the program instruction words (see Section 1.2.2).

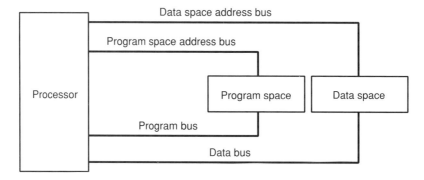

FIGURE 1–3
Harvard architecture.

- Writing programs for Harvard machines is somewhat more complicated than for Von Neumann architecture. Getting used to the sequencing problems created by the simultaneous fetching and execution of instructions takes a while.

To provide some flexibility in the memory space allocation, some processors allow special memory blocks to be reconfigured as program *or* data spaces. This allows the processor to accommodate long programs that have little need of data space and, conversely, short programs with a great appetite for data space.

As a final note, many signal processors use an *extended* Harvard architecture. These processors have access to multiple data spaces, each using its own address and data bus.

1.2.2 SINGLE-CYCLE PROGRAM INSTRUCTIONS

To get very high processing speeds, it is necessary to be able to fetch and execute the program instructions as fast as possible. There are two approaches to the design of a microprocessor instruction set to achieve this.

- *The very long instruction word (VLIW) approach.* This implements a large set of instructions that provides a special instruction to satisfy almost any situation. Using such a large set of instructions, the application requires fewer instructions to program; however, the code for each instruction typically requires many words in the program space. Consequently, fetching and executing a single VLIW instruction requires many clock cycles.
- *The reduced instruction set approach (RISC).* This implements a reduced instruction set that satisfies the basic requirements of the programming environment. The application requires more instructions to program; nevertheless, each instruction is coded using a single word in the program space. Fetching, decoding, and executing a RISC instruction can consequently be done within one clock cycle.

The typical DSP algorithm requires a small number of programmed steps. The speed at which these steps are processed is important. For this reason, signal processors typically use a RISC-like approach. The reduced instruction set of a signal processor, however, includes some very specialized DSP instructions.

Ideally, a signal processor should fetch, decode, and execute an instruction within one clock cycle. Some compromises are required to achieve such performance. One of the limiting factors is that the instruction words cannot be wider than the program bus. This sets an upper limit to the number of bits used to code the instruction words. Generally, the instruction words must define the operation and the location where this operation will take place. Figure 1–4 illustrates the coding of a program instruction.

FIGURE 1–4
Coding a program instruction word.

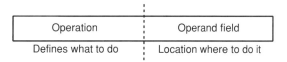

The number of bits in the *operation* part of the instruction word determines how many different instructions are possible. The *operand* part of the instruction may be defined using different addressing modes, each consuming a different number of bits in the instruction word. Because of the limited size of the instruction word, some familiar addressing modes may not be available on all signal processors. The following sections discuss different addressing modes and how they apply to the DSP environment.

Inherent Addressing

The operand specifies which part of the central processing unit (CPU) is operated on by the instruction. As an example, Figure 1–5 illustrates an instruction that operates on part of the CPU. The operand in this addressing mode consumes a small number of the instruction word bits.

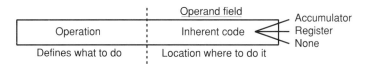

FIGURE 1–5
Inherent operand.

Direct Addressing

The operand directly specifies the address of the *memory location* being operated on by the instruction. Doing this requires many bits in the operand part of the instruction. To be able to fetch a complete instruction within a single clock cycle, the operation and operand part of the instruction must fit in a single code word. To achieve this, we can reduce the number of bits used in the operand by using a memory *page register,* which is preloaded with some of the most significant bits of the address (called the *page*). The operand contains the remaining least significant bits of the address. Combining the page register with the operand forms a complete direct address. As an example, Figure 1–6 illustrates the direct addressing of memory location 0107 with and without the use of a page register.

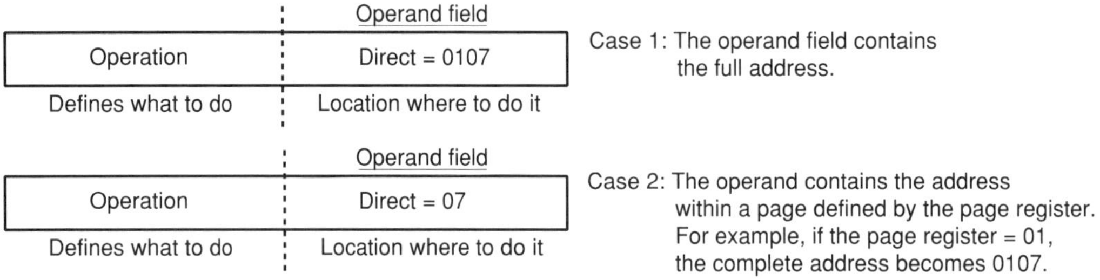

	Operand field	
Operation	Direct = 0107	Case 1: The operand field contains the full address.
Defines what to do	Location where to do it	

	Operand field	
Operation	Direct = 07	Case 2: The operand contains the address within a page defined by the page register. For example, if the page register = 01, the complete address becomes 0107.
Defines what to do	Location where to do it	

FIGURE 1–6
Direct operand.

In the figure, the operand in Case 1 uses a large number of bits in the instruction word; consequently, this addressing scheme is rarely implemented on signal processors. Case 2 reduces the number of operand bits; this addressing scheme is commonly found on signal processors.

Indirect Addressing

The operand specifies which CPU *address register* points to the memory location being operated on by the instruction. As an example, Figure 1–7 illustrates an instruction that operates on memory location 0107 whose address is contained in address register 1. The operand part of an indirect addressing mode instruction uses very few instruction word bits.

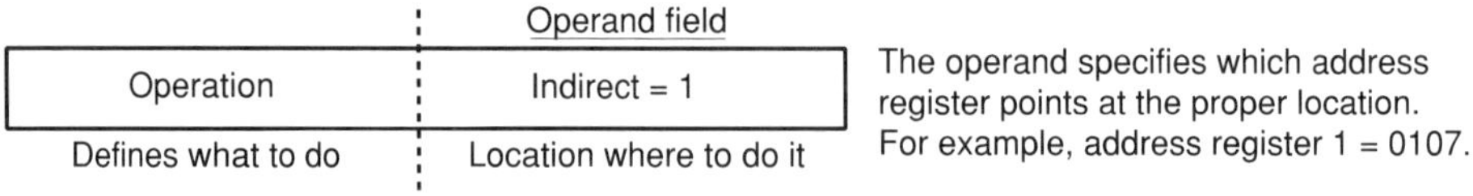

	Operand field	
Operation	Indirect = 1	The operand specifies which address register points at the proper location. For example, address register 1 = 0107.
Defines what to do	Location where to do it	

FIGURE 1–7
Indirect operand.

Indexed Addressing

The operand specifies an *offset* and a CPU *index register*. The memory location being operated on by the instruction is calculated as the sum of the index register and the offset value. As an example, Figure 1–8 illustrates an instruction that operates on the contents of the memory location whose address is obtained by adding index register 1 and an offset value of 3. The operand in this mode uses a fair number of instruction word bits.

	Operand field
Operation	Index = 1 Offset = 3
Defines what to do	Location where to do it

One of many index registers offset by a value. For example, if index 1 = 0104, resulting address = 0104 + 3 = 0107.

FIGURE 1–8
Indexed operand.

Circular Addressing

DSP makes extensive use of first-in, first-out (FIFO) buffers. To simplify the implementation of FIFOs, many signal processors include pointer registers that support the modulo addressing mode. In this mode, the pointer register contains an address that points within upper and lower boundaries defined by the programmer. When the modulo m pointer is incremented beyond a boundary, it automatically wraps around to the other boundary. The result is that the register always points within a fixed range of addresses. Figure 1–9 illustrates this circular addressing scheme.

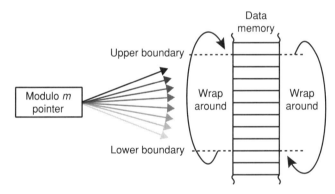

FIGURE 1–9
Circular addressing.

Other Addressing Modes

Some very popular DSP algorithms such as the fast Fourier transform (FFT), which is discussed in Chapter 5, require special addressing modes. These modes are quite different since they apply only to specialized DSP applications.

As an example, calculating an eight-point FFT operation results in an array containing elements whose order is scrambled. Unscrambling the elements requires a bit-reversed addressing scheme that enables the array elements to be fetched in the following binary order:

000   100   010   110   001   101   011   111

Because FFT applications are quite common, it makes sense to include special addressing modes that implement specialized addressing schemes such as the bit-reversed addressing mode. The availability of special addressing modes increases the efficiency of the programming, which results in faster processing.

1.2.3 MICROCONTROLLER-LIKE ARCHITECTURE

Signal processors include onboard devices that address the particular needs of the DSP environment. Typically, they include program memory (RAM and/or ROM), data memory, input/output (I/O) devices, and timers. Consequently, like microcontrollers, they incorporate on a single chip many of the components that define a computer.

This hardware structure is unavoidable because of the high-speed requirements of the DSP environment. Having the program memory, data memory, and processor embedded on the same chip of silicon allows the processor to simultaneously fetch, decode, and execute program instructions within one clock cycle. However, this severely limits the amount of memory that can be included on the processor chip. Fortunately, DSP applications typically execute programs that are quite short. Typically, a few kilo-words of program and data space provide plenty of memory to satisfy most applications.

Other devices typically used by signal processors are the analog-to-digital converter (ADC) and digital-to-analog converter (DAC). The ADC converts analog signals to the digital world of signal processing; the DAC outputs the processed digital signal back to the analog world. To reduce the number of I/O pins required for these conversion operations, the signal processor must implement one of the following two hardware solutions:

1. Incorporate the ADC and DAC onto the signal processor chip.
2. Provide special hardware that facilitates the interface to these devices.

In the first case, incorporating the ADC and DAC onto the signal processor chip consumes a good deal of space on the silicon chip. Signal processors that incorporate the ADC and DAC cannot usually include many other hardware features. Because of this, they are usually limited in performance and features.

In the second case, the signal processor includes a special *serial port* through which the converter data is exchanged. The use of a serial interface reduces the number of I/O pins required to exchange the data. This allows the signal processor chip to remain small while still being able to accommodate even the large converters that provide 16 or even 24 bits per sample. Special ADC/DAC integrated circuits are available to provide direct pin compatibility with the signal processor serial port interface.

The DSP environment calls for the precise timing of events such as ADC/DAC conversions. For this reason, many signal processors provide physical resources such as *event timers* to help control such events. The programmer typically uses the timers to generate periodic interrupts that trigger time-dependent segments of code.

1.2.4 PARALLEL PROCESSING

Parallel processing occurs when the signal processor performs many tasks simultaneously. The use of parallel techniques is one of the main innovations that allow signal processors to achieve the great processing speeds for which they are renowned. Parallel processing becomes possible in two separate cases.

1. When the operations in a programmed sequence are independent of each other, they may be executed in parallel. In a DSP environment, certain sequences of operations recur all the time. In this case, the signal processor may implement a single *specialized instruction* that performs all the required operations in parallel.

2. When programmed instruction segments are independent of each other, they may be executed in parallel. In this case, the signal processor may contain *many processing units* that work in parallel to execute the different instruction segments.

Figure 1–10 illustrates these two cases of parallel processing.

Parallel processing of operations

The multiply and accumulate instruction is executed by a processing unit to perform two operations in parallel:

• Accumulates the previous product.
• Multiplies two numbers.

Parallel processing of instruction segments

Multiple processing units (PUs) can execute many instruction segments in parallel:

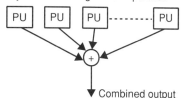

FIGURE 1–10
Parallel operation.

All signal processors implement special instructions such as the multiply and accumulate (MAC) instruction to speed the processing. Top-of-the-line processors implement specialized instructions that perform an incredible number of operations in parallel within a single clock cycle.

Some applications require the execution of billions of operations per second. Such speed can be achieved by using only processors that regroup many processing units through a parallel architecture. Unfortunately, these processors are expensive.

1.3 SPECIALIZED EMBEDDED HARDWARE

The hardware implementation of a function is typically a hundred times faster than its software equivalent. Signal processors contain many parts of specialized hardware that work in conjunction with the software to satisfy the high-speed requirements of DSP applications.

1.3.1 HARDWARE MULTIPLYING UNIT

Typical DSP applications require millions, if not billions, of multiplication and addition operations every second. For this reason, any processor that calls itself a *signal* processor requires a hardware multiplying unit capable of multiplying two numbers within one clock cycle. Building this multiplying unit, feeding it, and accumulating the results require a specialized embedded hardware structure.

The hardware that implements a multiplying unit on a silicon chip requires an enormous number of logic gates. This is the reason that most non-DSP processors include a product instruction instead of a multiplying unit. Two different versions of multiplying units exist:

1. *The fixed-point multiplying unit.* This is the cheapest and most commonly used multiplying unit. All the bits in the multiplied numbers are significant bits; therefore, these multipliers achieve great precision as well as great speed. Unfortunately, some software overhead is required to properly scale the fixed-point results (see Section 9.3).

2. *The floating-point multiplying unit.* This allows the use of exponentiated numbers such as 2.34567×2^9, which allow for the processing over a very large range of numbers. Storing the exponent bit in a reserved bit-field that is part of the multiplied numbers expands the range (see Section B.3.5). Because fewer bits remain to store the value of the number, floating-point numbers are not as precise as their fixed-point counterparts. The advantage with floating-point numbers is that the software is simpler to write and effectively requires fewer instructions since no scaling is required. Processors that include a floating-point multiplying unit are much more expensive, however, than their fixed-point counterparts.

All multiplying units operate on two separate numbers to provide a result. Consequently, two separate paths are required to feed the multiplying unit. The Harvard architecture (see Section 1.2.1) provides these two separate address and data paths. Figure 1–11 illustrates the paths that feed the multiplier.

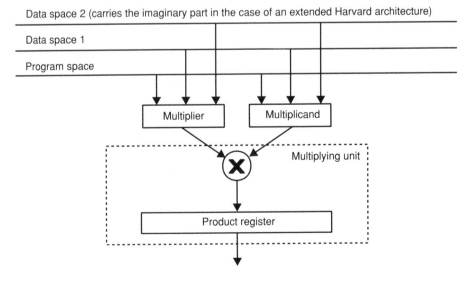

FIGURE 1–11
Feeding the multiplier.

Many signal processing algorithms involve complex numbers that require the manipulation of real and imaginary parts. As illustrated in Figure 1–11, the extended Harvard architecture allows some signal processors to provide separate data spaces for the real and imaginary parts.

1.3.2   SPECIAL ACCUMULATOR

The *accumulator* is the central component of any CPU. To avoid processing bottlenecks, it is important for the signal processor to implement some or all the following features:

- Dual accumulators
- Extra accumulator bits
- Overflow detect
- Overflow saturation

A *dual accumulator* structure simplifies the calculations that involve complex numbers. The real and imaginary parts may be accumulated without having to store intermediate results.

When calculations take place, the result may slightly overflow the number of bits provided by the accumulator. For this reason, some accumulators include a few extra accumulator bits to accommodate the extra-large results. These extra bits provide the headroom that allows some overflow during the execution of automatically repeated operations whose results are being accumulated.

When a CPU processes signals, overflow conditions are almost inevitable. An accumulator overflow can ruin the signal contents if it is not managed properly. For this reason, most signal processors provide special flags and overflow modes that minimize the problems.

Most DSP algorithms use 2's complement arithmetic, which allows the use of negative numbers. When a 2's complement number overflows, its value changes to the opposite polarity. This sharp polarity change ruins the frequency contents of the processed signal. To minimize the overflow effect, some processors implement a saturation mode that prevents abrupt changes of polarity during overflow conditions. Figure 1–12 illustrates the 2's complement overflow with and without saturation mode.

In the right side of Figure 1–12, the saturation mode implements the equivalent of hitting one of the voltage rails in analog electronic circuits. When compared to the 2's complement overflow, the saturation condition alters the signal contents to a lesser degree.

The accumulator has flags that record the overflow condition. The processing may then undertake some corrective action if necessary.

1.3.3   BARREL SHIFTERS

The signal processor often needs to scale the computed results. In the binary world, the shifting operation results in the scaling of numbers. For example, every shift left operation (while concatenating zeros) scales the shifted number by a factor of 2. The number of shift operations therefore scales the shifted number by the powers of 2:

Scaling factor: $2^{\text{\# of shift operations}}$

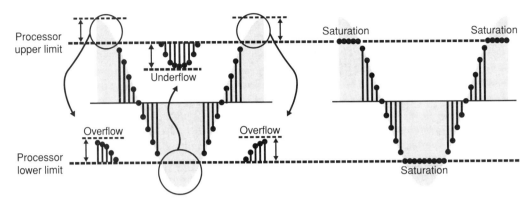

FIGURE 1–12
Saturation arithmetic.

Scaling a number by any factor larger than 2 requires more than one shift operation. In this case, the repeated shift operations consume much of the precious processor time. To solve this problem, most signal processors include a hardware version of the shift operation that we refer to as a *barrel shifter*.

A barrel shifter is actually a bank of selectors inserted between the source and the destination of the data. With a barrel shifter inserted, the transfer of data from a source to a destination must go through the bank of selectors. By choosing which input is selected, the bank of selectors effectively performs the equivalent of shifting operations. A barrel shifter does not alter the contents of the source; it is the destination that receives a shifted version of the source data. Figure 1–13 illustrates the upper half of an 8-bit barrel shifter.

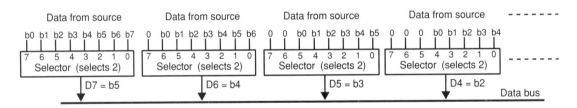

FIGURE 1–13
Portion of an 8-bit barrel shifter.

All of the selectors in a barrel shifter select the same input. For example, in Figure 1–13, the second input is selected and the data bus carries a version of the source data that has been shifted two positions to the left. The destination therefore receives the source data scaled by a factor of 4. Because the shifting occurs as the transfer takes place, the source data is not altered by the operation.

1.3.4 ZERO-OVERHEAD LOOPING

Signal processors use every possible trick to squeeze the most processing performance out of the processor. The great majority of DSP algorithms call for a continuous repetition of the same processing operations over a number of iterative passes. The normal programming approach is to preset some iteration counter to the number of required passes. At the end of every pass, the iteration counter is first decremented, then tested, and a conditional instruction determines whether an additional pass is required.

To speed processing, most signal processors include special hardware registers that allow the automatic iteration of segments of instructions. Before the program enters the loop, a sequence of instructions loads an iteration register with the required number of passes and identifies the end of the instruction segment. The segment of code then repeats automatically with no need for decrement, test, or branching instruction. The entire iterative process therefore runs as if it had been written as straight-line code.

1.3.5 HARDWARE STACK

The *stack* is an area that the processor typically uses to store the return addresses of subroutines and interrupts. Traditionally, the stack area resides either in the program or data memory space. Using the stack in a traditional manner therefore ties up either the program or the data bus during the stacking operations. The architecture of most signal processors includes a special, separate, small area of memory reserved for stack operations. Using this hardware approach to store and retrieve return addresses, the processor does not waste time pushing or popping addresses from the stack.

The use of a hardware stack allows for faster interrupt responses and minimizes the amount of overhead when the processor calls or returns from subroutines. The drawback is that the small area of the stack memory usually limits the level of nesting for subroutines and interrupts.

1.4 SPECIALIZED SOFTWARE

Signal processors are full of specialized instructions that maximize the processing efficiency of special operations. For example, it is so common to accumulate the results of a sequence of multiplication operations that all signal processors implement a MAC.

Shifting the contents of memory from location to location is another operation that the great majority of processors implement as a parallel operation.

The availability of special instructions make some signal processors better than others at certain types of applications. Data compression, image processing, filtering, FFTs—whatever the application, it is likely that the choice can be narrowed to a small number of signal processors. Those of interest will implement some special instruction(s) that will make programming easier and the processing more efficient.

1.5 OTHER SPECIAL FEATURES

DSP is a very wide field that seeks to manipulate signals that comply with a variety of environments. In some cases, the environment in which a signal exists has undesirable characteristics. For example, a signal stored on magnetic tape should be conditioned to overcome the noise levels present on this storage medium. Conversely, some environments have characteristics that we can exploit. For example, a video signal contains many repetitive parts that lend themselves to signal compression.

A large number of standards are used to encode signals. When a signal appears in coded form, it usually has to be decoded before the processing is applied. Once processed, the signal usually needs to be re-encoded so that it can be returned to its environment. The decoding and encoding of signals are often time-consuming tasks that are better addressed by specialized hardware.

Fortunately, most popular standard formats are well supported by hardware devices. Some of these hardware devices are integrated onto specialized signal processors. For example, a signal processor that specializes in handling telephone signals is likely to include the μ-Law and/or A-law compander circuits that are standards in the telephone industry.

1.6 DIGITAL SIGNAL PROCESSING USING THE PURE-HARDWARE APPROACH

The alternative to using a signal processor is to design a pure-hardware circuit. This approach can achieve DSP at speeds that exceed one hundred times the performance of software-controlled systems. The disadvantage is that the flexibility of programmed systems is lost. A pure-hardware circuit specializing in the implementation of a specific application is often more cost effective than the programmed approach. A typical example is a system that requires a simple digital filter. If this is all that is required, a specialized integrated circuit would certainly bring considerable cost saving over a full signal processor approach.

The following are three types of hardware circuits.

1. *Specialized integrated circuits* (IC) address specific applications such as filtering, data compression, data encoding, echo suppression, etc.

2. *Programmable logic devices* (PLDs) provide a more flexible approach. They are somewhat more expensive than the specialized IC but they feature programmability. PLDs provide a more flexible solution since they can be used to implement additional logic functions that may be essential to a system. Standard designs implementing typical DSP applications can often be found as macrofunctions. In fact, complete signal processors (the small ones) can be programmed onto large PLDs. Applications such as filters can be put together inside a PLD with minimum effort. Many PLDs are reprogrammable, which allows the system to be upgraded. For example, upgrading an already existing product to be compatible with a new standard becomes possible if this product is implemented using a reprogrammable PLD.

3. *Application-specific integrated circuits* (ASICs) provide more density and more cost effectiveness when products are manufactured in large quantities. An ASIC is a customized IC onto which a number of already existing parts can be regrouped. For example,

a small microcontroller, a digital filter, an ADC/DAC, and some memory could be integrated onto the same ASIC. Silicon foundries can build large quantities of these custom devices as a special order. Each custom part has a special number, and this helps to reduce design piracy. Usually ASICs become cost effective only when thousands of devices are manufactured.

SUMMARY

- Once a number is stored in a digital memory, its value never changes.
- The functionality of digital processing systems is defined entirely by the value of stored numbers.
- The operation of DSP systems is

  As predictable and precise as the numbers that define its operation.

  Completely unaffected by aging.

  Completely stable with respect to changes in the environmental conditions such as temperature or pressure.

  Unaffected by component or power supply limitations.
- A DSP application may be completely redefined by changing the value of a few numbers.
- DSP applications are limited by the speed of the processing.
- DSP can be done either by pure-hardware circuits or by signal processors in programmed systems.
- DSP systems using the programmed approach bring flexibility. A signal processor can perform different signal-processing applications and it provides the means to perform other tasks such as peripheral control.
- Pure-hardware DSP systems bring great processing speeds. In the case of simple applications, these systems are usually more cost effective than programmed systems.
- Software tools are available to perform most of the DSP mathematics for the designer.
- Signal processors are specialized microprocessors that specifically address the needs of the DSP environment.
- Most signal processors implement a Harvard architecture, which provides the high bus bandwidth required by DSP applications.
- Signal processors usually adopt a RISC-like approach that allows them to fetch and execute an instruction per clock cycle.
- Because of the necessary short instruction words, some addressing modes are limited on signal processors.
- Most signal processors implement special addressing modes such as circular addressing and bit-reversed addressing.
- Signal processors include specialized hardware, such as timers and ADC/DAC interfaces that allow the handling of digital signals.
- The parallel processing of operations requires special instructions.

- The parallel processing of program segments requires a special processor structure that includes many processing units.
- Specialized hardware, such as multiplying units and barrel shifters, are necessary to achieve the required processing speeds.
- Fixed-point processors are less expensive than their floating-point counterparts, but require additional programming steps to interpret and scale the results.
- Floating-point processors are more expensive than their fixed-point counterparts but yield simpler and faster programs.
- Many DSP algorithms require the use of complex numbers. Processors that implement an extended Harvard architecture address the need for separate real and imaginary memory spaces.
- Processing signals sometimes result in accumulator overflows. Signal processors contain special flags, extra accumulator bits, and an overflow saturation mode to help manage overflow conditions.
- Special instructions and registers allow the iterative looping of program segments with the same efficiency as straight-line code.
- A special, separate, small area of memory is reserved for stack operations on most signal processors. This speeds all stacking operations, making the execution of subroutines and interrupts much faster.
- All signal processors include special instructions that address the needs of specific DSP environments. Specialized instructions make some signal processors better than others at certain types of applications.
- Specialized signal processors may include special hardware used to decode or encode signals according to specific standards.

PRACTICE QUESTIONS

1-1. Why are DSP systems not affected by environmental conditions such as temperature or pressure?

1-2. Why are DSP systems not affected by component or power supply limitations?

1-3. What is the main factor that limits DSP applications?

1-4. Is it possible to perform DSP using *any* microprocessor?

1-5. What is special about the Harvard architecture?

1-6. Why are instructions coded as single words on digital signal processors?

1-7. What is a *barrel shifter*?

1-8. What is the *overflow saturation mode*?

2

BUILDING DIGITAL SIGNALS

Digital signal processing is about managing signals. This management includes their building, analysis, and modification. This chapter develops the techniques that allow signals to be built using a digital processor.

Our research begins by considering the nature of signals as we examine the two basic forms that they can adopt. We recognize that any signal generated using a digital processor is in a form incompatible with the continuous-time signals that exist around us. This leads us to examine the standard technique to translate the digitally generated signals into continuous ones. The translation technique immediately exposes the computing performance limits that the digital processor faces.

We then explore the theory that leads to the building of digital signals. This takes us back almost two centuries in time as we view a few of the nineteenth century achievements in signal analysis. We briefly examine some of the results of the brilliant mathematical work that provides the background necessary to further our investigation into the nature of signals. By examining these results, we uncover that most signals actually consist of a mix of fundamental mathematical entities called *sinusoids.*

We next introduce the basics of what is certainly the most popular practical technique used to build digital signals. By implementing this technique on a digital processor, we demonstrate the programming steps required to synthesize a basic cosine waveform. We then develop additional programming techniques to control the frequency, amplitude, and phase of the synthesized cosine waveform. Finally, we consider a number of ways to synthesize elaborate signals.

We close by examining the different ways of structuring the hardware and the software of the digital processing system. The illustrated system architectures provide the information necessary to select an approach that enables us to achieve the required processing speed and/or the desired system cost.

2.1 TYPES OF SIGNALS

Most dictionaries define *signal* as a sequence of shapes or symbols that carry information. When a continuously changing shape carries the information, the resulting waveform defines a *continuous-time signal*. Alternately, when a sequence of symbols, such as a series of binary numbers, carries the information, the signal is noncontinuous and we refer to it as a *digital signal*.

The great majority of signals that we can hear, see, or feel around us are continuous-time signals. Unfortunately, the nature of digital processors limits them to manipulating binary numbers; consequently, these processors cannot directly handle the continuous-time signals that surround us. The following sections will help us to understand the differences that separate analog and digital signals. This understanding will guide us in using a processor to generate a signal and will give us the background necessary to translate this digital signal into a continuous form.

2.1.1 CONTINUOUS-TIME SIGNALS

Continuous-time signals, also referred to as *analog signals,* have a value that can be measured at any point in time. A good example of an analog signal is music, which consists of a continuously changing wave along which the atmospheric pressure varies. When the wave travels by us, we hear the sounds of music when our eardrum vibrates as it reacts to the changes in pressure. We know from experience that music continuously reaches our ear. This suggests at the intuitive level that all the parts of a sound wave, even the most microscopic ones, should be continuous. Figure 2–1 illustrates that analog signals are continuous.

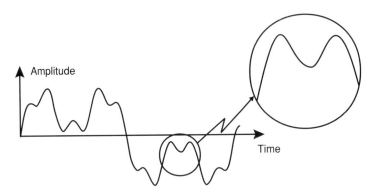

FIGURE 2–1
Analog signals are continuous.

Figure 2–1 indicates that an analog waveform (like a sound wave) has a particular amplitude value at all points along the time axis. Examining the enlarged part of the analog signal, we detect no evidence of discontinuity. This observation leads us to acknowledge that analog signals must consist of an infinite number of points that are chained together to shape the continuous-time signal. It follows that each of these infinite points must have its own distinct amplitude value. We must therefore conclude that an analog signal contains an infinite amount of information.

This conclusion presents a problem: A digital processor can store only a finite quantity of information in its semiconductor memory, and its processing speed limits it to process a finite amount of digital information. Therefore, a digital processor cannot handle the infinite amount of information contained in even the smallest part of any given analog signal.

2.1.2   USING DIGITAL SIGNALS

Digital signal processors are microprocessors whose architecture provides an optimized functionality to manipulate digitized signals. Some of these processors' main tasks include generating, analyzing, and/or modifying digitized signals. Since these devices manipulate binary numbers, it follows that a signal generated by a digital processor must consist of a sequence of binary numbers. This sequence of numbers defines the value of the signal amplitude at specific points in time.

Figure 2–2 illustrates a digital signal in numerical format and the corresponding graphical representation. The figure depicts a digital signal that has been obtained by plotting large dots at fixed intervals. The dots represent the magnitude of numbers that the digital processor generates sequentially. This illustrates that a digital processor can be used to digitally generate a signal by programming it to output, at fixed intervals, the numbers that correspond to points along the amplitude curve of the desired signal. When the signal exists as a sequence of numbers, we say that it exists in a *digitized format*.

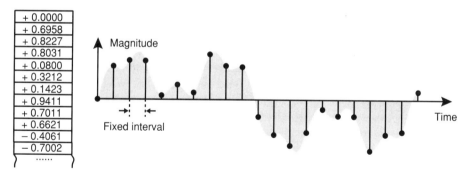

| + 0.0000 |
| + 0.6958 |
| + 0.8227 |
| + 0.8031 |
| + 0.0800 |
| + 0.3212 |
| + 0.1423 |
| + 0.9411 |
| + 0.7011 |
| + 0.6621 |
| − 0.4061 |
| − 0.7002 |

FIGURE 2–2
Digital signals consist of a sequence of numbers.

2.1.3   SENDING A DIGITIZED SIGNAL TO THE ANALOG WORLD

Most DSP systems require the conversion of the digitized signal to the analog world. As noted earlier, a *digital-to-analog converter* (DAC) performs this conversion.

The DAC translates digital values into voltages by linking every possible digital value to a different voltage amplitude. The voltage that the DAC outputs is designed to be directly proportional to the digital value at its input: the greater the numerical value, the higher the output voltage. Since the converter must associate every possible input value to a distinct output voltage level, we must ask the following question: How many different digital values does the DAC need to convert?

To answer this question, we must remember that a digital value consists of a group of bits. The number of bits that we choose to regroup to form a binary number determines the number of different digital values that are possible according to the following relationship:

Number of different digital values = 2^{Number of bits used}

This means that digital signals can adopt only certain values and that DACs can output only certain voltage amplitudes. For example, Figure 2–3 illustrates a hypothetical 4-bit DAC that can accept $2^4 = 16$ different digital values and translate these into a range of 16 distinct voltage levels.

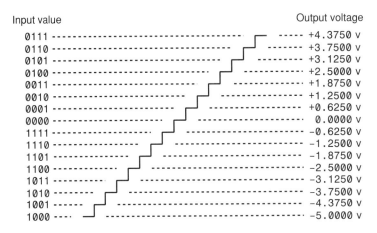

FIGURE 2–3
Digital values converted into voltage levels.

The DAC system consists of a latch section whose function is to remember the last digital value that the DSP provided. This latch feeds an analog section that converts the latched value into a voltage. Figure 2–4 illustrates the basic structure of a DAC.

A new digital-to-analog conversion starts when we use a clock event to transfer a digital value, which the processor data bus provides, into the latch. The latch feeds the stored value to a converter that continuously translates it into a voltage level. The analog output maintains this voltage level until we use the clock event to store a new digital value into the latch.

The DAC therefore refits the output voltage level following each clocking event. When producing signals, it is standard to make the clock events periodic to make the output voltage change at fixed time intervals. The duration of these time intervals is called the *sampling period* and is abbreviated T_S.

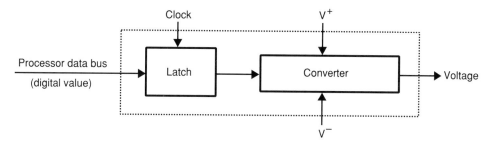

FIGURE 2–4
Basic structure of a DAC.

Since the output voltage level remains unchanged between clock events, the resulting analog waveform exhibits a staircaselike appearance. Figure 2–5 illustrates the typical appearance of a DAC's analog output.

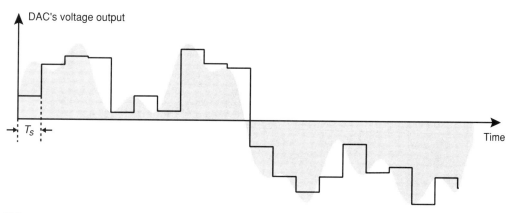

FIGURE 2–5
DAC stretches samples over time intervals.

As Figure 2–5 indicates, the converter provides a constant voltage output between the clocking events. The clocking events generate discontinuities (sudden changes of level) that result in a stepped output. This jagged output may be satisfactory for some applications, but in most cases, it is necessary to correct this situation by smoothing the output signal. Sections 6.4.5 and 6.4.6 illustrate and explain standard techniques that correct the output's appearance.

2.1.4 CONVERTING BINARY NUMBERS TO VOLTAGES

DSP mainly uses *signed* DACs. As discussed in Section 2.1.3, the DAC operation starts by latching a digital value that it receives from the processor. It then translates the latched value into a specific voltage that it selects from a range of possible output voltages. The number of different voltage steps that the DAC can output depends on the number of bits in the latched digital value.

You can adjust the range of voltages that the DAC covers, but the DAC always segments this range into the same number of voltage steps. Consider the examples illustrated in Figure 2–6. Note that the DACs in Figure 2–6 produce different output voltages although they receive exactly the same 8-bit digital value. Since there is no binary point in the digital numbers latched by the DAC, it can choose to interpret the values of these numbers using any format.

DACs always choose to interpret their input as signed pure fractional numbers.

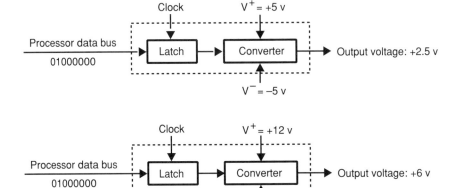

FIGURE 2–6
Changing the voltage range of a DAC.

Viewed this way, the digital input value dictates what fraction of the full output voltage must appear at the DAC output.

The following examples require familiarity with fractional binary systems. Refer to Appendix B if you need to review the Q format of fractional numbers. For example, in Figure 2–6, the DACs will interpret their 8-bit digital input as being a Q7 value:

$$01000000 \xrightarrow{\text{Q7}} + \frac{01000000}{2^7} \xrightarrow{\text{Decimal}} \frac{64}{128} = +0.5$$

The digital input therefore tells the DACs to output 0.5 of their full positive range.

It is important to note that the DAC interpretation and the processor interpretation can differ. For example, consider what happens when the processor outputs signed Q4 values to the DAC. To the processor, the digital value has three integer bits (I-bits) and four fractional bits (F-bits) but to the DAC, the number has seven F-bits!

The DAC effectively sees the binary point moved three locations to the left:

Processor interpretation: `SIII.FFFF` ... DAC interpretation: `S.FFFFFFF`

In this case, the DAC has actually moved the binary point three positions to the left and therefore sees the number as being eight times smaller. The DAC interpretation is actually scaling down the value of the processor output by a factor of 8 since it converts three I-bits into F-bits.

An understanding of these different interpretations is important when the processor's digital output is linked to the voltage produced by the DAC. For example, refer to the practical situation described on Figure 2–7.

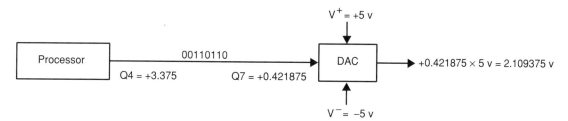

FIGURE 2–7
The DAC scales the digital output.

The processor Q4 output in Figure 2–7 corresponds to the following numerical value:

$$0011.0110 \xrightarrow{\text{Decimal}} + \frac{00110110}{2^4} = \frac{54}{16} = +3.375$$

The DAC interprets its input as a pure fractional Q7 number that has a value of

$$0.0110110 \xrightarrow{\text{Decimal}} + \frac{00110110}{2^7} = \frac{54}{128} = +0.421875$$

and, therefore, the output voltage is

$$+0.421875 \times 5 \text{ v} = +2.109375 \text{ v}$$

The processor output format and the DAC interpretation are constant, which means that the DAC always scales the digital value by the same factor when it moves the binary point. Considering this, we can calculate the output voltage in a single step:

Processor output × Constant DAC scaling × DAC voltage range = Output voltage

$$+3.375 \times \frac{1}{8} \times \pm 5 \text{ v} = +2.109375 \text{ v}$$

2.1.5 UNDERSTANDING THE CONCEPT OF DISCRETE TIME

In most DSP applications, the output sampling period is kept constant. This means that the processor must feed the DAC latch a new digital value at the end of every T_S interval. We call this digital value an *output sample*. The constant sampling period implies that the processor is producing output values periodically at a specific *output sample rate* (also referred to as the *sampling frequency*):

$$Output\ sample\ rate\ =\ \frac{1}{Output\ sample\ period}$$

also called

$$Sampling\ frequency:\ f_S\ =\ \frac{1}{T_S}$$

Since the processor must deliver a new digital value at the end of every sampling period, the processing necessary to generate every new value is constrained to execute in the time interval T_S. Changing the sample rate therefore changes the amount of processing time allocated to the calculation of an output value. We have the processor perform the calculation by programming it to execute a sequence of instructions. Once we know how many instructions are required and how much time is available to produce an output sample, we can determine the digital processor speed required to meet these constraints.

For example, suppose that the signal processing algorithm calls for 20 instructions to generate each of the output samples. Given an arbitrary sampling period of 10 μs, the processor needs

$$Minimum\ processor\ performance\ =\ \frac{20\ instructions}{10\ \mu s}$$

$$=\ 2 \times 10^6\ instructions/second$$

In practice, even the slowest digital signal processors should be able to perform at least 10 million instructions per second (MIPS).

It is important to note that the processor calculates the next output value *during* the present T_S interval. This means that the result of this calculation becomes valid only at the beginning of the next T_S interval. Consequently, the digitized signal is valid only at specific (discrete) instants in time. These points in time occur at integer multiples of T_S:

$$0T_S,\ 1T_S,\ 2T_S,\ 3T_S,\ 4T_S,\ 5T_S \ldots$$

or

$$nT_S$$

where

$$n = 0,\ 1,\ 2,\ 3,\ 4,\ 5 \ldots$$

We use the variable n, which we call the *discrete-time variable,* to refer to the discrete-time instants when the digitized signal has a value. Figure 2–8 illustrates the instants at which the discrete-time variable n exists.

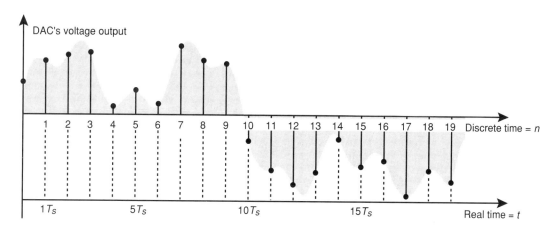

FIGURE 2–8
Time instants at which *n* exists.

It is important to reiterate that the digitized signal is valid only at discrete instants in time and that it does not exist at any other time since it is in the process of being computed. A digitized signal is therefore *not* continuous in time and, because of this, the discrete-time variable n can adopt only *integer* values. The variable t, which is normally used to describe continuous-time, is valid only at the discrete-time instants nT_S. A continuous-time expression can therefore be converted to a discrete-time expression by replacing the continuous-time variable t by nT_S. Once this is done, the signal exists only at the instants described by nT_S.

Consider the following examples that illustrate the change in the time variable when moving from continuous-time to discrete-time:

Continuous-Time Signal		Discrete-Time Signal
$x(t)$	$\Leftrightarrow$	$x(nT_S)$
$A\cos(\omega t)$	$\Leftrightarrow$	$A\cos(\omega nT_S)$

where $n = 0, 1, 2, 3, 4, \ldots$ and T_S is the sampling period

Refer to Figure 2–9 and consider the following important questions:

- What happens to the signal *between* the discrete-time instants?
- If all the samples in the sequence that describes a signal are shifted in position by the same small interval, will the signal still contain the same information?
- At what time interval should we sample a signal? What is the advantage of sampling more often?

These very important questions are investigated and answered in Chapter 3.

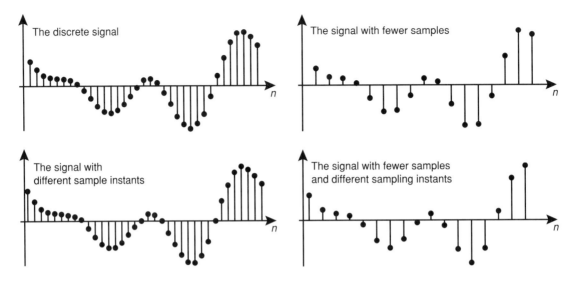

FIGURE 2–9
Do these samples represent the same signal?

2.2 INTERNAL STRUCTURE OF SIGNALS

For a long time, thinkers have wondered about the internal structure of signals. They had already noticed that nature usually produces solutions that are elegant and simple. Because of this, these thinkers suspected that the intricate signals that they could observe really consisted of a mix of some fundamental waveforms.

2.2.1 PERIODIC SIGNALS

There are many different classes of signals.

- Some signals are *pure real*, which means mathematically that they have no imaginary part.
- Some signals are *complex*, which in a mathematical sense means that they have both a real and an imaginary part.
- Some signals are *periodic*.
- Some signals are *not periodic*.

The special class that regroups real and periodic signals really fascinated the eighteenth and nineteenth century mathematicians and is the subject of most of this book. Real signals include those that we can observe and produce with our electronic instruments. Periodic signals consist of a single waveform segment that keeps repeating itself as time advances. Figure 2–10 illustrates examples of periodic waveforms.

The sine and the cosine waveforms are sinusoid in nature. At the beginning of the nineteenth century, Jean Baptiste Joseph Fourier (1768–1830) had the intuitive belief that

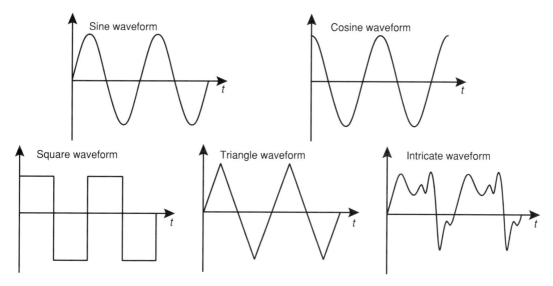

FIGURE 2–10
Examples of periodic signals.

sinusoids were the basic waveform from which all other real and periodic signals could be built. He proposed that any real and periodic signal could be constructed by combining a number of harmonically related sinusoids. Note that "harmonically related" means that the sinusoids need to oscillate at an integer multiple of a reference base frequency that we call the *fundamental frequency,* or simply the *fundamental.*

We refer to the many sinusoids that may be contained in a signal as being the *harmonics* of that signal. We identify each of the individual harmonics by using the particular integer number k. This number identifies a specific harmonic by describing its frequency in terms of the fundamental frequency.

For example, a periodic signal that oscillates at a frequency of 400 hertz (Hz), contains some of the following harmonically related sinusoid components:

Fundamental: 400 Hz
Harmonic 2: 800 Hz (2×400 Hz)
Harmonic 3: 1200 Hz (3×400 Hz)
Harmonic 4: 1600 Hz (4×400 Hz)
. . .
Harmonic k: $k \times 400$ Hz

Synthesis is the process of building a periodic signal from its component harmonics. As a typical example, we consider the partial synthesis of a square wave. It can be shown mathematically that this periodic signal, which looks simple in appearance, really consists of an infinite number of sinusoid components that are odd harmonics. As illustrated in Figure 2–11, we can add the odd harmonic components one at a time and, with each added harmonic, we see an improvement to the shape of our approximated square wave.

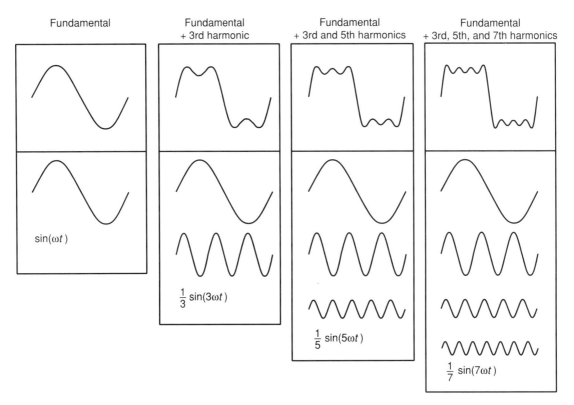

FIGURE 2–11

Partial synthesis of a square wave.

Some signals, such as the square wave, require an infinite number of harmonics. In practice, it is impossible for us to combine an infinite number of harmonics using a signal processor; fortunately, however, we can build most useful signals from a finite number of harmonics. Actually, most practical applications approximate the required signal by combining only the signal harmonics that carry the most information.

The idea of building signals by combining sinusoid harmonics fascinated a number of mathematicians and, following Fourier's insight, they came to accept that most periodic signals (with a few exceptions) could be built by adding harmonically related sinusoids. Peter Gustav Dirichlet (1805–1850) eventually provided the precise conditions that allow the representation of periodic signals by a series of sinusoids.

From this nineteenth century research eventually emerged the results showing how periodic signals can be built by mixing the harmonics of a fundamental frequency. These results took the form of two equations:

1. The *synthesis equation,* which describes the harmonic components required to build almost any periodic signal. This equation is developed in the following sections and will be used in practical exercises of signal synthesis.

2. The *analysis equation*, which is used to determine the sinusoid harmonic components of an existing periodic signal. Chapter 5 explores the analysis of signals.

These nineteenth century mathematicians probably did not expect the modern practical applications that would emerge from their discoveries. Nevertheless, they gave us the fundamental knowledge that would lead us to program our modern digital processors to perform the synthesis and the analysis of signals. The following sections of this chapter provide the basic information necessary to synthesize signals in practice.

2.2.2 SYNTHESIS EQUATION

Signal synthesis forms the basis for many practical applications in DSP. These include music synthesizers, modems, and cellular telephones.

The synthesis of a periodic signal requires mixing harmonically related sinusoids of different amplitudes and phase relationships. To synthesize a real digital signal $y(n)$, we need a list that individually describes the characteristics of all the harmonics that are contained in $y(n)$. This list begins with harmonic 0 (the direct current [DC] bias), proceeds to harmonic 1 (the fundamental frequency of the periodic signal annotated $f_{Fundamental}$), and continues with all the other harmonics (multiples of the fundamental):

$$
\begin{aligned}
y(n) = \ & A_0 \cos\left(2\pi\left[0 \times f_{Fundamental}\right]nT_S\right) && \text{(Harmonic 0 is the DC component)} \\
& + A_1 \cos\left(2\pi\left[1 \times f_{Fundamental}\right]nT_S + \phi_1\right) && \text{(Fundamental frequency)} \\
& + A_2 \cos\left(2\pi\left[2 \times f_{Fundamental}\right]nT_S + \phi_2\right) && \text{(Harmonic 2)} \\
& + A_3 \cos\left(2\pi\left[3 \times f_{Fundamental}\right]nT_S + \phi_3\right) && \text{(Harmonic 3)} \\
& + \ldots \\
& + A_k \cos\left(2\pi\left[k \times f_{Fundamental}\right]nT_S + \phi_k\right) && \text{(Other harmonics)}
\end{aligned}
$$

We express the discrete-time value of the kth harmonic as

$$
y_k(n) = A_k \cos\left(2\pi\left[k \times f_{Fundamental}\right]nT_S + \phi_k\right) \tag{2-1}
$$

Let's interpret the significance of each of the terms in Equation (2–1):

k is the harmonic number.
$f_{Fundamental}$ is the frequency of the signal being synthesized—harmonic 1 of $y(n)$.
nT_S is the time at which the nth sample is outputted.
2π is the factor used to convert cycles/second into radians/second.
A_k is the amplitude of the kth harmonic.
ϕ_k is the phase of the kth harmonic.

For example, consider that a signal $y(n)$ has its second harmonic defined arbitrarily as

$$
y_2(n) = 0.5 \cos\left(2\pi\left[2 \times f_{Fundamental}\right]nT_S + \frac{\pi}{2}\right)
$$

In this case, the harmonic component of the signal $y(n)$ consists of a cosine having an amplitude of 0.5 oscillating at twice the fundamental frequency and shifted by an angle of $\pi/2$ radians.

Consider that the fundamental of the signal $y(n)$ is a cosine of unit amplitude with no phase shift. Figure 2–12 illustrates the amplitude and phase relationship between the signal fundamental and its second harmonic. We can represent the sum of all the harmonics that make up the signal $y(n)$ using the series notation of Equation (2–2), which describes a general form of the *synthesis equation*:

$$y(n) = \sum_{k=0}^{\text{Number of harmonics}} A_k \cos\left(2\pi\left[k \times f_{\text{Fundamental}}\right] nT_S + \phi_k\right) \quad (2\text{–}2)$$

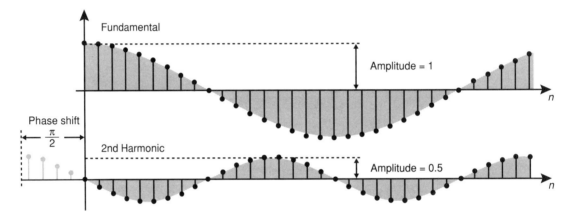

FIGURE 2–12
Fundamental and second harmonic.

We have defined the discrete-time signal $y(n)$ as *periodic,* which means that it consists of a set of N samples repeating itself as time advances. The discrete-time period of this signal is therefore the number of samples (N) in the periodic set. The duration of this periodic signal is the discrete-time period N, which multiplies T_S, the time interval between each of the samples.

Duration of one period of the signal $y(n)$: NT_S.

As an example, Figure 2–13 illustrates a signal that has a discrete-time period of 12 samples. The frequency of this signal must be the inverse of the duration of one period NT_S:

$$f_{\text{Fundamental}} = \frac{1}{NT_S}$$

This frequency corresponds to the rate at which the periodic segments repeat themselves, which is the lowest frequency contained in the signal. This is the fundamental frequency (harmonic 1) from which we determine the specific frequency of any of the other harmonics

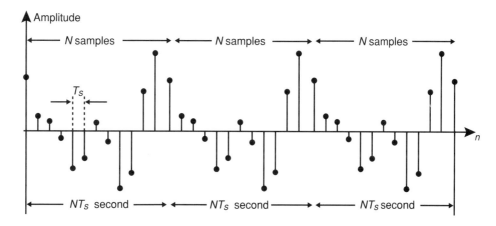

FIGURE 2–13
A periodic discrete-time signal.

making up this periodic signal. We can therefore substitute this expression for the fundamental frequency in the synthesis equation:

$$y(n) = \sum_{k=0}^{\text{Number of harmonics}} A_k \cos\left(2\pi\left[k \times \frac{1}{NT_S}\right]nT_S + \phi_k\right) \tag{2–3}$$

The fundamental, second, third, fourth, and so forth harmonics therefore have frequencies of

$$1 \times \frac{1}{NT_S}, \quad 2 \times \frac{1}{NT_S}, \quad 3 \times \frac{1}{NT_S}, \quad 4 \times \frac{1}{NT_S}, \quad \dots$$

Figure 2–14 illustrates the period of the first four harmonics of a 12-sample periodic signal. An examination of Figure 2–14 indicates that the number of samples decreases on the period of higher harmonics. For reasons that we will see in Section 3.4, there must be at least two samples on the period of any sinusoid. This sets an upper limit to the number of harmonics that can be used to synthesize a periodic signal.

For a signal with a period of N samples, the number of samples used to describe one period of each of the different harmonics is

Number of samples in the fundamental: $\dfrac{N}{1}$

Number of samples in a period of the second harmonic: $\dfrac{N}{2}$

Number of samples in a period of the third harmonic: $\dfrac{N}{3}$

. . .

Number of samples in a period of the kth harmonic: $\dfrac{N}{k}$

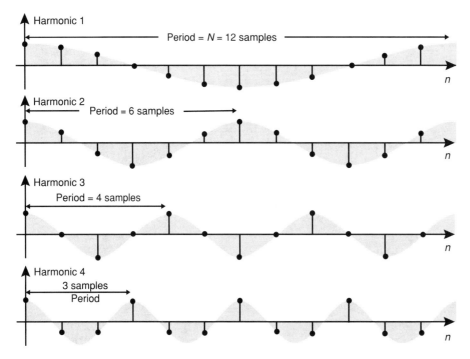

FIGURE 2–14
Harmonics of a signal that has a 12-sample period.

Since the number of samples must be at least two, we reach the highest possible harmonic when

$$\frac{N}{k} = 2$$

which means that the highest harmonic is

$$k = \frac{N}{2}$$

To accommodate the maximum number of harmonics contained in a real and periodic signal, we rewrite the synthesis equation:

$$y(n) = \sum_{k=0}^{N/2} A_k \cos\left(2\pi\left[k \times \frac{1}{NT_S}\right]nT_S + \phi_k\right) \tag{2–4}$$

For example, say that we need to synthesize a 1 kilohertz (kHz) periodic signal by repeating a 10-sample period. In this case, the signal is built of harmonics of a fundamental frequency of 1 kHz, and

The period of the fundamental is $\dfrac{1}{1\ \text{kHz}} = 1$ millisecond (ms)

Since our period consists of a 10-sample segment, the time interval between samples is

$$T_S = \frac{1\,\text{ms}}{10} = 0.1\,\text{ms}$$

Knowing that we need two samples to synthesize the highest harmonic, the period of the highest harmonic must be

$$2 \times 0.1\,\text{ms} = 0.2\,\text{ms}$$

and the frequency of the highest harmonic is the inverse of this period:

$$\frac{1}{2 \times 0.1\,\text{ms}} = 5\,\text{kHz}$$

In conclusion, we see that this 10-sample periodic signal can be synthesized using a maximum of five harmonic sinusoids: 1 kHz, 2 kHz, 3 kHz, 4 kHz, and 5 kHz.

2.3 PRACTICAL TECHNIQUES FOR SYNTHESIZING SINUSOIDS

A number of different techniques exist to synthesize sinusoids. The choice of a particular technique depends on the application. Some applications call for producing fixed tones, which can be generated using techniques described in Section 7.7. Other applications call for full control of the sinusoid's frequency, phase, and amplitude characteristics. This section describes a practical technique that generates a sinusoid while maintaining full control of all of these characteristics.

2.3.1 USING A COSINE TABLE TO SYNTHESIZE A SINUSOID

The most practical and flexible technique used to synthesize a sinusoid is to store a cosine table in the memory of the digital signal processor. The entries contained in Table 2–1 consist of a sequence of cosine values taken at fixed intervals from the complete period of a cosine function.

For example, let's assume that we arbitrarily choose to store a 36-entry table in memory. In this case, the individual entries correspond to cosine values that we calculate at 10-degree intervals to cover the full period of 360 degrees. Table 2–1 illustrates some of the values that are stored in this 36-entry table.

To synthesize a signal using such a cosine table, we must program a digital signal processor to output selected table entries to a DAC at every T_S interval.

The processor uses a *table pointer* to select the table entries outputted to the DAC. During every sample interval, the processor must compute the address of the next table entry by adding an *offset* value to the contents of the table pointer. For example, if we use an offset value of 1, the table pointer sequentially addresses each of the 36 table entries, and these will be outputted to cover a full period of the sinusoid. When the table pointer's content increases beyond the last table entry, it must wrap back to the top of the table.

TABLE 2–1
36-entry cosine table.

Entry 0:	$\cos(0°)$	$= +1.00000$
Entry 1:	$\cos(10°)$	$= +0.98481$
Entry 2:	$\cos(20°)$	$= +0.93969$
Entry 3:	$\cos(30°)$	$= +0.86603$
	$\cdots$	
Entry 35:	$\cos(350°)$	$= -0.98481$

Assume that we are outputting a new table value at intervals of $T_S = 100$ μs. The sinusoid that is synthesized when we use an offset of 1 displays the following characteristics:

Period of the waveform:

$$N T_S = \frac{36 \text{ samples}}{\text{Cycle}} \times \frac{100 \text{ μs}}{\text{Sample}} = \frac{3600 \text{ μs}}{\text{Cycle}}$$

Frequency of the waveform:

$$\frac{1 \text{ cycle}}{3600 \text{ μs}} = \frac{277.8 \text{ cycles}}{\text{Second}} = 277.8 \text{ Hz}$$

Figure 2–15 illustrates one period of this synthesized waveform.

Now assume that we change the processor instructions to modify the value of the offset. For example, if we use an offset value of 4, then every fourth table entry is addressed by the table pointer. Consequently, the processor outputs only one-quarter of the table entries to cover a full period of the output waveform. The synthesized sinusoid now displays the following timing characteristics:

Period of the waveform:

$$N T_S = \frac{\frac{36}{4} \text{ samples}}{\text{Cycle}} \times \frac{100 \text{ μs}}{\text{Sample}} = \frac{900 \text{ μs}}{\text{Cycle}}$$

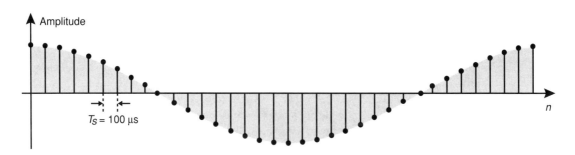

FIGURE 2–15
Outputting every table entry.

Frequency of the waveform:

$$\frac{1 \text{ cycle}}{900 \text{ μs}} = \frac{1111 \text{ cycles}}{\text{Second}} = 1111 \text{ Hz}$$

Figure 2–16 illustrates the new output samples that our processor transfers to the DAC.

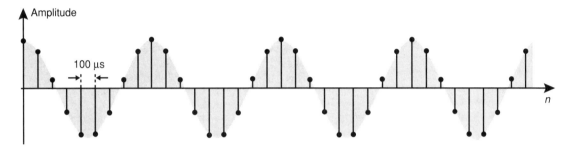

FIGURE 2–16
Outputting every fourth table entry.

An examination of Figure 2–16 indicates that the period of the synthesized waveform is reduced to nine samples. When we increase the value of the offset, the digital processor uses fewer samples to cover a complete sinusoid period. As we increase the value of the offset, the following questions relating to the quality of the synthesized sinusoid may come to mind:

- We use a DAC to generate voltage steps at the output. As we increase the offset value, fewer steps are outputted to cover the sinusoid period. What does this do to the quality of the sinusoid we are trying to synthesize? You will be pleasantly surprised to learn in Section 3.5.3 that the DAC steps are not a problem.
- Because a minimum of two samples is needed to describe the period of any sinusoid, the maximum offset value is one-half the number of entries in the table. What is the *minimum* offset value that we can use? The extent of these limits is addressed in Section 2.3.3.

2.3.2 CONTROLLING THE FREQUENCY OF SYNTHESIZED SINUSOIDS

By now, you understand how the processor uses the offset value and the table pointer to select the table entries that it must send to the DAC. The number of table entries required to cover a full sinusoid period is

$$\text{Number of table entries used in one period} = \frac{\text{Total number of table entries}}{\text{Offset value}}$$

The duration of the synthesized sinusoid period is therefore

$$\text{Sinusoid period} = \text{Number of table entries in one period} \times T_S$$

By inverting the sinusoid's period, we obtain the

$$\text{Synthesized sinusoid frequency} = \frac{\text{Offset value}}{\text{Total number of table entries}} \times \frac{1}{T_S}$$

As long as we use at least two table entries in each period, the offset value may be any number, even if it has a fractional part! The use of fractional numbers in binary computations is as is discussed in Appendix B. Let's arbitrarily choose an offset value of 2.7 to synthesize an output sinusoid. In this case, we calculate the output frequency of the synthesized sinusoid as follows:

$$\frac{2.7}{36} \times \frac{1}{100 \text{ μs}} = 750 \text{ Hz}$$

To generate a sinusoid using a particular offset value, we need to program the processor to compute the location of a new table entry at every sample interval T_S. We perform this computation by instructing the processor to:

1. Add the offset value to the contents of the *running offset* variable.
2. Copy the integer part of the running offset to a table pointer.
3. Copy the table entry that is addressed by the table pointer to the DAC latch.

Figure 2–17 depicts the programming steps required to output each sinusoid sample when a constant offset of 2.7 is used to synthesize a 750 Hz sinusoid. The 36-entry cosine table is stored in memory starting at address location 1000 and extending to include address 1035.

Table 2–2 depicts the processing operations required to generate the address of five of the output samples that will be sent to the DAC. The table pointer must always refer to one of the 36 addresses that correspond to the table entries. When the running offset value overflows to a value greater than/or equal to address 1036, it must wrap around to the start of the table. We wrap around the table by subtracting 36, the number of table entries, from the running offset. Doing this ensures that the running offset always contains values we can round down to an address corresponding to one of the 36 table entries.

The table address, which is held in the table pointer, is the integer part of the running offset. Consequently, when the running offset does not contain an exact integer number, the sinusoid sample that is outputted to the DAC is the table entry referred to by the *integer* part of the running offset. Truncating the fractional part creates an approximation that introduces small deviations in the cosine values required to synthesize an ideal sinusoid.

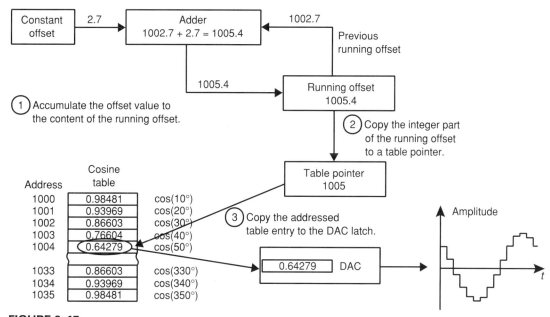

FIGURE 2–17
Mechanics of signal synthesis.

TABLE 2–2
Processing to generate the address of the table entry.

Time = nT_S	Running Offset Calculation	Table Pointer
$0 \times 100 \ \mu s = 0 \ \mu s$	1000	1000
$1 \times 100 \ \mu s = 100 \ \mu s$	$1000 + 2.7 = 1002.7$	1002
$2 \times 100 \ \mu s = 200 \ \mu s$	$1002.7 + 2.7 = 1005.4$	1005
. . .	. . .	. . .
$13 \times 100 \ \mu s = 1300 \ \mu s$	$1032.4 + 2.7 = 1035.1$	1035
$14 \times 100 \ \mu s = 1400 \ \mu s$	$1035.1 + 2.7 = 1037.8$ ≥ 1036 therefore wrap around $1037.8 - 36 = 1001.8$	1001

Figure 2–18 compares an ideal sinusoid with a sinusoid synthesized by extracting entries from a 36-entry table using an offset of 2.7. The small deviations illustrated in Figure 2–18 occur when the running offset contains a fractional part. These deviations create *harmonic distortion*, which occurs when the offset is not a whole integer. The presence of harmonic distortion indicates that the synthesized sinusoid is not ideal and that some of the signal power leaks to frequencies other than the one we are attempting to synthesize. The amount of harmonic distortion is therefore a measure of the amount of power leakage resulting from imperfections in the shape of the synthesized sinusoid.

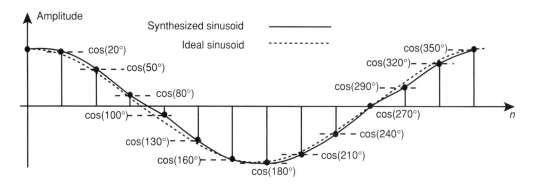

FIGURE 2–18
Sinusoid value is approximated.

2.3.3 CONTROLLING THE HARMONIC DISTORTION

We can control different factors to reduce the amount of harmonic distortion exhibited by the synthesized sinusoid. The first factor that we should consider is the size of the sinusoid table used to synthesize the signal. A larger table means that the running offset uses more integer bits to select the computed sample. Since there are more entries to pick from, the fetched sinusoid value is closer to the ideal value.

Using a larger sinusoid table produces less harmonic distortion.

A second factor that influences the amount of harmonic distortion is the precision with which we encode the value of the table entries. As reviewed in Appendix B, converting a fractional number into a binary number introduces a tolerance problem. To illustrate this, Table 2–3 compares the tolerance that results when we use 8, 16, or 20 bits to encode the table entries as signed numbers. As data in Table 2–3 indicates, using more bits to encode the table entries yields an improved tolerance. This brings the table values closer to ideal cosine values and, since the table contains more accurate sinusoid values, the synthesized samples are closer to an ideal cosine sequence. Consequently, using more bits to encode the table entries reduces the amount of harmonic distortion in the synthesized signal.

**Using more bits to encode the table entries
produces less harmonic distortion.**

Table 2–4 lists the worst-case harmonic distortion levels for tables that contain a different number of 16-bit entries. As an example, consider using a cosine table containing one hundred twenty-eight 16-bit entries. Using the round-down technique, the resulting harmonic distortion is below 0.03% (Table 2–4). This means that less than 0.03% of the total signal power spills to frequencies other than the intended one.

A third factor that we can control is the computational process used to obtain the sinusoid value. When we describe a curve using a sequence of digital values, we provide only a limited number of samples taken along that curve. No matter the number of entries

TABLE 2–3
Expressing a table value using a different number of bits.

Bits Used to Encode the Table Entries	Tolerance
1 sign, 1 integer, and 6 fraction bits	$\pm \dfrac{1}{128}$
1 sign, 1 integer, and 14 bits of precision	$\pm \dfrac{1}{32,768}$
1 sign, 1 integer, and 18 bits of precision	$\pm \dfrac{1}{524,288}$

TABLE 2–4
Table size (16-bit entries) versus harmonic distortion levels.

Table Size (16-bit entries)	Maximum Harmonic Distortion
32	Below 0.4%
64	Below 0.08%
128	Below 0.03%

we use in the table and the number of bits we use to encode the entries, the digitized curve is still undefined between the individual samples. It is possible to reduce the amount of harmonic distortion by estimating the value of the curve *between* samples. Since the curve is undefined between the samples, we can only guess at its value; fortunately, it is possible to take an educated guess at the shape of the curve between the samples. We call this educated guess *interpolation*.

There are many forms of interpolation; the simplest one called *linear interpolation* assumes that the curve is a straight line between each pair of samples. Linear interpolation can yield results that are very close to reality for sinusoids. For example, if enough samples are available, we can interpolate cosine values with great accuracy. Consider the linear interpolation of the cosine curves depicted in Figure 2–19. There we see that increasing the number of samples brings the interpolated cosine closer to a perfect cosine. In practice, using 128 samples provides ample accuracy for most applications.

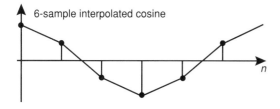
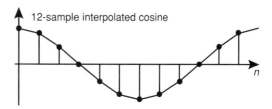

FIGURE 2–19
Linear interpolation of a cosine curve.

Interpolating a value between table entries requires processing fractional binary numbers. Appendix B presents a full review of the important concepts used to manipulate and process fixed-point binary numbers. To implement digital signal processing systems, a person must be thoroughly familiar with the content of Appendix B.

Let's illustrate the linear interpolation technique by using the example of a cosine curve stored in a 128-entry table. In this example, we assume that the table entries are 16-bit Q14 samples. Using this table to synthesize the harmonics of a signal, the processor extracts values from this table at a fixed offset as discussed throughout Section 2.3.2. The processor uses the contents of a running offset pointer to determine which table entry to extract. The linear interpolation technique is different in that it uses the fractional part of the running offset to adjust the table entry value.

Let's assume that we have a 16-bit processor and that the running offset pointer is an *unsigned* number that refers to a location in the 128-entry table. To point to 128 different entries, the pointer requires 7 integer bits; the leftover 9 bits are used to hold the fractional part of the offset:

Unsigned Q9 format of a 128-entry, 16-bit running offset: `IIIIIII.FFFFFFFFF`

Note that the running offset does not need to be a signed number. Let's continue the example by assuming that the running offset presently contains the following arbitrary Q9 value:

Contents of the running offset:

$$0000101.101101000 \xrightarrow{\quad \text{Decimal} \quad} 5.703125$$

The location 5.703125 does not actually exist in the table since entries exist only at whole integer locations. We therefore need to interpolate the value of the cosine at 5.703125 as adopting a value between table entries 5 and 6. As a reference, the values for the Q14 table entries 4, 5, and 6 are provided as hexadecimal (Hex) values:

Entry 4:

$$\cos\left(\frac{4}{128} \times 360°\right) = 0.980785$$

$$\xrightarrow{\quad \text{Q14} \quad} 0.980785 \times 2^{14} \xrightarrow{\quad \text{Hex} \quad} = 3\text{EC5}$$

Entry 5:

$$\cos\left(\frac{5}{128} \times 360°\right) = 0.970031$$

$$\xrightarrow{\quad \text{Q14} \quad} 0.970031 \times 2^{14} \xrightarrow{\quad \text{Hex} \quad} = 3\text{E14}$$

Entry 6:

$$\cos\left(\frac{6}{128} \times 360°\right) = 0.956940$$

$$\xrightarrow{\quad \text{Q14} \quad} 0.956940 \times 2^{14} \xrightarrow{\quad \text{Hex} \quad} = 3\text{D3E}$$

Figure 2–20 illustrates the calculations required to linearly interpolate between locations 5 and 6 to obtain a value for the nonexisting entry at 5.703125. Referring to Figure 2–20, we must program the system so that it executes the following steps:

- Fetch the Q14 table entry at the rounded-up running offset location:

$$\text{6th entry} = \text{3D3E}$$

- Fetch the Q14 table entry at the rounded-down running offset location:

$$\text{5th entry} = \text{3E14}$$

- Find the Q14 slope by subtracting the upper table entry from the lower one:

$$\text{3D3E} - \text{3E14} = \text{FF2A}$$

Note: The slope is negative because the waveform is going down.

- Zero the I-bits of the running offset to isolate the Q9 fractional part:

$$0000000.101101000 \xrightarrow{\text{Hex}} 0168 \xrightarrow{\text{Decimal}} 0.703125$$

- Find the correction amount by multiplying the fraction part of the running offset by the slope:

$$0168 \times \text{FF2A} = \text{FFFED310}$$

Note: This 32-bit answer is a $Q9 \times Q14 = Q23$ number.

- Add the 32-bit Q23 correction to the Q14 entry stored in the fifth table location. To perform this addition, we must align the two numbers in the 32-bit accumulator. We convert the fifth table entry from a 16-bit Q14 format to a 32-bit Q23 format by shifting it left nine times.

$$\text{3E14} \xrightarrow{\text{9 shift left}} 007\text{C}2800$$

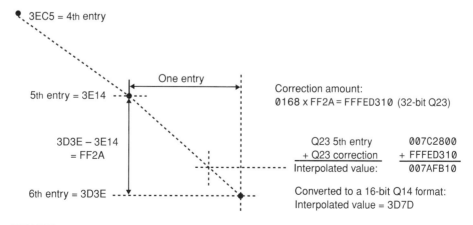

FIGURE 2–20
Processing interpolation.

- Interpolate by adding the correction to the lower table entry:

$$007C2800 + FFFED310 = 007AFB10$$

- Convert the Q23 result to a Q14 value by discarding the nine least significant fractional bits:

$$007AFB10 \xrightarrow{\text{Q23 to Q14}} 3D7D$$

We can check the accuracy of this Q14 result by converting it to a decimal value and then comparing it to the ideal cosine value:

$$\frac{3D7D}{2^{14}} \xrightarrow{\text{Decimal}} 0.960754$$

compared to:

$$\cos\left(\frac{5.703125}{128} \times 360°\right) = 0.961069$$

This yields a percentage error of

$$\frac{0.961069 - 0.960754}{0.961069} \times 100 = \frac{0.0002536}{0.961069} \approx 0.03\%$$

Interpolating therefore takes us very close to the result without storing a large table. The drawback is that interpolating requires processing time. The interpolation processing time can be reduced by precalculating the value of the slope and storing these values in a second table. This eliminates the processing time required to calculate the slope. Table 2–5 lists the worst-case harmonic distortion levels when linear interpolation is used with tables containing 16-bit entries. Compare with Table 2–4.

TABLE 2–5
Interpolated 16-bit values versus harmonic distortion levels.

Table Size (16-bit entries)	Maximum Harmonic Distortion Using Interpolation
32	Below 0.2%
64	Below 0.02%
128	Below 0.0002%

2.3.4 CONTROLLING THE PHASE OF SYNTHESIZED SINUSOIDS

One primary technique used to carry information in data communication applications is *phase control*. It consists of transmitting a sinusoid of fixed frequency and periodically shifting its phase. The information is contained in the amount of phase shift. For example, consider the simple example of a phase control system generating four different phase shifts. Since four (2^2) phase changes are possible, each of the changes can be made to carry 2 bits of information (see Table 2–6).

TABLE 2–6
Phase shifts that carry information.

Amount of Phase Change	Binary Information
45°	0 0
135°	0 1
225°	1 0
315°	1 1

The present phase of a synthesized sample is directly related to the table entry we are outputting to the DAC. Each of the table entries corresponds to a particular phase position along a full period of the sinusoid. Figure 2–21 illustrates the relationship between the phase status and the entries in a 36-entry cosine table.

In Section 2.3.2, we saw that the running offset contains the ideal table entry we should be using at specific discrete-time instants. Figure 2–21 shows that there is a direct relationship between the table entries and the phase. The running offset must therefore correspond to the current phase position of the synthesized sinusoid. Controlling the value of the running offset consequently gives direct control over the phase position of the sinusoid.

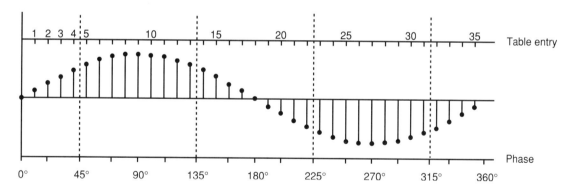

FIGURE 2–21
Phase of a synthesized sinusoid.

For example, we can see (Figure 2–21) that a 45-degree phase shift corresponds to moving forward in the table by 4.5 entries. It follows that if we suddenly increase the contents of the running offset by 4.5, the sinusoid's phase jumps forward by 45 degrees. Note that the amount by which the phase jumps is independent of the current sinusoid phase position.

The amount by which the running offset changes corresponds to the desired phase jump.

Producing periodic phase jumps in the synthesized sinusoid is therefore a matter of changing the content of the running offset at fixed time intervals. Figure 2–22 illustrates a synthesized sinusoid having its phase changed at periodic intervals to carry data.

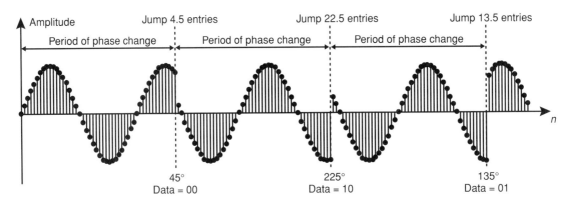

FIGURE 2–22
Periodic phase changes.

2.3.5 CONTROLLING THE AMPLITUDE OF SYNTHESIZED SINUSOIDS

Amplitude control is another technique used to shape the characteristics of a synthesized sinusoid. Applications that use this technique include amplitude modulation (used in radio applications), modems (quadrature amplitude modulation technique), and the synthesis of intricate signals, as we will see in Section 2–4.

We can control the amplitude of the synthesized sinusoid by scaling the extracted table entries before we send them to the DAC. In practice, we do this by instructing the processor to multiply the extracted table entry by a *magnitude factor*. For example, if the processor multiplies all the extracted values by a magnitude factor of 0.5, the values outputted to the DAC are reduced to one-half of their original value. Consequently, the synthesized sinusoid will be generated with one-half of its full amplitude. Figure 2–23 illustrates the extra processing step required to control the amplitude using the value of a magnitude factor.

Multiplying by fractional numbers requires using some of the binary manipulation techniques outlined in Appendix B. For example, assume that we want to synthesize a signal that contains up to 20 harmonics. Each harmonic sample consists of a cosine table entry multiplied by a magnitude factor:

$$\text{Harmonic sample} = \text{Table entry} \times \text{Magnitude factor}$$

Remember that the Q format of the table entry is one of the factors that determine the amount of harmonic distortion. Similarly, the Q format of the magnitude factor determines the precision with which we can control the amplitude.

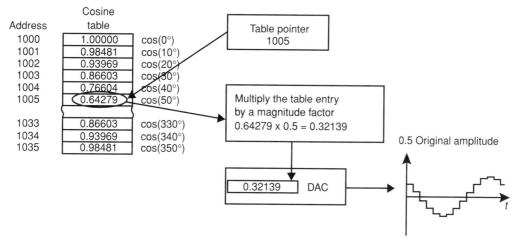

FIGURE 2–23
Controlling the amplitude.

Let's arbitrarily assume that we are using a cosine table (other types of tables are possible as shown in Section 2.4.4) and that the value of the magnitude factors cannot exceed a value of 10. This immediately sets the following constraints on some of the bit-formats:

- The cosine table entries require at least one S-bit (sign) and one I-bit (integer) to accommodate the cosine range of ±1.
- The magnitude factors are unsigned and require four I-bits to accommodate numbers ranging between 0 and 10.

Let's assume that we are using a 16-bit processor. In this case, the cosine table entries and the magnitude factors are 16-bit entities. The processing requires the multiplication of 16-bit numbers; this means that the processor product register, arithmetic logic unit, and accumulator must include 32 bits to multiply and accumulate the results. The processor computes the value of the harmonic samples by multiplying cosine table entries by magnitude factors. The Q format of a product (see Appendix B) is the sum of the Q formats of the multiplier and multiplicand. Consequently, the Q format of the harmonic samples is as follows:

$$
\begin{array}{r}
\text{Q format of table entry} \\
+ \quad \underline{\text{Q format of magnitude factor}} \\
\text{Q format of harmonic sample}
\end{array}
$$

In this example, the magnitude factors are limited to a value of 10 and the largest cosine table entry has a value of 1. The largest computed harmonic sample in the product register therefore has a maximum value of $10 \times 1 = 10$. Since the complete signal calls for accumulating 20 such harmonics, the signal samples have a maximum value of $20 \times 10 = 200$. These maximum values drive the choice for the Q factors of the table entries and magnitude factors.

Let's examine the case in which the processor directly accumulates the products. The 32-bit accumulator requires at least one S-bit and eight I-bits to accommodate the maximum sample value of 200. Consequently, every accumulated signal sample has the following Q23 format:

Signal sample in accumulator: SIII IIII IFFF FFFF FFFF FFFF FFFF FFFF

Each signal sample in the accumulator is the result of directly accumulating 20 products:

$$\text{Signal sample} = \text{Harmonic sample 1} + \text{Harmonic sample 2}$$
$$+ \ldots + \text{Harmonic sample 20}$$

Since the processor computes the signal sample by *directly adding* the harmonic samples, the Q format of the signal sample is the same as that of the harmonic samples (see Appendix B). Consequently, both the accumulator and the product register hold a Q23 format. This puts additional constraints on the Q format of the cosine table entries and on the magnitude factors. Remember that the 16-bit cosine table entries cannot exceed a Q14 format (±1 range requires at least one S-bit and one I-bit) and the unsigned 16-bit magnitude factors cannot exceed a Q12 format (value of 10 requires at least four I-bits). This yields four possibilities that accommodate the required Q23 format:

Table Entries		Magnitude factors		Product and Accumulator Samples
Q14	×	Q9	=	Q23 (least harmonic distortion)
Q13	×	Q10	=	Q23
Q12	×	Q11	=	Q23
Q11	×	Q12	=	Q23 (most precise magnitude control)

Any of these four possibilities works; the decision of which one to select depends on the harmonic distortion and magnitude control that are acceptable to the application.

The constraints on the cosine table entries and on the magnitude factors are imposed by their individual range requirements. In the last example, we considered these requirements separately; however, it is possible to combine these requirements to improve the performance of the system. This is possible because the processing always requires that we compute the *product* of these two numbers.

For example, the cosine table entries cover the range of ±1 and, because of this, we had to use one I-bit to accommodate this range. Note that a fixed-point number that contains one I-bit actually covers the $\pm 2^-$ range (see Appendix B); therefore, approximately one-half of the range is wasted. In the case of the magnitude factors, four I-bits are required to accommodate the ±10 range, but again, this is a waste since four integers can cover the much broader range of $\pm 16^-$.

To eliminate some of this waste, we can scale one of the numbers *down* by a certain factor and scale the other number *up* by exactly the same factor. Since the two numbers are being multiplied, the scaling cancels and the result is unchanged:

$$\text{Output sample} = \left[\text{Table entry} \times \frac{1}{\text{Scaling}} \right] \times \left[\text{Magnitude factor} \times \text{Scaling} \right]$$

$$= \text{Table entry} \times \text{Magnitude factor}$$

The advantage lies in the fact that this decreases the range of one number while it increases the range of the other. For example, we can arbitrarily use a factor of 0.75, which yields the following results:

New scaled magnitude factor range: $10 \times \dfrac{1}{0.75} = 13.3$

New scaled cosine table entries range: $\pm 1 \times 0.75 = \pm 0.75$

The new scaled magnitude factor still requires four I-bits, but the new scaled cosine entries no longer require an integer bit. This opens the possibility of using a Q15 format for the cosine table entries (and Q8 for the magnitude factor), which yields less harmonic distortion.

Alternately, we could reverse the scaling to yield

New magnitude factor entries range: $10 \times 0.75 = 7.5$

New cosine table range: $\pm 1 \times \dfrac{1}{0.75} = \pm 1.33$

In this case, the magnitude factors may now be coded with a Q13 format (and Q10 for the cosine entries), which yields better amplitude control.

2.4 SYNTHESIS OF ELABORATE SIGNALS

We can use intricate signals in practice to synthesize music, to generate telephone-dialing tones, to synthesize voice or sounds, and so on. As Section 2.2.2 indicated, when we developed the synthesis equation, intricate periodic signals actually consist of a mix of cosine harmonics characterized by their amplitude and phase. The synthesis equation (2–4) for a periodic signal containing N samples is repeated here for convenience:

$$y(n) = \sum_{k=0}^{N/2} A_k \cos\left(2\pi \left[k \times \frac{1}{N\,T_S} \right] n\,T_S + \phi_k \right)$$

2.4.1 INTRODUCING THE NORMALIZED SYNTHESIS EQUATION

Examining the synthesis equation, we notice that the sampling period T_S cancels out to yield the following version, which is independent of the sample interval:

$$y(n) = \sum_{k=0}^{N/2} A_k \cos\left(2\pi \left[k \times \frac{1}{N} \right] n + \phi_k \right) \tag{2-5}$$

This means that the discrete-time samples $y(n)$, making up the signal, remain the same no matter how fast we choose to output them. Since the signal samples are independent of the sample rate, we call this equation the *normalized synthesis equation,* which uses a number of samples to describe the *shape* of the required signal.

The sample rate is a function of the choice of the DAC clocking frequency. If we increase the sampling frequency (make the DAC clock faster), the output samples of the signal $y(n)$ are delivered faster, resulting in a signal of higher frequency. If we reduce the sampling frequency, the output samples are generated at a slower rate and the signal contains lower frequency components.

Changing the output sample rate effectively changes the *frequency* of the synthesized signal, not its *shape* or *harmonic content*.

This has a practical application for music synthesizers if we need to adjust the pitch of the sounds we are synthesizing.

2.4.2   COMPUTING THE SIGNAL SAMPLES

The number of sinusoids to be mixed depends on the signal we are trying to build. To build a specific periodic signal, we need a description of its harmonic constituents. This is basically a recipe that describes the amplitude and phase of each of the signal harmonic ingredients. Once we know the recipe, we can build the complete signal by individually synthesizing the required harmonics and by mixing the results. For example, consider a signal described by the following recipe:

$$A_0 = 0.0 \qquad A_1 = 1.0 \qquad A_2 = 0.8 \qquad A_3 = 0.7$$
$$\phi_0 = 0° \qquad \phi_1 = 0° \qquad \phi_2 = 20° \qquad \phi_3 = 120°$$

When we examine the signal recipe, we notice that it has no DC component since $A_0 = 0.0$. To build this signal, we need to synthesize the fundamental, the second, and the third harmonics. Because this signal requires only the three first harmonics, it has a minimum period of six samples (the highest harmonic requires a minimum of two samples; the fundamental requires three times as many). The signal period must be at least six samples long, but it can be made longer. Making the period longer gives the practical advantage of being able to add extra higher harmonics to create a more elaborate signal.

For example, assume that we arbitrarily choose to make the signal period 36 samples long. In this case, five out of six samples are superfluous; nevertheless, it provides room to expand the signal to a maximum of 18 harmonics if this need occurs.

Let's examine the samples that belong to the second harmonic ($k = 2$). Since $A_2 = 0.8$ and $\phi_2 = 20°$, the samples are represented by:

$$y_2(n) = 0.8 \cos\left(360°\left[2 \times \frac{1}{36}\right] n + 20°\right) \text{ (in degrees)}$$

$$= 0.8 \cos\left(2\pi\left[2 \times \frac{1}{36}\right] n + \frac{20°}{180°}\pi\right) \text{ (in radians)}$$

We can evaluate the value of some of the second harmonic samples at different discrete instants by substituting values for the discrete-time variable $n = 0, 1, 2, \ldots$:

$$y_2(0) = 0.8 \cos\left(360°\left[2 \times \frac{1}{36}\right] 0 + 20°\right) = 0.8 \cos(20°) = 0.75175$$

$$y_2(1) = 0.8 \cos\left(360°\left[2 \times \frac{1}{36}\right] 1 + 20°\right) = 0.8 \cos(40°) = 0.61284$$

$$y_2(2) = 0.8 \cos\left(360°\left[2 \times \frac{1}{36}\right] 2 + 20°\right) = 0.8 \cos(60°) = 0.40000$$

Doing this on a signal processor requires a very fast coprocessor to evaluate the required cosine values. It should be obvious that storing a cosine table speeds up the processing and substantially reduces the cost and the complexity of the system by avoiding the need for a coprocessor.

No matter which technique was selected to evaluate the cosine values, the resulting signal and its three harmonics components are individually graphed in Figure 2–24.

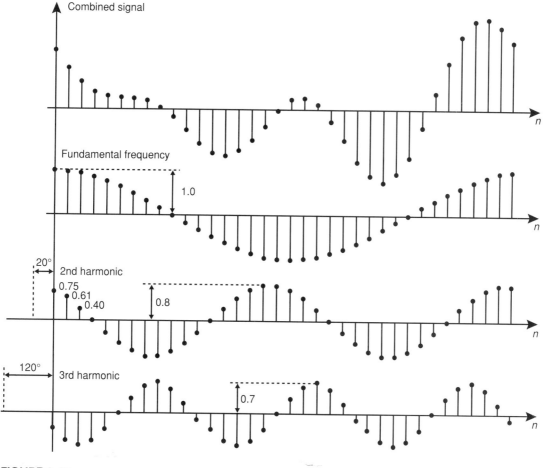

FIGURE 2–24
Mixing harmonics to build a signal.

The values of A_k and ϕ_k used in any signal recipe call for precise control of the amplitude and of the phase shift of the different harmonics contained in a signal. We can set the amount of phase shift by selecting the initial value we store in the running offset. For example, Figure 2–25 illustrates the sinusoids that result when the running offset is initialized with two different initial conditions.

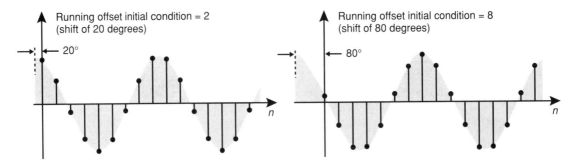

FIGURE 2–25
Controlling the phase with the running offset initial value.

The fact that intricate signals consist of a combination of many sinusoid harmonics provides the opportunity to simplify the signal synthesis approach. Breaking a large complex problem into a number of smaller and easier-to-manage problems has always been the favored engineering approach. As we will soon see, we can synthesize any elaborate signal by generating its individual harmonic components separately.

2.4.3   SEGMENTING THE PROCESSING INTO SECTIONS

When we output a synthesized signal to a DAC, we are actually outputting a sequence of numbers that represent a combination of sinusoids. To keep the hardware and the software modular, it is necessary to break the total required processing into smaller separate subsections. Use of this approach causes each subsection to contribute the samples of a single harmonic to the overall output. We generate the final signal by combining the outputs of the individual subsections. This combination may be done using either serial or parallel system architectures. Figures 2–26 and 2–27 illustrate how to regroup the system subsections using either a parallel or a cascaded serial architecture.

In the parallel architecture illustrated in Figure 2–26, the individual subsections *simultaneously* synthesize harmonic components, and the adder effectively combines all results to produce the complete signal. This method has the advantage of being fast since the many processors work in parallel to perform the required processing. The disadvantage associated with this approach is the high hardware cost of this multiprocessor architecture.

In the cascaded processing method illustrated in Figure 2–27, the processing proceeds one subsection at a time and the output signal slowly accumulates the required har-

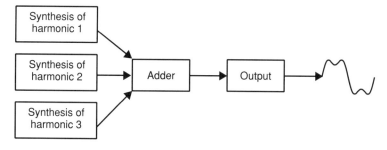

FIGURE 2–26
Parallel processing architecture.

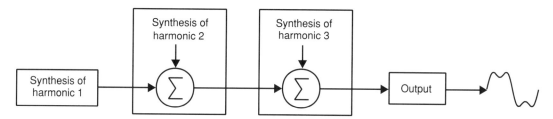

FIGURE 2–27
Cascaded serial processing architecture.

monics as each subsection completes its job. Although it is much slower than using a parallel architecture, the serial approach is by far the more popular in practical applications because of its low cost.

2.4.4 STORING AN INTRICATE SIGNAL INTO THE TABLE

If a serial architecture is used, the processor must repeat the computing required to generate a specific harmonic sample in every subsection. Each subsection therefore consumes a part of the total interval T_S, which the processor uses to compute the complete signal sample. The more harmonics the signal contains, the less time each subsection has to compute a specific harmonic sample. For example, consider that the samples of a signal are being outputted every $T_S = 125$ μs and that the signal consists of a combination of 50 harmonics. The maximum computing time for the subprocessing of one of the harmonics is

$$\text{Time to process one harmonic} = \frac{\text{Sample interval}}{\text{Number of harmonics}}$$

$$= \frac{125\ \mu s}{50} = 2.5\ \mu s$$

If we have a slow processor limited to an instruction-processing rate of 10 instructions/μs, this processor must compute each of the subsections using no more than

$$
\begin{aligned}
\text{Number of instructions} &= \text{Available computing time} \times \text{Instruction processing rate} \\
&= 2.5 \ \mu s \times 10 \ \text{instructions}/\mu s \\
&= 25 \ \text{instructions}
\end{aligned}
$$

Obviously, increasing the number of harmonics decreases the overall processing time, which reduces the number of instructions that can be included in a subsection. If the required number of harmonics is pushing the processor beyond its processing limits, we can use a parallel processing architecture (which is expensive) or store the entire signal in a table using the technique described in the following section.

Setting Up a Signal Table

An alternative to synthesizing the harmonics of a signal separately is to store the complete signal in the table. In other words, the table contains all the harmonics, which are already premixed with the right phases and amplitudes. For example, let's use the signal described by the following recipe:

$$
\begin{aligned}
A_0 &= 0.0 & A_1 &= 1.0 & A_2 &= 0.8 & A_3 &= 0.7 \\
\phi_0 &= 0° & \phi_1 &= 0° & \phi_2 &= 20° & \phi_3 &= 120°
\end{aligned}
$$

This signal corresponds to the following equation:

$$
\begin{aligned}
y(n) = \ &\cos\left(2\pi\left[1\times\frac{1}{N}\right]n\right) + 0.8\cos\left(2\pi\left[2\times\frac{1}{N}\right]n + \frac{20°}{180°}\pi\right) \\
&+ 0.7\cos\left(2\pi\left[3\times\frac{1}{N}\right]n + \frac{120°}{180°}\pi\right)
\end{aligned}
$$

Let's create a table that represents this signal. The first step is to select a value for N, the length of the table. We know that each harmonic must contain at least two table entries. Since this signal contains three harmonics, it must contain at least six entries. Unfortunately, as was shown in Section 2.3.3, using only two entries results in a very large harmonic distortion. We must therefore use a much larger number of entries to keep the harmonic distortion within acceptable levels. For example, according to the data in Table 2–4, using thirty-two 16-bit entries results in less than 0.4% of harmonic distortion. If we arbitrarily choose to use thirty-two entries to encode the third harmonic (the highest frequency), encoding the fundamental (the lowest frequency) requires three times that many entries. In this case, the complete table requires $N = 96$ entries:

$$
\begin{aligned}
y(n) = \ &\cos\left(2\pi\left[\frac{1}{96}\right]n\right) + 0.8\cos\left(2\pi\left[\frac{2}{96}\right]n + \frac{20°}{180°}\pi\right) + \\
&0.7\cos\left(2\pi\left[\frac{3}{96}\right]n + \frac{120°}{180°}\pi\right)
\end{aligned}
$$

In this case, the table entries are as follows:

$y(0) = 1.402$	$y(4) = 0.804$	. . .	$y(88) = 2.085$	$y(92) = 1.935$
$y(1) = 1.246$	$y(5) = 0.678$	. . .	$y(89) = 2.098$	$y(93) = 1.826$
$y(2) = 1.091$	$y(6) = 0.568$	. . .	$y(90) = 2.075$	$y(94) = 1.697$
$y(3) = 0.943$	$y(7) = 0.475$	. . .	$y(91) = 2.020$	$y(95) = 1.554$

When we periodically fetch entries from this table at a fixed offset, we synthesize all three harmonics *simultaneously*. Each sample contains the precomputed contribution from all three harmonics. This technique therefore saves much processing time at the cost of using a longer table, which consumes a large amount of the very limited memory of our digital signal processor. Note that the table size grows as we increase the number of signal harmonics or if we aim for lower harmonic distortion levels.

Let's consider another example in which we synthesize a particular signal that requires 18 harmonics (chosen arbitrarily). Assume that we must use 128 samples to keep the harmonic distortion within acceptable levels for this application. In this case, a full period of the fundamental occupies $18 \times 128 = 2304$ samples; since the fundamental is the lowest harmonic, 2304 is the required signal table size. Generally, the number of entries required in the table may be calculated as follows:

$$\frac{\text{Number of harmonics required in a signal} \times \text{Number of samples required to meet the harmonic distortion specification}}{\text{Number of entries required in the signal table}}$$

Obviously, we need to allocate a significant amount of memory to store the signal table. The memory requirements grow significantly for applications that call for the synthesis of a large number of different signals.

Generating Different Signals Using Signal Tables

Some applications call for the synthesis of a large quantity of different signals. Examples of such applications include the synthesis of voice and music. Consider, for example, a piano synthesizer that must contain a separate table for each of the piano keys. We can build the tables that describe different signals by sampling real signals (such as a piano key), or we can synthesize a particular signal shape by mixing harmonics. No matter which technique we use to build the samples of the signal tables, it is important to understand that the size of the signal table and the number of harmonics it contains ultimately determine the quality of the synthesized signal.

Once all required signals are stored in their respective tables, we can "play" with this information to

- Change the frequency, phase, and amplitude of each signal (for example, to obtain different musical notes, sounds, etc.).
- Mix many signals (for example, to create musical chords).
- Change the contents of the tables to create new signals (for example, to invent new musical sounds).

Regrouping all of these signals into a signal output can be done using either the serial or the parallel architecture discussed in Section 2.4.3.

SUMMARY

- Digital signals consist of a sequence of numbers.
- A DAC generates voltage steps that correspond to the numbers contained in the digital signal sequence at periodic intervals of T_S.
- A digital signal is valid only at specific instants defined by the discrete-time variable n, which must be an integer.
- The processing *performance* required of the processor(s) depends on the time interval between samples and on the complexity of the processing.
- Any real and periodic signal can be built by adding sinusoids that are harmonics of the signal's fundamental frequency.
- The synthesis equation describes a signal in terms of the individual harmonic components. Each of the sinusoid harmonic components is characterized by a particular frequency, phase, and amplitude.
- The highest harmonic component contained in a discrete-time signal cannot exceed a frequency of one-half the output sample rate: $f_S/2$.
- If the period of a signal is N samples, the maximum number of harmonics that can be contained in that signal is $N/2$.
- Sinusoids can be synthesized using a cosine table stored in the digital signal processor memory.
- The harmonic distortion may be reduced by using a larger sinusoid table, by using more bits to encode the table entries, and by interpolating between table entries.
- Processing the cosine table entries allows us to control the frequency, phase, and amplitude characteristics of the sinusoid.
- Intricate signals can be built by individually synthesizing the required harmonics and by combining the results.
- Serial processing uses a single processor to synthesize the signal harmonics, one after the other, while accumulating the individual results until all harmonics have been generated.
- Parallel processing requires the use of a number of processors to compute the different signal harmonics simultaneously and then to combine the results.
- We can reduce the amount of processing required to build an intricate signal by including all the signal harmonics in a single table, which requires much memory.

PRACTICE QUESTIONS

2-1. Why can digital signal processors not handle analog signals directly?

2-2. How many distinct voltage levels can a 14-bit DAC output?

2-3. A DAC latches and converts a new sinusoid sample every 125 μs. What is the absolute maximum frequency of a sinusoid produced by this DAC?

2-4. A digital signal processor requires 4 μs to execute all instructions necessary to extract a sinusoid sample from a cosine table. If the output sample rate is 8 kHz, how many sinusoids can this processor mix?

2-5. A digital signal processor extracts values from a 128-entry cosine table to generate a new sinusoid sample every 125 μs. At what offset should the table entries be extracted to generate a 2.32 kHz sinusoid?

2-6. A digital signal processor extracts values from a 128-entry cosine table to generate a 2.32 kHz sinusoid sample every 125 μs. What must the processor do to generate a sudden +22° phase change in the output sinusoid?

2-7. A frequency synthesizer extracts values from a 256-entry cosine table and transfers them to a DAC to generate tones. The samples are produced at a rate of 20 kHz, and the processor extracts entries at a fixed offset of 6.8. What is the fundamental frequency of the sinusoid at the output of the DAC?

2-8. A digital signal processor has a 128-entry cosine table stored in its memory starting at location 3400. What is the value stored at memory location 3500?

2-9. A digital signal processor has a 128-entry cosine table stored in its memory starting at location 3400. The running offset presently contains a value of 42.4, and an offset value of 29.2 is used to generate a sinusoid sample every 10 μs.

  (a) What is the frequency of the sinusoid being generated?

  (b) What is the present phase of the synthesized sinusoid?

  (c) Plot the next eight samples of this synthesized sinusoid.

2-10. A cosine table is to be stored in 16-bit memory locations as 64 signed Q14 entries.

  (a) What are the hexadecimal values of table entries 21 and 22?

  (b) Use linear interpolation to determine the cosine sample value when the running offset pointer accesses entry 21.7.

2-11. A discrete-time signal has a period of 42 samples, and the signal is being generated at a rate of 8 k-samples/second.

  (a) What is the frequency of the seventh harmonic of this periodic signal?

  (b) What is the frequency of the highest harmonic that this periodic signal could contain?

2-12. A total of 4 μs of processing is required to extract a sinusoid sample from a cosine table. If 8 k-samples/second must be sent to the DAC, is it possible to generate a *periodic* signal that requires mixing 40 harmonics? Explain.

2-13. A periodic discrete-time signal is generated at a rate of 8 k-samples/second. If this signal requires the mixing of 12 harmonics, what is the highest possible fundamental frequency of the signal?

2-14. A periodic signal contains four harmonic sinusoids that have the following characteristics:

Fundamental	Harmonic 2	Harmonic 3	Harmonic 4
$A_1 = 0.3$	$A_2 = 1.2$	$A_3 = 0.7$	$A_4 = 0.2$
$\phi_1 = 0°$	$\phi_2 = 0°$	$\phi_3 = 32°$	$\phi_4 = 120°$

This signal is generated by extracting 8 k-samples/second from a 128-entry cosine table.

(a) If the fundamental frequency is 500 Hz, what are the frequencies of the other harmonics?

(b) What is the offset value used to generate each harmonic?

(c) How many running offset pointer(s) are required to program the synthesis of this signal?

(d) Find the initial value of each running offset pointer used to generate the harmonics?

(e) What is the value of the first sample that the processor will be sending to the DAC?

3

PROPERTIES OF DIGITAL SIGNALS

Digital signals can come from different sources. As noted in Chapter 2, one way is to create the digital signals inside the processor by synthesizing them from a sinusoid table. Another way is to convert existing analog signals to a digitized format. This chapter begins by showing that an analog-to-digital converter (ADC) can be used to convert an analog signal to a digital format.

Translating a continuous-time signal to discrete time introduces fundamental effects on the signal. This chapter concentrates on investigating the fundamental properties that separate continuous-time signals from discrete-time signals.

The investigation begins by discussing how to represent sinusoids in the time domain and in the frequency domain. We follow by investigating and comparing the periodicity properties of continuous- and discrete-time sinusoids. This leads to the discovery that continuous-time and discrete-time signals have very different frequency contents.

Further investigation reveals that the digitization process results in effects that have both an advantage and a disadvantage. Through a number of examples, we learn that the key to controlling these effects is proper selection of the sampling rate. Playing with the sampling rate, we observe that DSP can handle only a limited range of frequencies. We develop the use of an antialiasing filter to maximize the size of this useful range of frequencies.

Inspecting the frequency spectrum of a digitized signal, we recognize that a reconstruction filter allows the conversion of digital signal back to the analog format.

3.1 ACQUIRING DIGITAL SIGNALS

As discussed in Section 2.1, the signals that surround us exist as continuous analog signals. To process these signals with a digital processor, it is necessary to convert the continuous signals to a digital format. We use an ADC to perform this conversion.

A typical ADC periodically samples the amplitude of its analog input and converts these samples into a sequence of digital numbers. During the conversion process, the value of the analog sample is kept constant by a sample-and-hold circuit, which can be either part of or separate from the ADC. Figure 3–1 illustrates this process.

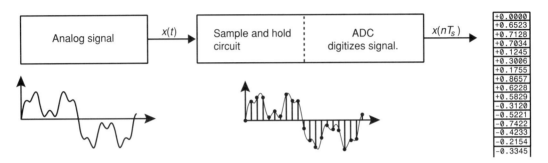

FIGURE 3–1
ADC sampling process.

Since the ADC takes samples only at periodic intervals of T_S, the value of the signal between the samples is completely lost. This means that the resulting sequence of digital numbers describes the signal only at the discrete-time instants nT_S (see Section 2.1.5 for a description of discrete time).

Sampling an analog signal $x(t)$ at periodic intervals of T_S yields a discrete-time signal $x(nT_S)$.

The following are important questions:

- How much information is lost in the sampling process?
- Can we completely recover the analog signal from the sequence of samples?

To answer these questions, we must look into the nature of the sinusoids from which the signals are built.

3.2 REPRESENTING SINUSOIDS

To understand the following sections, the reader is encouraged to become familiar with the content of Appendix A, which reviews vectorial representation in detail. This book has intentionally broken from the standard pure mathematical presentation normally used to explain DSP principles. Although some mathematics is inevitable, all important results are illustrated graphically using vectors. This has the advantage of giving the reader a visual representation of signals, which helps in acquiring a "feel" for the nature of signals even if the reader does not fully understand the mathematics. It also puts the material within the reach of almost any audience who has had exposure to graphs and vectorial representations.

Mathematicians have known for a long time that sinusoids can be broken into "elementary" parts. Euler showed that both the periodic sine and cosine functions consist of a pair of complex exponentials:

$$\sin(\omega t) = \frac{e^{+j\omega t} - e^{-j\omega t}}{2j} \quad \text{and} \quad \cos(\omega t) = \frac{e^{+j\omega t} + e^{-j\omega t}}{2}$$

The analysis of real signals requires that the signal be broken into sinusoid components and that each sinusoid be broken further into a pair of complex exponentials. Using this approach, the analysis of intricate real signals reduces itself to the analysis of pairs of time-varying complex exponentials.

3.2.1   TIME DOMAIN REPRESENTATION

We can examine the oscillations of electrical waveforms using a device called an *oscilloscope*. This device displays the amplitude fluctuations of the waveform as time increases. Figure 3–2 shows an oscilloscope's display of two cosine waveforms. The cosines displayed in Figure 3–2 are real sinusoids that contain no imaginary part.

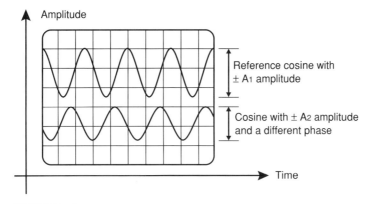

FIGURE 3–2
The time-varying amplitude of cosine waveforms.

It takes three parameters to define a sinusoid:
1. The *period,* which defines the amount of time required to go through one oscillation.
2. The *amplitude,* which defines the maximum and minimum of the oscillation.
3. The *phase,* which defines the starting position of the oscillation.

Euler also showed that we can build any *real* sinusoid by combining the following conjugate complex exponentials:

$$A e^{+j\omega t} = A\cos(\omega t) + j\,A\sin(\omega t)$$
$$A e^{-j\omega t} = A\cos(\omega t) - j\,A\sin(\omega t)$$

Since they are conjugates of each other, when we combine these two complex exponentials by adding them, the imaginary parts cancel to yield a pure real cosine. The complex exponentials that correspond to the elementary parts of a periodic cosine are time varying since they include the time variable t in the exponent. In the continuous-time domain, as time increases from $t = 0$, the value of the exponent $j\omega t$ keeps increasing monotonously. If $A = 1$, the complex exponential $e^{+j\omega t}$ is actually a vector of length 1 and its phase angle is the ωt part of the argument. This means that the phase angle keeps increasing as time increases. This results in a periodic rotation of the vector as the phase angle increases through the full circle values of 2π, 4π, 6π, and so on.

Since the rotating complex exponential takes on imaginary values, an oscilloscope cannot be used to display it. An Argand diagram is used to graph imaginary values, and the complex exponential $e^{+j\omega t}$ corresponds to a vector that rotates continuously about the origin. Figure 3–3 illustrates the rotating vector.

FIGURE 3–3
$e^{+j\omega t}$ time-varying complex exponential.

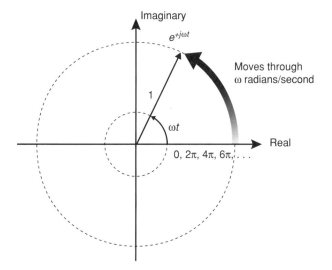

If we want to picture the complex exponential parts of a real cosine, we need to plot two rotating vectors. Since the two complex exponentials that make up a cosine are conjugates of each other, one has a positive exponent and the other has a negative exponent. This means that a real cosine consists of two vectors that rotate in exactly opposite directions. Figure 3–4 illustrates the two vectors as they rotate in opposite directions.

When inspecting Figure 3–4, you should try to visualize the two vectors as rotating along the dotted line of the circle. The two vectors always stay exactly opposite each other as they periodically go through all the possible positions along the dotted circle.

FIGURE 3–4
Two components of a real cosine.

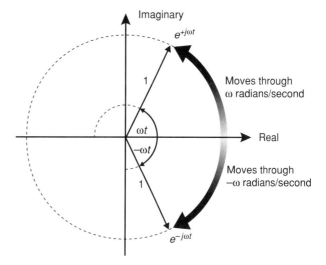

3.2.2 FREQUENCY DOMAIN REPRESENTATION

The vector associated with $e^{-j\omega t}$ can be rewritten as $e^{j(-\omega)t}$, where $-\omega$ corresponds to a *negative* frequency. The concept of a negative frequency may seem weird at first, but—as we will soon see—it will help us to visualize real signals in the frequency domain. Picturing sinusoids in various ways will help us to grasp the nature of this fundamental waveform.

We have seen two ways to represent a real sinusoidal in the time domain:

1. As a waveform that periodically repeats the same amplitude oscillation.
2. As two conjugate vectors that continuously rotate through the real/imaginary plane of an Argand diagram.

Both representations of a real sinusoid result in a picture that constantly changes as the time variable t increases in value. This has the unfortunate consequence of creating a depiction that is difficult to follow. To simplify the representation, we must illustrate the sinusoidal waveforms in a domain that produces a nice static (nonmoving) picture that we can easily inspect.

Let's develop an alternative to the time-domain representation of a sinusoid. To define a sinusoid, we need a representation that illustrates its frequency, amplitude, and phase. If we use frequency as a reference, we can use two graphs to plot the amplitude and the phase as functions of frequency.

As a first example, let's produce a frequency domain picture for a real periodic sinusoid that has an amplitude of $\pm A_1$, oscillates at ω_1 radians/second, and is phase shifted by ϕ_1 radians:

$$x_1(t) = A_1\cos(\omega_1 t + \phi_1)$$

Euler showed that this function can be broken into the following pair of conjugate complex exponentials:

$$x_1(t) = A_1\cos\left(\omega_1 t + \phi_1\right) = \frac{A_1 e^{+j(\omega_1 t + \phi_1)} + A_1 e^{-j(\omega_1 t + \phi_1)}}{2}$$

This real sinusoid contains two conjugate complex exponentials oscillating at angular frequencies of $+\omega_1$ and $-\omega_1$. These complex exponentials have a magnitude of A_1 and are phase shifted by respective amounts of $+\phi_1$ and $-\phi_1$.

Figure 3–5 illustrates two graphs that provide a general picture of the magnitude and phase of the two conjugate complex exponentials as a function of the angular frequency ω. Note the standard notation used in Figure 3–5. Since the signal is a function of ω, we say that it is a *frequency-domain representation* of the signal. We use capital letters to annotate a signal that is expressed as a function of frequency $X(\omega)$:

$$\text{Time domain} \quad \leftrightarrow \quad \text{Frequency domain}$$
$$x(t) \quad \leftrightarrow \quad X(\omega)$$

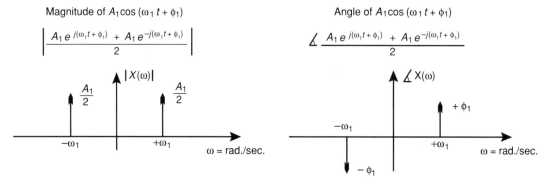

FIGURE 3–5
Frequency domain representation of $A_1\cos(\omega_1 t + \phi_1)$.

Let's define a complex exponential C_1 that represents both the magnitude and the phase of the complex exponential:

$$C_1 = \frac{A_1}{2}e^{+j\phi_1} \quad \text{and} \quad \overline{C_1} = \frac{A_1}{2}e^{-j\phi_1}$$

Substituting C_1 in Euler's identity, we rewrite the cosine as

$$x_1(t) = A_1 \cos(\omega_1 + \phi_1) = \frac{A_1}{2} e^{+j\phi_1} \times e^{+j\omega_1 t} + \frac{A_1}{2} e^{-j\phi_1} \times e^{-j\omega_1 t}$$

$$= C_1 e^{+j\omega_1 t} + \overline{C}_1 e^{-j\omega_1 t}$$

Using C_1 allows us to reduce the frequency-domain representation to a single graph as illustrated in Figure 3–6.

FIGURE 3–6
Single graph for a complete frequency domain representation.

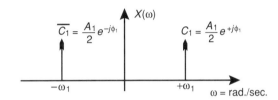

As a second example, consider the partial synthesis of a 1000 radians/second square wave that uses only three harmonics to approximate:

$$\text{Square wave approximation} = x(t) = \cos(1000\,t) + \frac{1}{3}\cos(3000\,t) + \frac{1}{5}\cos(5000\,t)$$

When we break $x(t)$ into its complex exponential constituents, we find three pairs of conjugate complex exponentials. These six complex exponentials correspond to six spikes in the frequency domain. Let's use C_1, C_3, and C_5 to describe the magnitude and phase characteristics of these three harmonics. Figure 3–7 illustrates a time-domain representation and a frequency-domain representation for comparison.

Note in Figure 3–7 that the frequency axis is represented using two types of units. To convert from one unit to the other, we remember that there are 2π radians in each cycle:

$$2\pi \times \text{Cycles/s} = \text{Radians/s} \qquad \frac{\text{Radians/s}}{2\pi} = \text{Cycles/s}$$

The cycles/second units are also called hertz (as noted earlier, abbreviated Hz) after Heinrich Hertz (1857–1894), who built resonators and oscillators and who linked electromagnetic waves to the propagation of light.

As a third example, let's consider the representation of the human voice. Human beings can produce a wide variety of sounds. We know that these sounds are signals that consist of a mix of sinusoids of different frequencies. When a source is able to generate a signal that covers a range of different frequencies, there is too much information to represent the infinite number of frequencies contained in the range. In this case, we show how the signal power is distributed over a range of frequencies by drawing the power *spectrum*. The range of frequencies over

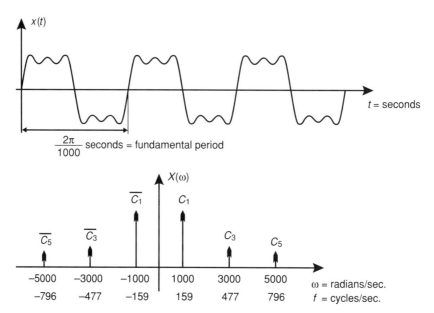

FIGURE 3–7
Partial synthesis of a square wave in the time and frequency domains.

which the signal carries power is called the *bandwidth*. This bandwidth is often subdivided into bands that carry different parts of the information. For example, the larynx, which produces the human voice, has been engineered by nature to produce a signal that carries most of its power over a bandwidth ranging approximately from 300 to 3000 Hz. This bandwidth can be subdivided into low-frequency, medium-frequency, and high-frequency bands. Figure 3–8 illustrates the power spectrum produced by a typical human larynx.

It is important to notice in Figure 3–8 that the power spectrum at negative frequencies exactly mirrors the power spectrum at positive frequencies. This is because *real* sinusoids, at any frequency, consist of two *conjugate* complex exponentials: $e^{+j\omega t}$ and $e^{-j\omega t}$.

Further inspection of Figure 3–8 indicates that the human voice does carry a small amount of power over the high-frequency band lying outside the 300 to 3000 Hz range. In

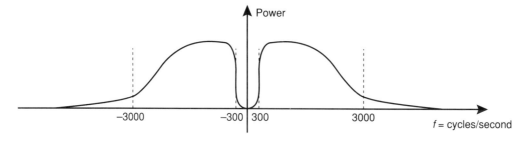

FIGURE 3–8
Power spectrum of a typical human voice.

practice, this small amount of power is not essential to interpret the information carried by voice. The telephone industry takes advantage of this fact to reduce the bandwidth of the frequencies that need to be transmitted. The telephone central office uses filters to remove some frequencies from the voice spectrum so that only the frequency band between 300 and 3000 Hz is transmitted. Figure 3–9 illustrates a voice spectrum that has been truncated by the filters of a telephone central office.

The fact that the telephone central office filters out some frequencies slightly alters the way the voice sounds. We know this from experience since we can easily recognize the

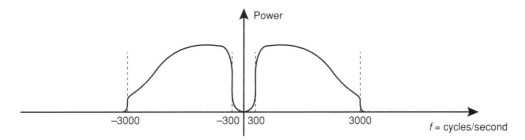

FIGURE 3–9
Human voice truncated at 3000 Hz.

intonations of a "telephone" voice. Even if some frequencies are truncated, enough information is left in the signal to identify the speaker's voice and, obviously, the contents of the conversation. This small concession in voice quality results in an efficient use of the bandwidth of telephone lines since a smaller part of the total voice bandwidth needs to be transmitted. When we do signal processing, we often need to truncate parts of the input signal spectrum. To simplify some problems, you should always identify the essential and the redundant parts of the signal spectrum.

To help you examine the spectrum of a signal, you can use a *spectrum analyzer*. This device displays a frequency domain representation of real signals. The spectrum analyzer displays the power over only the positive range of frequencies since the negative frequencies exactly mirror the positive part. Good-quality spectrum analyzers are capable of determining the signal power carried over very narrow bands of frequencies. Poor spectrum analyzers can be found, for example, on some audio amplifiers. They use a vertical bar of light-emitting diodes (LEDs) to produce a display that seems to jump up and down as music is being played. This moving display gives a rough real-time portrayal of the amount of music power carried over a number of frequency bands. Chapter 5 explains how these devices work.

3.3 ANALYZING THE PROPERTIES OF SINUSOIDS

We have seen that the complex exponential $e^{+j\omega t}$ is the fundamental component from which continuous signals are built. Consequently, the properties of this complex exponential reflects the properties of the complete signal.

The following sections explore some of the very important properties of complex exponentials. The following analysis is introductory and is limited to exploring the periodicity properties of complex exponentials. Still, even the very fundamental properties discussed in the following sections can be exploited and will lead us to discover some of the many incredibly useful practical applications presented in the remainder of this book.

3.3.1 ANALYZING CONTINUOUS-TIME COMPLEX EXPONENTIALS

Our analysis of complex exponentials begins with *continuous-time signals* since we are already intuitively familiar with their properties. We already know that time-domain periodic signals repeat themselves monotonously forever. The analysis of these signals presents the mathematical procedure used to investigate periodicity properties in signals.

The mathematical presentations that follow can be boring to some and appealing to others. We must present them here because they form the basis of all the signal-processing techniques described in this book. If you are not interested in the mathematical development, you need to examine only the results of each section presented in bold type. If you are one of the curious who finds this mathematical analysis appealing, you will find a number of books that explore the many other properties of complex exponentials, which lead to other fascinating applications.

Investigating the Periodicity in the Continuous-Time Domain

The complex exponential entity analyzed in the following sections is the fundamental component from which all sinusoids are built. For convenience, Figure 3–10 regroups the mathematical and vectorial representation of both $e^{+j\omega t}$ and $e^{-j\omega t}$.

We want to determine whether the continuous-time complex exponential function is periodic in time. Expressed in other terms, we wonder whether a continuous-time complex exponential repeats the same set of values as time t increases. If it does, over what interval of time does it repeat?

If the function does repeat itself in time, it must do so at fixed intervals of time, which we define as the period T. If periodicity exists with a period of T, the following mathematical relation should exist:

$$e^{+j\omega t} = e^{+j\omega(t+T)}$$
$$= e^{+j\omega t} \times e^{+j\omega T}$$

An interpretation of this equation is that if periodicity exists, the value of the complex exponential at time t must repeat itself at time $t + T$. Notice that the expression $e^{+j\omega t}$ appears on both sides of the equation. The only way that this equation can hold is if $e^{+j\omega T} = 1$, which satisfies $e^{+j\omega t} = e^{+j\omega t} \times e^{+j\omega T}$.

We know that $e^{+j\phi} = \cos(\phi) + j\sin(\phi)$. The expression $e^{+j\phi}$ therefore contains both a real and an imaginary part.

Since $\phi = \omega T$, we must determine the value(s) of ωT that will satisfy $e^{+j\omega T} = 1$. Examining Figure 3–11 helps us to determine when this condition is satisfied. Examining the

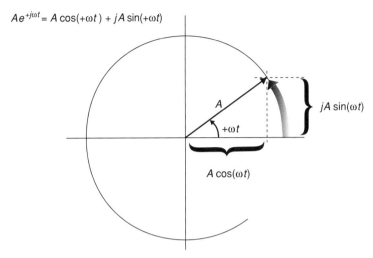

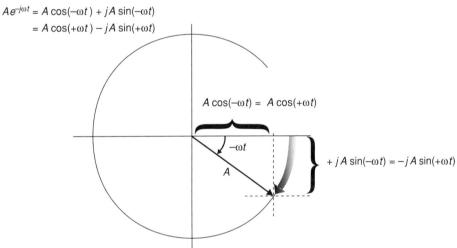

FIGURE 3–10
Time-varying complex exponentials.

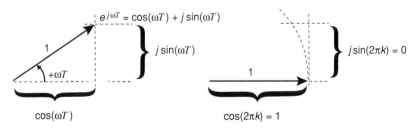

FIGURE 3–11
Complex vector $e^{+j\omega T}$.

left part of Figure 3–11, we see that meeting the condition requires that we adjust the angle ωT so that the imaginary part of the vector has a value of 0.

$$e^{+j\omega T} = \cos(\omega T) + j\sin(\omega T)$$
$$= 1 + 0j$$

Examining the right part of Figure 3–11, we see that the vector meets the requirements when it is horizontal and points right. This happens when the value of the angle ωT is a multiple of 2π since

$$\cos(0) = \cos(2\pi) = \cos(4\pi) = \cos(6\pi) = \cos(8\pi) = \ldots = \cos(2\pi k) = 1$$

where $k = 0, 1, 2, 3, \ldots$ and since

$$\sin(2\pi) = \sin(4\pi) = \sin(6\pi) = \sin(8\pi) = \ldots = \sin(2\pi k) = 0$$

where $k = 0, 1, 2, 3 \ldots$ Therefore, the values of ωT that satisfy $e^{+j\omega T} = 1$ are

$$\omega T = 2\pi k$$

where $k = 0, 1, 2, 3, \ldots$

Manipulating, we solve for the value of the period:

$$T = \frac{2\pi k}{\omega}$$

Since there is an infinite number of values for k, there is an infinite number of possible periods. Choosing $k = 1$, we can solve for the smallest possible (nonzero) period:

$$T = \frac{2\pi}{\omega}$$

To interpret this result in familiar terms, we must remember that ω is the *angular* frequency expressed in *radians/second* and that there are 2π radians in one cycle of a signal oscillation at a frequency f.

$$\omega = 2\pi f$$

Substituting $2\pi f$ for ω, we obtain the following expression for the smallest period:

$$T = \frac{2\pi}{2\pi f} = \frac{1}{f}$$

This yields the familiar result showing that the period is the inverse of the frequency.

Periodic continuous-time domain signals have a period of $T = 1/f$.

We did the preceding proof to illustrate the method used to investigate periodicity in complex exponentials. We will reuse this procedure a number of times in the following sections.

As previously noted, a single continuous-time complex exponential contains both real and imaginary parts. The waveforms that we use in real life have no imaginary part. To create a pure *real* waveform such as the function $A\cos(\omega t)$, we need to combine two conjugate complex exponentials. Since these two complex exponentials have exactly the same period, it follows that the complete function $A\cos(\omega t)$ is also periodic with exactly the same period:

$$\text{Period of } e^{+j\omega t} = \text{Period of } e^{-j\omega t} = \text{Period of } \cos(\omega t) = T = \frac{1}{f}$$

Figure 3–12 illustrates the periodicity of the $A\cos(\omega t)$ function.

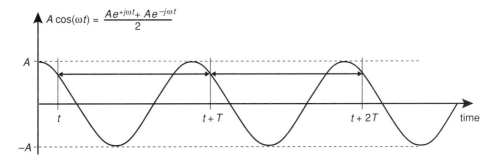

FIGURE 3–12
Periodicity of continuous-time complex exponentials.

Investigating Periodicity in the Frequency Domain

Let us now investigate whether our continuous-time-complex exponential exhibits periodicity in the *frequency* domain. If it does, this implies that the signal power manifests itself at periodic frequency intervals. If the complex exponential function repeats itself in frequency, it must do so at fixed frequency intervals, which we define as Ψ. If periodicity exists with a period of Ψ, the following mathematical relation should hold true:

$$e^{+j\omega t} = e^{+j(\omega + \Psi)t}$$
$$= e^{+j\omega t} \times e^{+j\Psi t}$$

According to this equation, if periodicity exists, the value of the complex exponential at frequency ω must repeat itself at frequency $\omega + \Psi$.

Notice that the expression $e^{+j\omega t}$ appears on both sides of the equation. The only way this equation can hold is if $e^{+j\Psi t} = 1$, which satisfies $e^{+j\omega t} = e^{+j\omega t} \times e^{+j\Psi t}$.

As we found in the previous case, this is satisfied only for values of Ψt that are integer multiples of 2π:

$$\Psi t = 2\pi k$$

For the smallest possible period, we choose $k = 1$, and we can manipulate the relation to isolate the period Ψ:

$$\Psi t = 2\pi$$

or that

$$\Psi = \frac{2\pi}{t}$$

Now since t represents the continuously increasing time variable, the preceding relationship cannot yield a constant value for the frequency interval Ψ. This means that there is no periodicity in the frequency domain; we expected this result since we already knew that a particular continuous sinusoid carries all of its power only at the $\pm\omega_0$ frequencies. Figure 3–13 illustrates the signal $A\cos(\omega_0 t)$ in the frequency domain.

Continuous-time domain signals are *not* frequency periodic.

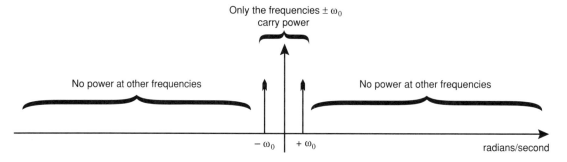

FIGURE 3–13
Continuous-time complex exponentials are *not* frequency periodic.

3.3.2 ANALYZING DISCRETE-TIME COMPLEX EXPONENTIALS

As Section 3.1 indicated, continuous-time domain signals must be digitized using an ADC before they can be processed by a digital computer. Since the digitization process loses the signal value between the samples, this process has serious consequences that must be thoroughly understood. The following sections investigate what happens to the contents of a signal when it is stored in digital memory as a sequence of numbers.

Representation of Discrete-Time Domain Complex Exponentials

The ADC delivers a new sample at the end of each T_S interval. We refer to these instants using the discrete-time variable n. For example, if the ADC takes samples at a frequency $f_{Sampling} = f_S$, this corresponds to taking samples at periodic intervals of

$$T_S = \frac{1}{f_S} \text{ second}$$

The sampling process transforms the continuous-time signal into a signal that exists only at the periodic instants nT_S:

	Time	Discrete Time
The first sample is acquired at:	$t = 0$	$n = 0$
The second sample is acquired at:	$t = 1T_S$	$n = 1$
The third sample is acquired at:	$t = 2T_S$	$n = 2$
and so on	. . .	. . .

Figure 3–14 illustrates the sampling process and shows how it modifies the continuous-time signal.

As explained in Section 2.1.5, when we mathematically express a sampled signal, we replace the continuous-time variable t by the discrete-time instants nT_S. Before being sampled, the original analog cosine consists of the following mix of complex exponentials:

$$\cos(\omega t) = \frac{e^{+j\omega t} + e^{-j\omega t}}{2}$$

Once sampled, the continuous-time variable t has been replaced by the discrete-time instants nT_S:

$$\cos(\omega n T_S) = \frac{e^{+j\omega n T_S} + e^{-j\omega n T_S}}{2}$$

Notice that the complex exponentials are now discrete. This means that the complex exponentials exist only at the sampling instants nT_S. Figure 3–15 illustrates a complex exponential that has been sampled a little more than 10 times over one full period of 2π radians. Obviously, the samples represent a set of vectors separated by an angular distance of ωT_S radians (indicated by small circles in Figure 3–15). Let's examine the terms of this angular distance:

The angular frequency of the sampled signal: $\omega = 2\pi f_{Input}$

The sampling interval: $T_S = \dfrac{1}{f_{Sampling}}$

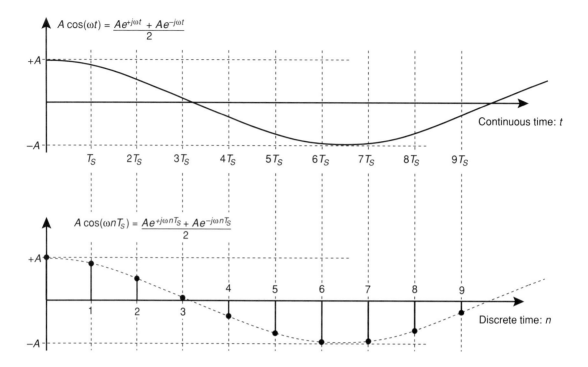

FIGURE 3–14
Sampling of a continuous-time signal.

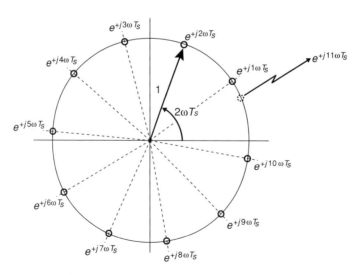

FIGURE 3–15
Digitized complex exponential $e^{+jn\omega T_s}$.

If the frequency of the signal being sampled is *increased,* the value of the angle between the samples ωT_S becomes greater and fewer samples are taken along a full period of 2π radians (the full circle). As we can see, ω and T_S are closely linked together; in fact, they become inseparable in the discrete-time domain.

When you examine Figure 3–15, notice that the 11th sample starts a new set of samples, which will not repeat the first set of 10 values. This next set of 10 samples produces 10 new vector positions. Oddly enough, even if the sampled cosine is periodic, the samples in the example are not producing a periodic set of samples! This phenomenon is examined further in the next section.

Investigating Periodicity in the Discrete-Time Domain

We found in the preceding section that continuous cosines and their complex exponential components are periodic in time but not in frequency. Now that we've sampled these signals, we want to find out whether these properties are maintained for discrete-time signals. If periodicity exists, the ADC should output a repeating set of numbers.

If a digitized signal function repeats the same set of values, it must do so at fixed intervals of discrete time, which we define as N. This means that a periodic digitized signal should repeat the same value after every interval of N samples. If we apply this to a discrete-time complex exponential, it yields the following relation:

$$e^{+jn\omega T_S} = e^{+j(n+N)\omega T_S}$$
$$= e^{+jn\omega T_S} \times e^{+jN\omega T_S}$$

An interpretation of this relationship is that if periodicity exists, the value of the complex exponential at time n must be the same as its value at time $n + N$. Notice that the expression $e^{+jn\omega T_S}$ appears on both sides of the equation. The only way this equation can hold is if $e^{+jN\omega T_S} = 1$, which satisfies $e^{+jn\omega T_S} = e^{+jn\omega T_S} \times e^{+jN\omega T_S}$. As we found in the previous periodicity case studies, this is satisfied only for values of $N\omega T_S$ that are multiples of 2π, which means that

$$N\omega T_S = 2\pi k$$

We can therefore solve for the discrete-time period N:

$$N = \frac{2\pi k}{\omega T_S}$$

Remember that ω is the *angular* frequency of the sampled complex exponential expressed in *radians/second* and that there are 2π radians in one cycle so that $\omega = 2\pi f$. Substituting $2\pi f$ for ω, we obtain

$$N = \frac{2\pi k}{2\pi f \times T_S} = \frac{k}{f \times T_S}$$

We can manipulate this further since T_S is the sampling period, which is the inverse of the sampling frequency f_S.

Substituting $1/f_S$ for T_S yields the following:

$$N = \frac{k}{f \times \dfrac{1}{f_S}} = \frac{f_S}{f} k \tag{3–1}$$

Since we are analyzing a *discrete*-time complex exponential, the value of N must be an integer. Because the value of k can take the value of any integer, we can try to adjust it so that the expression for N becomes an integer. If we can do this, the discrete-time signal will be periodic. This simple fact yields a very special condition for the function to be periodic. A digitized signal is periodic only if the sampling frequency f_S is a ratio of integers of the signal frequency f. This means that discrete-time functions can become periodic only under very special conditions! For example, if the sampling frequency f_S is 21/4 (a ratio of integers) of the input frequency f being sampled, then k can be adjusted to adopt a value of 4, which yields

$$N = \frac{k \times f_S}{f} = \frac{4 \times \dfrac{21}{4} f}{f} = 21$$

This means that the set of numbers being produced by the ADC will repeat itself every 21 samples. In practice, the signal that the ADC samples is likely to come from a "world" that is external to our digital processing system and is likely to cover a large spectrum of frequencies.

For the ADC samples to be periodic, all frequencies contained in the sampled signal must be integer ratios of the sampling frequency. This can be the case only if all frequencies contained in the input signal are synchronous with the sampling rate. Since the signal comes from the outside world and since it contains a large number of frequencies, it is very unlikely that the conditions necessary to obtain periodicity will be met.

A discrete-time signal can be periodic only if all of its frequency components are synchronous with the sampling frequency of the ADC. This is unlikely to happen in practical applications.

There is, however, an important special technique by which we can force a set of samples to become periodic. This allows us to perform spectral analysis on signals. We call this technique the *discrete Fourier transform*. It is explained in Chapter 5. Figure 3–16 illustrates a discrete-time comparison of what happens when a signal is synchronous with the sampling rate and when it is not.

Investigating Periodicity in the Frequency Domain

Let us now investigate whether a discrete-time signal exhibits periodicity in the frequency domain. *Frequency periodicity* implies that the power carried by the signal manifests itself at multiples of a certain base frequency.

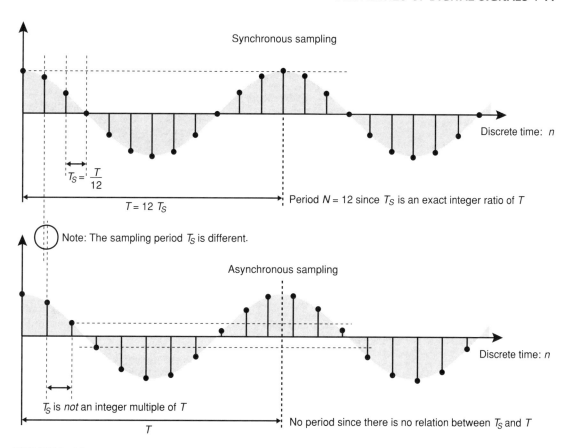

FIGURE 3–16
Time periodicity of discrete-time signals.

If a signal is periodic in frequency, its spectrum repeats itself at fixed frequency intervals, which we define as Φ. In this case, the following mathematical relation should exist:

$$e^{+j\omega T_s n} = e^{+j(\omega + \Phi)T_s n}$$
$$= e^{+j\omega T_s n} \times e^{+j\Phi T_s n}$$

According to this relationship, if periodicity exists, the value of the discrete-time complex exponential at frequency ω must be the same as its value at frequency $\omega + \Phi$.

Notice that the expression $e^{+j\omega T_s n}$ appears on both sides of the equation. The only way that the relationship holds is if $e^{+j\Phi T_s n} = 1$, which satisfies $e^{+j\omega T_s n} = e^{+j\omega T_s n} \times e^{+j\Phi T_s n}$. As we found in the previous case studies, $e^{j\Phi T_s n} = 1$ is satisfied only for values of $\Phi T_s n$ that are multiples of 2π. This condition means that

$$\Phi \, T_S n = 2\pi k$$

and consequently

$$\Phi = \frac{2\pi k}{T_S n}$$

For the smallest possible period, we choose $k/n = 1$, and we solve for the period Φ:

$$\Phi = \frac{2\pi}{T_S} = \frac{2\pi}{1/f_S} = 2\pi f_S = \omega_S \text{ radians}$$

A discrete-time complex exponential is periodic in the frequency domain: The period is the value of the sampling frequency.

Simply stated, the power carried by the digital signal manifests itself at an infinite number of frequency locations! This means that all digital signals contain an infinite number of copies of a base spectrum spanning a frequency range of f_S Hz. Since all periods replicate exactly the same information, only the base spectrum from $-f_S/2$ to $+f_S/2$ is important.

The periodic frequency property inherent to all discrete-time signals is probably the single most important property of digital signals. The importance of this property cannot be overemphasized. The reader must gain a thorough understanding of the implications of this property to understand DSP.

Figure 3–17 illustrates the periodicity properties of complex signals. Both the time and frequency domains are illustrated before and after the sampling of an input signal. Take the time to study this diagram in detail to ensure that you appreciate what happens to the frequency contents of a sampled signal.

The periodic frequency property of discrete-time signals warrants additional case studies to ensure that we fully appreciate the implications of digitization.

Case 1 This first case considers a system that digitizes an input signal that consists of the pure sinusoid tone: $\cos(2500t)$. Let's examine what happens in the frequency domain as this signal is digitized at a rate of $f_S = 1000$ samples/second (1000 Hz). We must ensure that we are using the same frequency units everywhere. We have a choice of working either in cycles/second (Hz) or in angular frequency (radians/second). Let's use the more intuitive Hertzian units.

The sampling rate of 1000 samples/second is already expressed using the proper units and corresponds to $f_S = 1000$ Hz. The $\cos(2500t)$ oscillates at a rate of 2500 radians/second. We convert this rate into cycles/second simply by dividing by a factor of 2π radians/cycle:

$$f_{IN} = \frac{\omega_{IN}}{2\pi} = \frac{2500 \text{ radians/second}}{2\pi \text{ radians/cycle}} \cong 398 \text{ cycles / second (Hz)}$$

Figure 3–18 illustrates some of the frequency replicates generated by the digitization process. There we can observe that the original spectrum of the $\cos(2500t)$ input is located at ±398 Hz and that these frequencies are replicated around multiples of the sampling

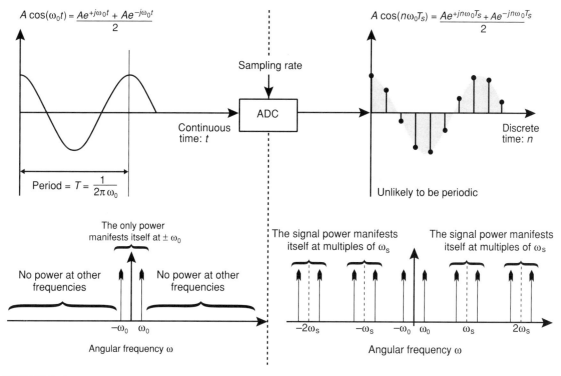

FIGURE 3–17
Periodicity properties of discrete-time signals.

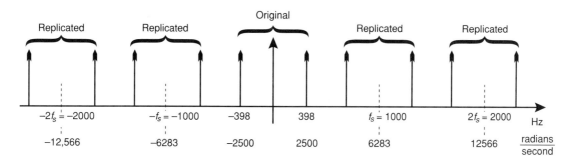

FIGURE 3–18
Some of the frequencies replicated by digitization.

frequency. You should understand that there are an infinite number of replicates and that the diagram shows only a small number of them.

Let's calculate the exact frequencies at which the digitized cosine signal power manifests itself. Table 3–1 shows the calculations for the first four sets of replicates around

TABLE 3–1
Some replicated frequencies.

±1000 Hz ± 398 Hz = ±602 Hz and ±1398 Hz
±2000 Hz ± 398 Hz = ±1602 Hz and ±2398 Hz
±3000 Hz ± 398 Hz = ±2602 Hz and ±3398 Hz
±4000 Hz ± 398 Hz = ±3602 Hz and ±4398 Hz
etc.

±1000 Hz, ±2000 Hz, ±3000 Hz, and ±4000 Hz. The cosine power manifests itself at ±398 Hz around an infinite number of sampling rate multiples.

Notice that the replicated frequencies in Table 3–1 occur in pairs at the following frequencies: ±602 Hz, ±1398 Hz, ±1602 Hz, ±2398 Hz, and so on. We must remember that an infinite number of other pairs are not listed.

The fact that the cosine signal power occurs at positive and negative frequencies illustrates that pairs of complex exponentials regroup to create pure-real functions as Euler showed:

$$\cos(\omega t) = \frac{e^{+j\omega t} + e^{-j\omega t}}{2}$$

since

$$
\begin{aligned}
e^{+j\omega t} &= \cos(\omega t) + j\sin(\omega t) \\
+ \; e^{-j\omega t} &= \cos(\omega t) - j\sin(\omega t) \\
\hline
e^{+j\omega t} + e^{-j\omega t} &= 2\cos(\omega t) + 0j
\end{aligned}
$$

Figure 3–19 illustrates how some of the complex conjugate pairs regroup to form pure real cosines.

You should realize at this point that the numerical samples stored in memory actually contain all replicated frequencies. The infinite number of frequencies contained in a digitized

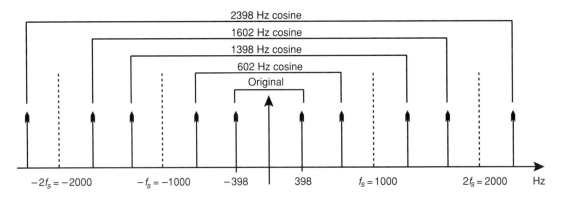

FIGURE 3–19
Replicated frequencies paired into pure real waveforms.

signal are real and will get in the way when we translate the numerical samples back to the continuous-time domain with the system DAC.

Case 2   This second case considers what happens when the system digitizes a signal that contains a spectrum of frequencies spread over a limited bandwidth. For example, Figure 3–20 illustrates the spectrum of music, which spans the audible frequency range of 0 to 20 kHz. We know that pure-real signals are composed of pairs of conjugate complex exponentials. Because of this, for pure real inputs, the negative frequency spectrum is always a mirror image of the positive side.

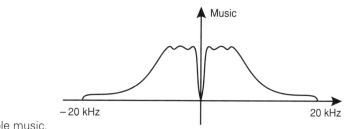

FIGURE 3–20
Spectrum of audible music.

Let's sample the music at a compact disc (CD) rate of 44.1 kilo-samples/second. Because discrete signals are periodic in the frequency domain, the sampling replicates the original music spectrum at multiples of the sampling rate. It is important to realize that once the samples are stored on a CD, every individual frequency contained in the music is replicated around multiples of the sampling rate. Figure 3–21 illustrates the spectrum of audible music that has been stored as numerical samples on a CD.

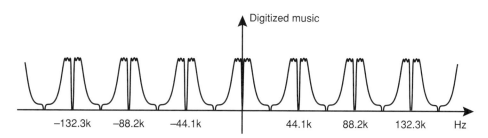

FIGURE 3–21
Spectrum of digitized music.

The original music spectrum located in the ±20 kHz range is centered on the origin, and it is the only part of interest. The power manifestation at all other frequencies comes from replicates of the original signal spectrum and contains no additional information. The frequencies outside the ±20 kHz range do not belong to the original signal and will distort the music unless we choose to do something about it.

We can remove the unwanted frequencies by outputting the digitized signal through a low-pass filter. This filter has a passband that matches the original bandwidth of the music. Such a filter removes all the unwanted frequencies outside the ±20 kHz range! We call this filter a *reconstruction filter* because eliminating the spectrum replicates effectively rebuilds the original music spectrum.

Figure 3–22 illustrates the position of the low-pass reconstruction filter in the digital system configuration and shows the frequency content of the signals as it travels through the various parts of the system.

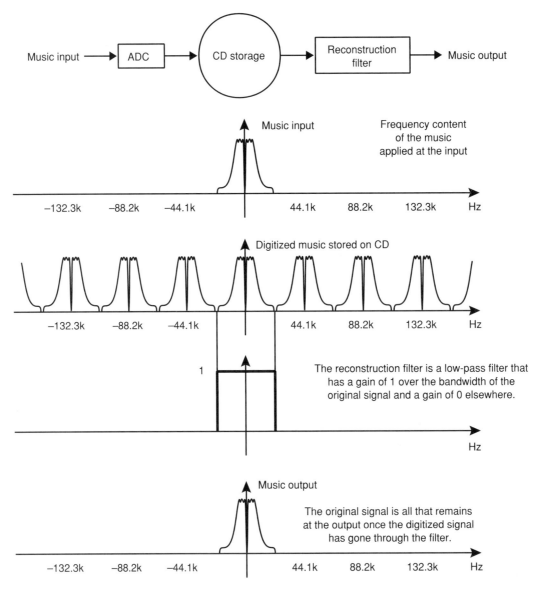

FIGURE 3–22
Frequencies going through an idealized digital system.

3.4 CHOOSING A SAMPLING RATE

When an ADC digitizes a signal, we can do nothing to prevent the input spectrum from being replicated around multiples of the sampling frequency. What we can control is the sampling rate used by the ADC; therefore, we control where the replicates will appear. If we are careful, we can select a sampling rate that positions the spectrum replicates where they do not disturb the processing.

The case studies described in the following sections demonstrate that the choice of the sampling rate sets the limits of the frequency range that the system will be able to process.

3.4.1 OVERSAMPLING

Let's oversample some music by doubling the standard rate used for digital music. The new sampling rate becomes 88.2 k-samples/second, which creates replicates that are spaced farther apart. Figure 3–23 illustrates the effect that oversampling has on the digitized signal spectrum.

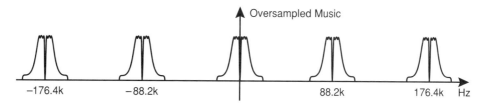

FIGURE 3–23
Spectrum of oversampled music.

As Figure 3–23 indicates, the original spectrum and its replicates are positioned at a comfortable distance from each other.

3.4.2 UNDERSAMPLING

We now choose to undersample the music at 30 kHz, which is well below the standard used for digital music. In this case, as illustrated on Figure 3–24, the replicated spectra overlap each other. Figure 3–24 indicates that the overlap causes the replicates to interfere with each other, and this distorts the shape of the signal spectrum. Once the original spectrum is corrupted, it becomes literally impossible to recover the original shape. The damage in the overlapping zones is permanent. There is no way to recover the original signal.

Figure 3–25 illustrates the damaged spectrum of an undersampled input signal. The overlapped sections add to create a distorted version of the original spectrum. This means that the music contained in that signal will never sound the same again. Essentially, our undersampling of the input signal has badly distorted the music.

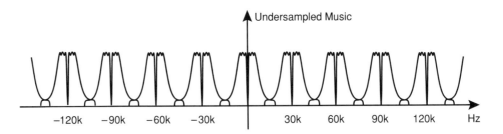

FIGURE 3–24
Spectrum of undersampled music.

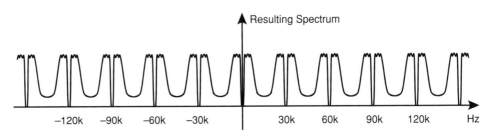

FIGURE 3–25
Overlapping replicates result in permanent spectral damage.

3.5 LINKING THE SAMPLING RATE TO THE INPUT SPECTRUM

This section presents the relationship between the sampling rate and the frequency contents of the sampled signal. We can investigate this relationship using the two approaches detailed in the next two sections.

3.5.1 USING FIXED SAMPLING RATES

Let's examine a situation in which we keep the sampling rate constant as we vary the frequency of an input cosine. We arbitrarily choose to sample the cosine waveform at a fixed rate of 10 kHz in this example.

Figure 3–26 illustrates what happens to the frequency components contained in the digitized samples when the cosine frequency is increased starting from a value of 2 kHz. As the input cosine frequency increases, so does the distance separating the components that make up each frequency pair. Notice that the individual components are actually moving away from their respective multiple of f_S.

As the input cosine frequency increases, the range separating the replicated frequency pairs decreases. When the cosine frequency reaches 4 kHz, the replicated components are getting dangerously close to each other. Figure 3–27 illustrates the distance that separates the replicated frequency pairs.

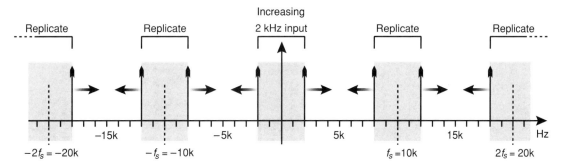

FIGURE 3–26
Input frequency is increasing.

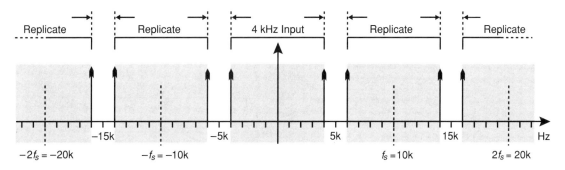

FIGURE 3–27
Replicated frequency pairs are getting close to each other.

If we increase the input cosine frequency beyond 5 kHz (note that this breaks the $f_S/2$ barrier), the replicates start overlapping each other's frequency range (see Figure 3–28). When overlapping occurs, the replicated parts overlap the original signal frequency range. This causes irreparable damage to the original signal spectrum. The replicates created by the sampling process can be pictured as mirror images reflected across odd multiples of the $f_S/2$ barrier. Figure 3–29 illustrates this mirroring of frequencies. The input frequency should not exceed $f_S/2$ because this would push the replicated frequencies to corrupt the original signal spectrum.

The input frequency should not exceed $f_S/2$.

3.5.2   USING VARIABLE SAMPLING RATES

To appreciate what happens when we sample an input signal at different rates, we will keep the input cosine at a constant frequency of 5 kHz. Let's see what happens to its spectrum when we reduce the sampling rate starting from a rate of 15 k-samples/second.

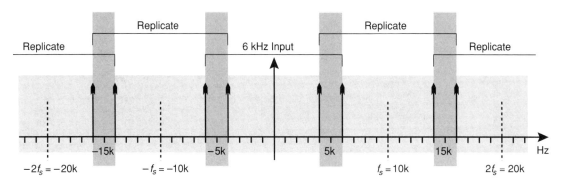

FIGURE 3–28
Replicates overlap each other's frequency range.

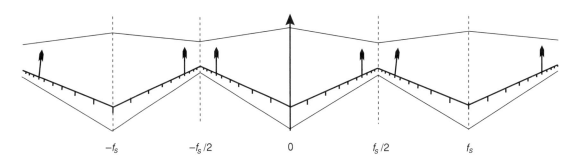

FIGURE 3–29
Mirroring of frequencies across the $f_S/2$ barrier.

Figure 3–30 illustrates that a reduction of the sampling rate results in a movement of the replicated parts toward the original frequency range of the input signal. Note that reducing the sampling rate moves *all* replicated parts toward the original spectrum of the input signal. Figure 3–31 illustrates the situation when the sampling rate is reduced to 12 k-samples/ second. Note that the replicated parts are getting relatively close to the original input spectrum.

If we reduce the sampling rate to less than twice the input frequency, the input cosine is undersampled and the replicated spectra start overlapping each other. When overlapping occurs, as in Figure 3–32, some of the replicated frequencies invade the original signal spectrum.

Notice that the original cosine frequency covers a bandwidth of 5 kHz. Referring to Figure 3–32, where the sampling rate is at 9 kHz, we can see that a frequency component exists at ±4 kHz. This indicates that a real 4 kHz cosine signal has entered the original signal frequency range. This 4 kHz cosine is created by the sampling process when the original 5 kHz input signal is replicated around f_S. Since this 4 kHz cosine is really the 5 kHz input

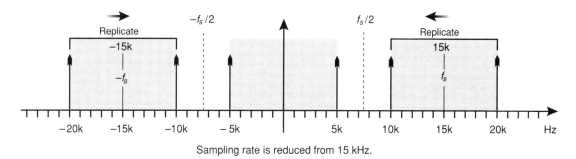

Sampling rate is reduced from 15 kHz.

FIGURE 3–30
Decreasing f_S.

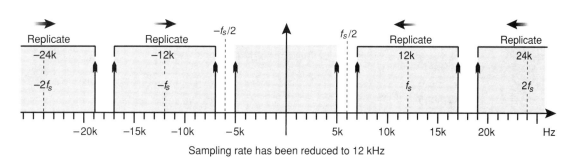

Sampling rate has been reduced to 12 kHz

FIGURE 3–31
Replicates are getting close to the input spectrum.

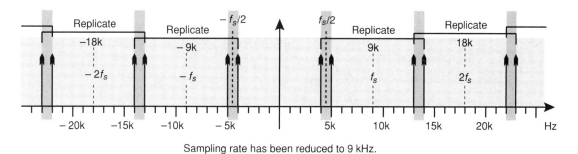

Sampling rate has been reduced to 9 kHz.

FIGURE 3–32
Replicates invade the original input frequency range.

coming back under a new "name", it is referred to as an *aliased frequency*. To prevent aliased frequencies from being generated, we must ensure that the ADC samples the input signal at a rate that is at least twice the highest frequency component contained in the input signal.

Undersampling may generate some aliased frequencies.

An intuitive way to look into the nature of aliased frequencies is to examine the actual samples taken in the time domain. Examine Figure 3–33 in which the sampling of two sinusoids is overlaid.

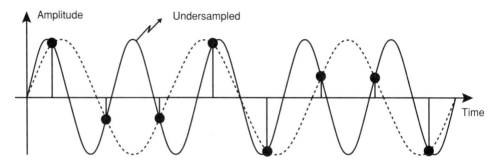

FIGURE 3–33
Creation of an aliased frequency in the time domain.

The sampling rate in Figure 3–33 is 8 k-samples/second. One sinusoid has a frequency of 3 kHz and the other 5 kHz. The 3 kHz sinusoid is sampled more than twice per period, but the 5 kHz sinusoid is undersampled.

As Figure 3–33 indicates, if we sample the sinusoids separately, we will acquire two *identical* sets of digitized samples. This means that sampling a 3 kHz or a 5 kHz sinusoid at 8 k-samples/second results in the acquisition of exactly the same numerical sequence of values. Because of this, once they are stored in memory, the two digitized sinusoids become indistinguishable. What is actually happening is that the undersampling of the 5 kHz sinusoid results in a 3 kHz frequency being mirrored across the $f_S/2$ barrier; in other words, undersampling the 5 kHz sinusoid creates a 3 kHz aliased frequency.

In practice, an infinite set of higher frequency sinusoids would result in exactly the same set of digitized values:

$$f_S - 3 \text{ kHz} = 5 \text{ kHz} \qquad f_S + 3 \text{ kHz} = 11 \text{ kHz}$$
$$2f_S - 3 \text{ kHz} = 13 \text{ kHz} \qquad 2f_S + 3 \text{ kHz} = 19 \text{ kHz}$$
$$3f_S - 3 \text{ kHz} = 21 \text{ kHz} \qquad 3f_S + 3 \text{ kHz} = 27 \text{ kHz}$$

$\cdots$

This result explains why the spectrum of a sampled sinusoid actually contains an infinite number of frequency components. There is no way to know which frequency the ADC actually acquired since the same sequence of values represents them all. Practically speaking, the digitized signal spectrum contains all of these frequencies!

If you watch western movies, you have seen cowboys in chuckwagons race away from danger. You may have noticed that the wagon wheels seem to rotate unusually slowly compared to the chuckwagon's speed. Sometimes the wheels even seem to rotate in the wrong direction! The filming process creates this unusual effect because it actually samples the scene. The film consists of a series of pictures (samples) of the wagon wheels as they rotate. Since the wheels rotate at high speed, the rate at which pictures are taken is insufficient to cover the required minimum of two pictures per wheel rotation. What results is an aliased rotation speed that is slower than the actual speed of the wagon wheel. When the wheels rotate in the wrong direction, the aliased rotation has an inverted phase.

A DSP system starts by digitizing the input signal so that it can process the acquired samples to manipulate the signal contents. The processing, which is the heart of the DSP system, usually extracts some information from the frequencies contained in the digitized signal. When the sampling generates aliased frequencies, the input signal spectrum is invaded by aliased frequencies that have no business being there. Since the aliased frequencies are located inside the original signal bandwidth, the signal processor has no way to know whether these frequencies belong to the original input signal. Because of this, the DSP system processes the aliased frequencies as if they were part of the original input signal. Aliased frequencies are therefore undesirable intruders from which we must protect the system.

Once aliased frequencies enter the signal spectrum, there is no way to distinguish them from frequencies that belong to the original input signal. Consequently, there is no way to remove these intruders. In practice, the only thing that we can do is to *prevent* the aliased frequencies from being generated so that the problem never needs to be addressed.

3.5.3   FORMULATING THE SAMPLING THEOREM

We have seen that increasing the sampling rate results in the replicated spectra moving away from each other. To avoid aliased frequencies, we must ensure that the sampling rate is high enough to keep the replicated spectra away from the original copy of the input spectrum. Sampling the input at a rate that is greater than twice the highest input frequency achieves this.

Figure 3–34 illustrates a digitized signal spectrum for different ADC sampling rates. The system illustrated in Figure 3–34 attempts to remove the spectrum replicates by forcing the output signal through a low-pass reconstruction filter. The box illustrated in the figure frames the range of frequencies that make it to the system output. As we can see, the oversampled and the critically sampled signals get through the system with no loss of information, but the undersampled signal suffers from a corrupted spectrum.

A close examination of Figure 3–34 reveals that the low-pass reconstruction filter has a cutoff frequency of $f_S/2$. Therefore, all input frequencies that cannot generate aliased frequencies make it to the output. As the third case of Figure 3–34 illustrates, any input frequency above the $f_S/2$ barrier produces a spectra overlap that the reconstruction filter cannot repair.

The sampling theorem:
If the highest frequency component
of an analog signal does not exceed $f_S/2$, this signal may be
digitized into a sequence of numerical samples and converted
back to the analog domain with no loss of information.

This amazing result means that the reconstruction filter rebuilds the analog signal by filling in all information missing *between* the samples! For example, a reconstruction filter can rebuild a complete cosine waveform as long as we provide it with at least two samples

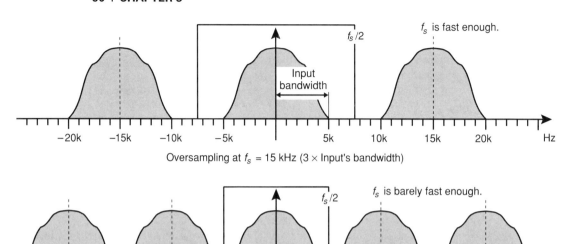

Oversampling at $f_s = 15$ kHz ($3 \times$ Input's bandwidth)

Critical sampling at $f_s = 10$ kHz ($2 \times$ Input's bandwidth)

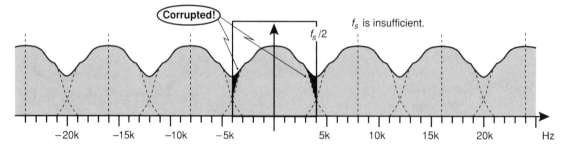

Undersampling at $f_s = 8$ kHz (smaller than $2 \times$ Input's bandwidth)

FIGURE 3–34
Sampling an input at different rates.

on each cosine period. The DAC steps are effectively a manifestation of frequencies that exist above the $f_s/2$ limit. Removing these frequencies using a reconstruction filter effectively smoothes out the DAC steps.

The sampling theorem sets practical limits on the frequency range that a DSP system can process.

3.5.4   FILTERING THE INPUT

The previous section has shown us that the sampling rate must be at least twice the bandwidth of the input signal to prevent aliased frequencies.

**Digital signal processing systems can accommodate
frequencies ranging from 0 to $f_s/2$.**

If we choose the sampling rate correctly, we can accommodate the bandwidth of any signal. In practice, as we increase the sampling rate, the cost of the ADC goes up, the amount of required memory increases, and more performance is required of logic circuits. Since there are limits to our budget, the size of circuits, and the speed of the available technology, we cannot use just any sampling rate. Our budget and the available parts impose practical limits on the sampling rate. What do we do if the signal we need to process exceeds the $f_S/2$ limit imposed by practical considerations?

Most practical signals cover a relatively large bandwidth but, fortunately, a limited range of frequencies usually carries the bulk of the information. For example, Section 3.2.2 indicated how telephone companies limit the bandwidth of voice for efficient use of the capacity of the transmission line.

Since we cannot allow aliased frequencies to be created during the sampling process, we need to limit the input signal bandwidth to $f_S/2$ *before* the sampling takes place. The required low-pass filter is called an *anti-aliasing filter*. It ensures that the input signal applied to the ADC contains no frequencies that can result in an aliasing effect.

Figure 3–35 illustrates how the frequency contents of an input signal can be frequency limited *before* the sampling process to prevent the generation of aliased frequencies.

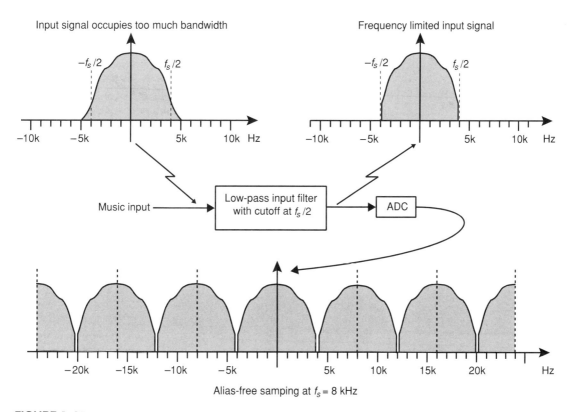

FIGURE 3–35
Anti-aliasing filter at the analog input.

With an antialiasing filter in place, the system can acquire and process a spectrum that is band limited from zero to $f_S/2$. We can achieve this performance only at the cost of truncating, from the input signal, the frequencies that exceed the $f_S/2$ barrier.

3.5.5   USING ALIASED FREQUENCIES

The previous section examined cases in which it was necessary to prevent aliasing so that the input spectrum remains unaltered. There are cases, however, in which the creation of aliased frequencies can be used to our advantage. This section deals with some applications in which we create aliased frequencies *on purpose* to provide elegant solutions to problems that at first glance seem rather difficult to solve.

The most familiar example is a strobe light. In this case, short bursts of light are flashed periodically on a moving object. The flashes of light actually produce a sampling of the object's position as it moves. When a rotating object receives strobe flashes at a rate of less than two samples per rotation, an aliasing effect results. The aliasing appears to slow the rotation of the object to speeds less than half the strobe rate. Mechanics use strobe lights in shops to observe objects that rotate too fast for direct observation. A strobe light makes dancers on a dance floor seem to move in slow motion.

The next example, concerning a science fiction movie plot, is provided to keep things interesting. Imagine that we receive a visit from friendly batlike beings from outer space and that we want to learn their language. Unfortunately, these beings communicate by speaking a language that uses ultrasounds (like real bats) that humans cannot hear. We must therefore design a device that shifts the ultrasounds that these beings emit to a range of frequencies that we can hear. This will allow linguists to work on a translation of their language.

Figure 3–36 displays these beings' voice spectrum and illustrates what we are trying to accomplish. Notice in this figure that the beings' spectrum is larger than our audible range. This means that even if we manage to move the spectrum, some parts of it will still overflow our audible range. We hope that this handicap will not result in the loss of much information. Note that we will need to move both the negative and the positive frequency parts to create a real signal that we can hear.

Imagine now that we purposely undersample the beings' spectrum at a sampling rate of 44 kHz. This sampling replicates the beings' spectrum around ±44 k, ±88 k, ±132 k, and at infinite multiples of our sampling rate. Figure 3–37 examines the two spectrum replicates closest to the origin.

Inspecting Figure 3–37, we notice that a pair of spectrum replicates falls inside our audible range! These consist of aliased frequencies, but, fortunately, they are occupying a range of frequencies that was void of information. Because of this, the aliases are not interfering with any parts of the beings' signal. The only minor problem is the small amount of power that spills over narrow bands close to the origin. This spilling occurs because the beings' spectrum is wider than $f_S/2$. To prevent this annoying effect, we can trim the beings' spectrum *before* we sample it. Using a band-pass input filter to limit the spectrum to a bandwidth of $f_S/2$ removes the problem. We hope that the little power that is chopped off does not contain much information. Figure 3–38 illustrates the complete system.

Although the human race has not had to face these bat beings yet, the technique that we have just described has practical applications. For example, the communications industry

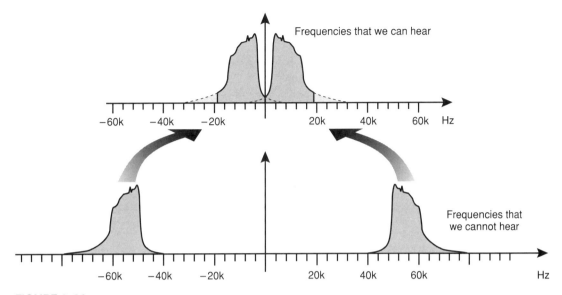

FIGURE 3–36
Repositioning a spectrum.

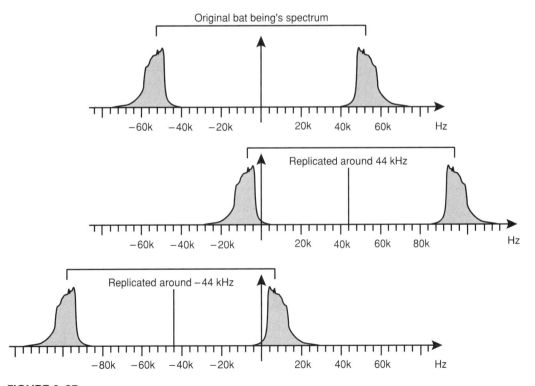

FIGURE 3–37
Purposely undersampling a spectrum.

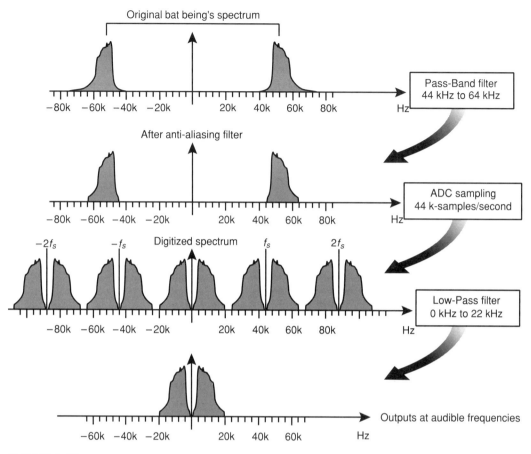

FIGURE 3–38
System to move a whole spectrum of frequencies.

uses this technique to demodulate signals that have been shifted to higher frequencies. The details of how to implement digital demodulation are beyond the scope of this introductory level book, but you should now understand the basic principle, which will allow you to research the subject further.

3.6 NORMALIZING THE SAMPLING RATE

The previous sections emphasized that the input frequency ω_{Input} is closely related to the sampling interval T_S. This is so because the sampling process converts the continuous-time variable t into a discrete-time expression: $T_S n$ (refer to Section 2.1.5). Consequently, when we sample a continuous-time signal, the expression $\omega_{\text{Input}} t$ becomes $\omega_{\text{Input}} T_S n$.

A sampled signal therefore has its input frequency linked to the sampling period. Let's analyze the $\omega_{Input}T_S$ expression to interpret its significance. Looking at units, we note first that $\omega_{Input}T_S$ is the angular distance covered by the signal between the ADC sampling instants:

$$\omega_{Input}\ T_S\ =\ 2\pi\ f_{Input}\ \text{radians/second} \times T_S\ \text{second/sample}$$

$$=\ 2\pi\ \frac{T_S}{T_{Input}}\ \text{radians/sample}$$

Second, note that this angular distance is the ratio T_S/T_{Input} of the signal's full period 2π.

Interpreting this, if the input period T_{Input} becomes smaller (the input frequency increases), the ADC takes samples over a larger angular distance. This means that the ADC takes fewer samples over a full period of the input signal. Figure 3–39 illustrates this situation with input sinusoids of different frequencies being sampled at T_S intervals. As Figure 3–39 indicates, an increase in the input frequency means that the ADC takes fewer and fewer samples on each period of the input signal. The last case illustrates a frequency of $f_S/2$ at which the ADC takes only two samples on each signal period. This input frequency touches the theoretical sampling limit that avoids the generation of aliased frequencies.

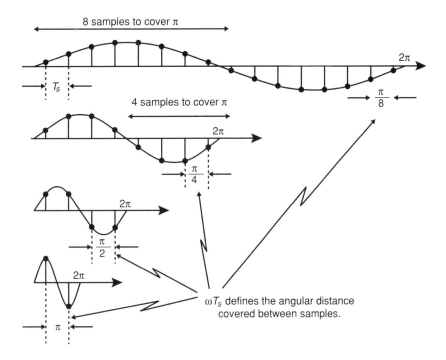

FIGURE 3–39
Significance of ωT_S.

When f_{Input} reaches a maximum of $f_S/2$, then

$$\text{Maximum } \omega_{Input} \, T_S = 2\pi \, \frac{f_S}{2} \times \frac{1}{f_S} = \pi \text{ radians/sample}$$

Since DSP systems usually deal with input frequencies that range between zero and $f_S/2$, the value of $\omega_{Input} T_S$ in this useful range of frequencies corresponds to values that lie between zero and π.

In DSP, since the input frequency is closely linked to the sampling frequency, we mainly refer to the input frequency in terms of the sampling frequency. To simplify this reference, we manipulate the $\omega_{Input} T_S$ expression to a more intuitive form:

$$\omega_{Input} \, T_S = 2\pi \, f_{Input} \times \frac{1}{f_S} \text{ radians/sample}$$

$$= 2\pi \, \frac{f_{Input}}{f_S} \text{ radians/sample}$$

Expressed as a ratio of frequencies, we can see that $\omega_{Input} T_S$ links the input frequency f_{Input} to the sampling frequency f_S. The value of $\omega_{Input} T_S$ therefore expresses the input frequency in terms of the sampling frequency; we call it the *normalized input frequency* Ω:

$$\omega_{Input} \, T_S = 2\pi \, \frac{f_{Input}}{f_S} = \Omega \text{ radians/sample}$$

where $0 \leq \Omega \leq \pi$.

As an example of using the normalized notation, let's consider a modem tone that is being sampled at the telephone network sampling rate of $f_S = 8$ kHz. If the modem tone has a frequency of $f_{Tone} = 1270$ Hz, it corresponds to a normalized input frequency of

$$\Omega_{Tone} = \omega_{Tone} \, T_S$$

$$= 2\pi \times f_{Tone} \times \frac{1}{f_S}$$

$$= 2\pi \times 1270 \text{ Hz} \times \frac{1}{8000 \text{ Hz}} = 0.9975 \text{ radians/sample}$$

From the normalized frequency of the modem tone, we can calculate the number of samples that the ADC takes on each 2π period of the input signal:

$$\frac{2\pi}{0.9975} \cong 6.3 \text{ samples}$$

Figure 3–40 shows a frequency domain graph that uses three different frequency units to display the DSP useful range of frequencies.

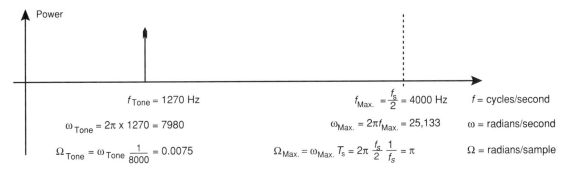

FIGURE 3–40
Three units used to express frequency value.

SUMMARY

- The ADC converts an analog signal into a discrete signal by replacing the continuous-time variable t with the discrete-time expression nT_S.
- A sinusoid is defined by three parameters: frequency, amplitude, and phase.
- The time domain plot of a real sinusoid results in a continuously changing picture.
- Plotting a sinusoid in the frequency domain provides a static picture.
- The *power spectrum* of a signal illustrates the power that the signal carries over a range of frequencies.
- Removing some frequencies that carry little power often can be done without changing much of the information contained in the signal.
- Continuous-time domain signals are time periodic with period: $T = 1/f$.
- Continuous-time domain signals are *not* frequency periodic.
- A digitized real signal consists of a pair of rotating complex conjugate vectors.
- A discrete-time signal can be periodic only if all of its frequency components are synchronous with the sampling frequency of the ADC. This is unlikely to happen in practical applications.
- A discrete-time complex exponential *is* periodic in the frequency domain. The period is the value of the sampling frequency.
- A digitized signal can be translated to the analog world by sending it through a DAC followed by a *reconstruction filter*.
- The input signal should be oversampled to ensure that its spectrum is not corrupted.
- Undersampling can generate some aliased frequencies.
- The sampling theorem states that if the highest frequency component of an analog signal does not exceed $f_S/2$, this signal may be digitized into a sequence of numerical samples and converted back to the analog domain with no loss of information.
- DSP systems have a useful range that covers frequencies from zero to $f_S/2$.
- Sending the input signal through an anti-aliasing filter prevents the generation of aliased frequencies.
- Aliasing can be used to move a frequency spectrum.

- The normalized frequency is the angular distance covered by the signal between the ADC sampling instants.
- The frequency can be expressed using three different units:

Angular frequency units expressed in radians/second, where one oscillation is measured to cover 2π radians.

Cycles/second units expressed in Hertz, where one oscillation is measured to cover one cycle.

Normalized frequency units expressed in radians/sample, where one oscillation is measured to cover $2\pi/\Omega$ samples.

PRACTICE QUESTIONS

3-1. Consider the expression $\cos(1000t)$.
    (a) Give the two complex exponentials that correspond to the expression.
    (b) At $t = 1$, give the Cartesian coordinates for the two complex exponentials of part (a).
    (c) What is the frequency of the cosine expressed in Hertz?
    (d) How much time will the expression take to complete one cycle?
    (e) Plot the first two cycles of the expression as a function of time.

3-2. For the expression $\cos(1000t + \pi/4)$ sampled 1000 times/second.
    (a) Give a frequency domain representation of the undersampled expression.
    (b) Give a frequency domain representation of the sampled expression.

3-3. At time $t = 0$, the signal $\cos(2\pi \times 300t)$ is sampled at a rate of 1000 samples/second.
    (a) Expressed as decimal numbers, what are the numerical values of the samples at discrete times $n = 0, 1, 2, 3, 4, 5, 6, 7, 8, 9$? Plot these 10 samples.
    (b) Are the samples periodic? Explain.
    (c) Would other sampling rates yield a discrete-time period of $N = 10$ samples?
    (d) Repeat part (a) for the signal $\cos(2\pi \times 1300t)$.

3-4. An ADC samples an unfiltered 600 Hz cosine at a rate of 1000 samples/second. The samples are sent directly to a DAC, and its output goes through a reconstruction filter.
    (a) In a rough drawing, illustrate the shape of the DAC output.
    (b) In a rough drawing, illustrate the shape of the reconstruction filter output.

3-5. An ADC samples an unfiltered analog signal at a rate of 1000 samples/second.
    (a) In each of the following cases, specify whether the sampling results in the creation of an aliased frequency; if so, specify the aliased frequency:
        1. Case 1: Analog input is a cosine of 300 Hz.
        2. Case 2: Analog input is a cosine of 700 Hz.
        3. Case 3: Analog input is a cosine of 1300 Hz.
    (b) For each of the three cases in part (a), give the frequencies in the range of 0 to 2000 Hz contained in the sampled signal.

3-6. An ADC samples an unfiltered signal that has a bandwidth of 20,000 Hz.
    (a) If the ADC samples at a rate of 30 k-samples/second, how much of the spectrum can be recovered unchanged?
    (b) If the ADC sampling rate is increased to 35 k-samples/second, how much of the spectrum can be recovered unchanged?

3-7. An ADC samples an unfiltered analog signal at a rate of 1000 samples/second.
    (a) What range of frequencies can this DSP system process?
    (b) What is the cutoff frequency of the antialiasing filter?
    (c) What is the cutoff frequency of the reconstruction filter?

3-8. An analog signal occupies a spectrum that ranges from 540 kHz to 550 kHz. If this signal is undersampled at 50 k-samples/second, show where this spectrum is replicated in the 0 to 100 kHz range.

3-9. A 1000 Hz cosine is being sampled at a rate of 4000 samples/second.
    (a) What is the angular frequency of this cosine?
    (b) What is the normalized frequency of this cosine?
    (c) How many samples are taken on each cosine cycle?

3-10. A sampled signal stored in memory has a normalized frequency of $\pi/5$. The samples are sent to a DAC whose output goes through a reconstruction filter.
    (a) The DAC receives the samples at a rate of 5000 samples/second. Expressed in hertz, what is the frequency at the output of the filter?
    (b) The DAC now receives the samples at a rate of 10,000 samples/second. Expressed in hertz, what is the frequency at the output of the filter?

4

PROCESSING SIGNALS

Digital signal processing (DSP) systems must be able to input external signals. To do this, the digital system includes an analog-to-digital converter (ADC), which converts the analog input signal into a sequence of samples. Once the signal is reduced to a digital format, the digital system processes the samples to compute the output samples.

This chapter investigates the nature of the processing. What operations can a digital system perform on the input samples? How does the processor use these operations to produce the output signal? Answering these questions provides the necessary knowledge to do signal processing.

We begin by stepping back and looking at the system as a box that has an input and an output. We then set rules about the insides of the box so that we can investigate the operation of a specific class of systems. Most engineering systems consist of linear and time-invariant systems. They are *time invariant* because the mechanics of the box are not allowed to change during the processing. They are *linear* because they can be broken into a number of subsystems that can be sequenced in any order.

The analysis of the linear time-invariant class of systems leads to the discovery that their operation is deceptively simple. Actually, digital systems use a certain amount of memory to store past input samples, and they process these samples to compute output samples. The operations used in the processing are mainly limited to addition and multiplication operations. From this, we describe the processing steps using a *difference equation*.

Examining the structure of the difference equation, we realize that a simple list of constant numbers entirely describes the system operation. This provides an incredibly simple way to modify the system functionality! We develop techniques to extract this list of numbers from systems and thus develop tools to verify the system operation.

The amount of memory a system holds determines the level of sophistication of its processing. We discover that using feedback creates an infinite amount of virtual memory. This provides the key to practically implementing increased levels of sophistication. As we investigate the limits of feedback systems, we discover the conditions necessary to maintain system stability.

We close by developing the *convolution operation,* which implements the mechanics necessary to solve the difference equation. The simulation and practical implementation of systems rely extensively on this operation. Other chapters also use the convolution operation to develop the tools that allow the design of the system operation.

4.1 DIGITAL SIGNALS

According to Section 3.1, an ADC can be used to convert a continuous-time signal $x(t)$ into a discrete-time signal $x(nT_S)$. The ADC therefore provides an endless sequence of binary numbers and this sequence represents the signal. What is the binary format used to express the ADC sample values?

Section 2.1.4 showed that a DAC always chooses to interpret its digital input as a signed fixed-point pure fractional number. This number dictates what fraction of the full output voltage range the DAC should produce.

An ADC follows the same principle of operation. The ADC translates each sampled input voltage into a numerical sample whose value is a signed fraction of the input voltage range. The value of signed ADC samples therefore covers the pure fractional range of $\pm 1^-$ (see Appendix B for the range of signed fractional numbers).

Figure 4–1 illustrates this for two 8-bit ADCs that happen to be sampling an input voltage positioned halfway in different positive ranges; both produce an output sample value of +0.5.

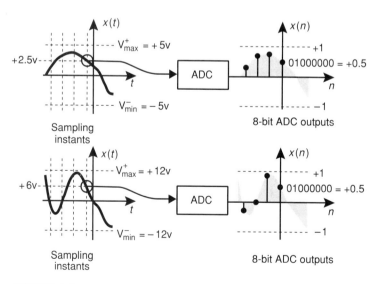

FIGURE 4–1
The ADC samples are pure signed fractions.

4.2 DIGITAL SIGNAL PROCESSING

DSP occurs when the binary numbers that correspond to one signal are manipulated to create a new sequence, which corresponds to the processed signal. We call the rules that define how to manipulate a digital signal the *processing algorithm*. It consists of mathematical operations that are applied to a sequence of binary numbers.

DSP is usually performed on samples taken from an input signal provided by an ADC. For every sample coming in, the processing algorithm delivers an output sample. The sequence of output samples creates a new digital signal, which becomes the output of the system. Figure 4–2 illustrates that the processing system outputs a sample for every input sample coming in.

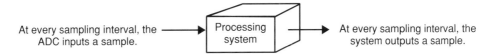

At every sampling interval, the ADC inputs a sample. → Processing system → At every sampling interval, the system outputs a sample.

FIGURE 4–2
System defined as a box.

What operations does the processing system box implement to perform filtering, spectral analysis, and so forth? The answer to this question cannot be specific for any particular processing algorithm. The answer must apply to any form of processing and to a broad class of systems.

In general, systems are categorized under *classes*. This book limits itself to a broad class of systems called *linear/time-invariant* (LTI) systems. This class of systems covers a very broad range of engineering applications including the popular digital filtering and spectral analysis techniques.

4.2.1 TIME-INVARIANT SYSTEMS

Let's examine the significance of the term *time invariant*. When setting up the digital system box, we instruct it to apply a specific processing algorithm. Once this is accomplished, the digital system knows exactly how to process the input samples to produce the output samples. A time-invariant system uses a fixed processing algorithm that always uses exactly the same operations to compute the output samples. A fixed processing algorithm means that the digital system is incapable of modifying its own instructions. In practice, if we wait a minute, a day, a year, or forever, a time-invariant system will still be processing the input samples in exactly the same way.

Time-invariant systems are easy to achieve in digital systems. The whole functionality of the system can be defined with numbers whose values remain exactly the same as the system ages and as environmental conditions such as pressure and temperature change. Compare this to analog systems; their functionality depends on a multitude of components such as resistors and capacitors whose values can fluctuate with age, temperature, and so on. Note that we can only *approximate* analog systems as being time invariant when we consider their operation over a relatively short, stable period of time. A digital system, as opposed to an analog system, can truly be time invariant since its operation depends on numbers whose value never changes.

Note that advanced digital systems can be made to modify their own behavior to adapt to changing environmental conditions. In such sophisticated systems, the procedures used to produce the output actually change over time. When the processing needs to adapt to a changing environment, the system uses sophisticated procedures to modify the numbers that define the processing algorithm. Such advanced systems are not time invariant and are discussed in more advanced DSP books.

4.2.2 LINEAR SYSTEMS

Most scientists accept that large complex problems are easier to solve when they are broken into a number of smaller, easier-to-manage, subproblems. Following this sound approach, the preferred engineering technique is to divide the system into a number of subsystems. This simplifies the engineering job since each subsystem contributes only a small portion of the total system functionality. Figure 4–3 shows such a system being broken into a number of subsystems.

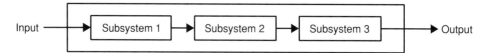

FIGURE 4–3
A system broken into subsystems.

In *linear* systems, the operation of the overall system does not change if we rearrange the sequencing of the subsystems. In Figure 4–4, even if the ordering of the subsystems is shuffled, the overall linear system output remains exactly the same.

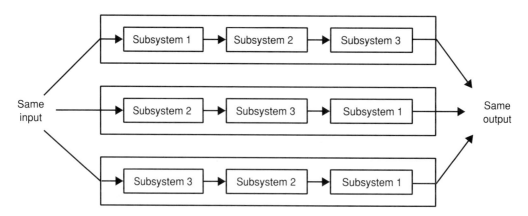

FIGURE 4–4
Order of subsystems changed.

Mathematically, linear systems are systems that implement commutative functions. *Commutativity* is defined as the property of being able to combine elements in such a manner that the result is independent of the order in which the elements are taken. For example, addition and multiplication are commutative operations:

$$a + b + c = c + b + a$$
$$a \times b \times c = c \times b \times a$$

For instance, the following equation shows two forms of a commutative function that uses only addition and multiplication operations:

$$b_0 \times x(0) + b_1 \times x(1) + b_2 \times x(2) = x(2) \times b_2 + x(1) \times b_1 + x(0) \times b_0$$

DSP makes wide use of the linear property of addition and multiplication operations to simplify the structure of many systems (see Chapter 9).

4.2.3 LOOKING INSIDE THE BOX

One must ask; how does a processing system manipulate the information contained in the input signal to produce the output? Answering this question requires playing a detective game of investigating, gathering facts, and drawing conclusions. We begin by reflecting on the obvious conditions that surround the system box.

The first thing that we must recognize is that the input signal $x(n)$ and the output signal $y(n)$ are both functions of discrete time. These signals sequence digital values at a rate of one sample per discrete-time interval (see Figure 4–5). If the input and output were not functions of time, they would never change their value, and this useless system would be receiving an unchanging ADC sample value while producing constant DAC output voltage.

FIGURE 4–5
Digital dignals are a function of discrete time.

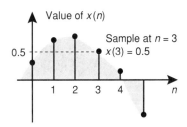

For example, if the ADC samples the analog input at a rate of 10 k-samples/second, then

$$T_S = \frac{1}{10 \text{ k - samples} / \text{second}} = 100 \text{ μs} / \text{sample}$$

This means that the discrete-time variable n increments every 100 μs. For instance, discrete instant $n = 15$ occurs 15×100 μs $= 1500$ μs after the time $n = 0$.

The second thing that we must realize is that the ouput signal depends on the presence of the input signal. If this were not the case, the system would produce an output whether there is an input or not.

We must conclude that the output signal is a function of the input signal. This implies that we should be able to perform a series of mathematical operations (a processing function) on the input signal $x(n)$ to compute the output signal $y(n)$:

$$y(n) = function\ \{x(n)\}$$

Finally, the third thing that we must realize is that the box could remember some of the past input samples. A system that has memory can use the past input samples to build the value of the *present* output sample. The system processing could therefore use a combination of present and past inputs to compute the present output value.

Figure 4–6 illustrates the way that the input values are remembered (every sampling interval T_S) by shifting them through a series of registers. This allows the processor to remember the value of some of the past input samples. Referring to Figure 4–6 we can see that the present and past input samples are labeled $x(n)$, $x(n-1)$, $x(n-2)$, $x(n-3)$, and so on. For example, $x(n-3)$ represents the value of the input three discrete-time units *before* time n. If time n is the present time, then $x(n-3)$ represents the input three time units in the past.

FIGURE 4–6
System with memory.

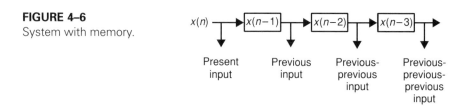

Figure 4–7 illustrates the linear processing stages inside a digital system that implements a processing algorithm that is limited to using four multiplication/addition operations.

4.2.4 DEFINING THE SYSTEM DIFFERENCE EQUATION

Consider a digital system that combines the values of the present and past input samples to compute the present value of the output sample. This system uses a processing algorithm to define the specific operations used to linearly combine the input samples. We can define the algorithm in a way that applies to all LTI systems.

We call the linear processing algorithm the *difference equation* of the system. The difference equation relates the value of the present output sample to the values of the present and past input samples. The following relationship between the input and the output values gives a general form for the difference equation for a linear system:

$$y(n) = b_0 x(n) + b_1 x(n-1) + b_2 x(n-2) + b_3 x(n-3) + b_4 x(n-4) + \ \ldots$$

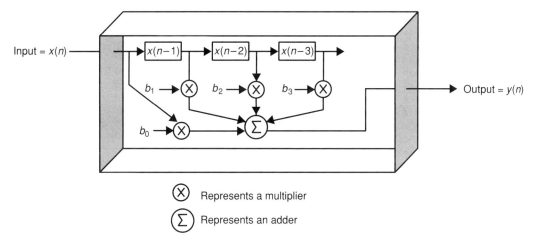

FIGURE 4–7
Output is a function of present and past inputs.

First note that this general difference equation relates the output $y(n)$ to a series of *weighted* present and past inputs. This means that the present output value $y(n)$ is computed by adding a portion (weighting factor b_0) of the present input $x(n)$, a portion (weighting factor b_1) of the previous input $x(n-1)$, a portion (weighting factor b_2) of the previous-previous input $x(n-2)$, and so forth.

Note these facts about the difference equation:

- It is linear because we can change the order in which the terms are added.
- It is time invariant because the values of the weighting factors b_0, b_1, b_2, b_3, b_4, and so forth, are all constants.

As we can see, the difference equation could consist of a very long series of operations. Computing the difference equation requires the memory of a number of past input events. Actually, if the series is very long, it could include the weighting of some very old input values. The amount of memory required to store the past inputs depends on the terms contained in the difference equation. The oldest input event required by the difference equation determines the amount of memory that the particular system requires.

We call the weighting factors the *coefficients* of the system. The values of these coefficients are the only parameters that we control when we design the system.

The values that we assign to the difference equation coefficients totally define the functionality of an LTI system.

This means that we can control the processing of a digital system simply by changing the values of the weighting factors b_0, b_1, b_2, b_3, b_4, and so on.

This fact is one of the main reasons that DSP is so versatile. The same digital system can be used to implement different processing functions simply by changing the value of a few constants. For example, changing the value of the weighting factors in the difference equation can convert a high-pass filter into a low-pass filter.

As an example of setting up a DSP algorithm, let's examine a system that implements the following difference equation:

$$y(n) = 1.2x(n) - 1.32x(n-1) + 1.188x(n-2) - 0.1944x(n-3)$$

The following coefficient values totally define the functionality of this particular system:

$b_0 = +1.2$ Weighting factor applied to the present input
$b_1 = -1.32$ Weighting factor applied to the previous input
$b_2 = +1.188$ Weighting factor applied to the previous-previous input
$b_3 = -0.1944$ Weighting factor applied to the previous-previous-previous input
$b_4 = b_5 = b_6 = \ldots = 0$

This particular system needs to remember only the last three inputs: $x(n-1)$, $x(n-2)$, and the oldest input known to this system: $x(n-3)$. Since older inputs are not required, the other coefficients b_4, b_5, b_6, . . . all carry a weight of zero.

When implementing this difference equation on a digital system, we must do the following:

- Allocate four memory locations to store the values of the present input and the past three inputs.
- Program the processor to generate the individual terms of the difference equation series. We do this by multiplying the stored input samples by the value of the appropriate weighting factor.
- Combine the terms of the difference equation to compute the value of the present output sample $y(n)$. We do this by summing the results of the preceding multiplication operations.

Note that a system that has memory of the past three inputs requires a fourth memory location and a fourth coefficient to accommodate $b_0x(n)$, which relates to the weighted present input. Figure 4–8 illustrates this particular system.

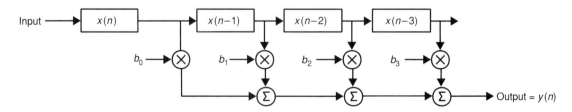

FIGURE 4–8
System that weights four input terms.

The number of terms included in the difference equation series can vary considerably from system to system. Because of this, it is better to use a more compact mathematical notation to express the difference equation that describes an LTI system:

$$y(n) = \sum_{k=0}^{\infty} b_k \, x(n - k) \qquad (4\text{--}1)$$

Equation (4–1) generates all terms that make up the difference equation defining an LTI system. The summation operator generates the terms one by one as the iteration variable k successively takes on the consecutive values that range from zero to infinity.

When implementing practical systems, the processor computes each term in the $y(n)$ series by iterating k to the next value so that reference is made to the next weighting factor b_k and to the next past input $x(n - k)$. As k is iterated, the processor must accumulate the intermediate result to the terms that have already been computed. Because of this, we cannot practically iterate k through an infinite number of values since the digital system would have to iterate through an infinite number of terms to compute a single output value.

In practice, the number of terms in the difference equation must therefore be finite. The system described in Figure 4–8 is an example of a system that requires the iteration of k through only four values to cover all terms of the difference equation. We can therefore implement this system in practice by programming the digital system to perform four multiplication/addition operations. Imagine how intense the processing of a system with hundreds of weighting factors would be like.

4.3 IMPULSE RESPONSE

Section 4.2.4 discussed the fact that the functionality of an LTI system is completely defined by the values of its difference equation coefficients. Reversing this statement, if we know the system coefficient values, we know what the system does and how it works. In other words, once we know the coefficient values, we know all that there is to know about the processing done by that particular system.

We can apply a special function to the system input to determine the coefficient values. We call this function a discrete-time *impulse function.* Figure 4–9 illustrates what it looks like. As Figure 4–9 indicates, the discrete-time impulse function has a value of 0 everywhere except at discrete time $n = 0$, where the function equals the value of 1.

FIGURE 4–9
Discrete-time impulse function.

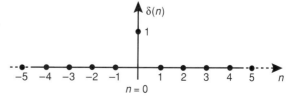

Let's determine what output signal we obtain when we apply this function to the input of a system. We must remember that the output values are determined by computing the difference equation, whose general format we repeat here for convenience:

$$y(n) = b_0 x(n) + b_1 x(n-1) + b_2 x(n-2) + b_3 x(n-3) + b_4 x(n-4) + \ldots$$

The response of the system starts at discrete time $n = 0$ when we apply the impulse function to the input. At that time, the impulse input value is $x(0) = \delta(0) = 1$. Since the impulse input has a value of 0 at all other discrete times, the other input values such as $x(1)$, $x(2)$, $x(3)$, ... all equal 0.

Let's compute the value of the output sample at time $n = 0$:

$$y(0) = b_0 x(0) + b_1 x(0-1) + b_2 x(0-2) + b_3 x(0-3) + b_4 x(0-4) + \ldots$$
$$y(0) = [b_0 \times 1] + [b_1 \times 0] + [b_2 \times 0] + [b_3 \times 0] + [b_4 \times 0] + \ldots$$
$$y(0) = b_0$$

At time $n = 0$, the system output therefore reflects the value of coefficient b_0. Let's increment to $n = 1$, which is one unit of discrete time after applying the impulse function to the input. The calculation of the value of the output sample at time $n = 1$ yields

$$y(1) = b_0 x(1) + b_1 x(1-1) + b_2 x(1-2) + b_3 x(1-3) + b_4 x(1-4) + \ldots$$
$$y(1) = b_0 x(1) + b_1 x(0) + b_2 x(-1) + b_3 x(-2) + b_4 x(-3) + \ldots$$
$$y(1) = [b_0 \times 0] + [b_1 \times 1] + [b_2 \times 0] + [b_3 \times 0] + [b_4 \times 0] + \ldots$$
$$y(1) = b_1$$

At time $n = 1$, the system output therefore reflects the value of coefficient b_1.

If we now go to $n = 2$, which is two units of discrete time after the impulse is applied, we compute the output value to be

$$y(2) = b_0 x(2) + b_1 x(2-1) + b_2 x(2-2) + b_3 x(2-3) + b_4 x(2-4) + \ldots$$
$$y(2) = b_0 x(2) + b_1 x(1) + b_2 x(0) + b_3 x(-1) + b_4 x(-2) + \ldots$$
$$y(2) = [b_0 \times 0] + [b_1 \times 0] + [b_2 \times 1] + [b_3 \times 0] + [b_4 \times 0] + \ldots$$
$$y(2) = b_2$$

At time $n = 2$, the system output therefore reflects the value of coefficient b_2. The system is sequentially outputting the value of its coefficients.

Applying an impulse function to a system input forces the sequential output of the coefficient values that define that system.

Because the system impulse response reveals so much about the system, we refer to it with the special name $h(n)$. Figure 4–10 illustrates the output of a system when we apply an impulse function stimulus. The figure shows that the system output remains at zero until the impulse is applied to the input at time $n = 0$. Following the impulse, the system consecutively outputs the coefficient values that define its difference equation.

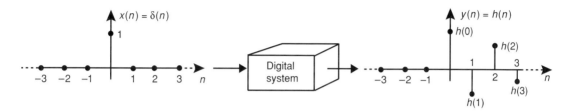

FIGURE 4–10
Impulse response $h(n)$.

One cannot help but admire the usefulness of the impulse function. One little pulse applied at the input of a system forces that system to reveal its entire identity. The impulse function is therefore a very important type of input used in practical engineering applications to test and analyze systems. Sections 4.3.2 and 4.3.3 put the impulse function to practical use to test and analyze systems.

4.3.1 FINITE IMPULSE RESPONSE

As discussed, the amount of memory required to create a particular system is directly related to the oldest input that the system must remember. The oldest input remembered by a system can be defined as being in the system for k_{Max} discrete-time units. This oldest sample drops from memory when the system acquires the sample number $[k_{Max} + 1]$. This means that the samples stay in memory for

$$[k_{Max} + 1] \times T_S \text{ second}$$

where T_S is the sampling interval.

If a system remembers the last k_{Max} past inputs, that system requires $[k_{Max} + 1]$ memory locations to accommodate the present input. This means that this system requires $[k_{Max} + 1]$ nonzero weighting coefficients. The other coefficients must have a value of 0 since we do not need to use them in the calculation of the system output.

As an example, imagine a system that samples its analog input every 100 μs and implements the following difference equation:

$$y(n) = 1.2x(n) - 1.32x(n-1) + 1.188x(n-2) - 0.1944x(n-3)$$

In this system, $k_{Max} = 3$, and the system requires four memory locations to store the input samples. This system has an impulse response $h(n)$ that lasts for

$$[k_{Max} + 1] \times 100 \text{ μs} = 4 \times 100 \text{ μs} = 400 \text{ μs}$$

After that interval, the impulse response remains at zero. Figure 4–11 illustrates the impulse response of this practical system. Note that the impulse response $h(n)$ goes back to a flat zero and stays there after the system runs out of nonzero coefficients. Because of this, a system that is limited to remembering the last k_{Max} inputs has an impulse response of *finite duration*. Systems that only make use of a limited number of past inputs are called *finite impulse response* (FIR) systems.

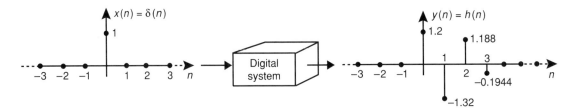

FIGURE 4–11
Example of an impulse response.

4.3.2 TESTING SYSTEMS

Most practical DSP systems acquire an analog input signal by using an ADC to digitize that input. To apply a discrete-time impulse function to such system, we must apply a pulse to the analog input of the ADC. When the ADC translates this pulse to the discrete-time domain, it becomes $\delta(n)$ and is transferred to the digital processing section. Figure 4–12 illustrates an analog pulse that the ADC digitizes into a discrete-time impulse.

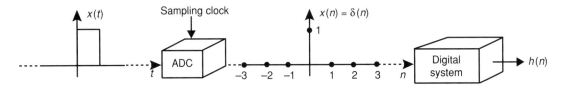

FIGURE 4–12
ADC digitizes an analog pulse.

To successfully translate the analog pulse to a discrete-time impulse, we must adjust the pulse duration so that the ADC samples it only once. The duration of the pulse ideally should equal the sampling interval T_S. If the analog pulse is shorter than T_S, the ADC sampling could miss it. If it is longer than T_S, it could be sampled more than once. Figure 4–13 illustrates the different cases.

The only way to ensure that the analog pulse is sampled exactly once is to synchronize its generation with the sampling clock of the ADC. To do this, the testing circuitry that generates the analog pulse must "steal" the clock from the system being tested and use it to adjust the duration of the analog pulse to last exactly T_S time units.

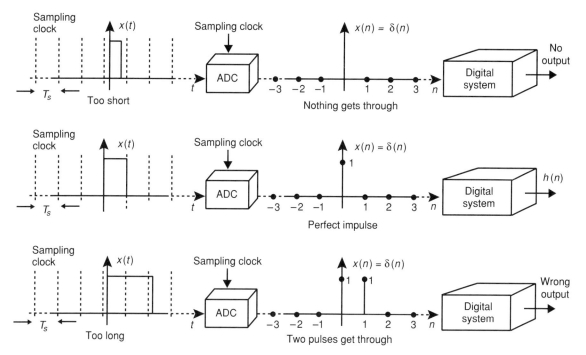

FIGURE 4–13
Duration of an analog impulse function.

For example, consider a system that is defined by the following difference equation:

$$y(n) = 1.2x(n) - 1.32x(n-1) + 1.188x(n-2) - 0.1944x(n-3)$$

When we apply a pulse $x(t)$ with a duration of T_S, the system responds by outputting its impulse response.

Figure 4–14 illustrates the testing circuitry necessary to generate the system impulse response. The figure indicates that the impulse response consists of a staircase response because the output is produced in practice by a DAC. The DAC translates each output sample to a voltage that is maintained over T_S time units. This staircase response still reflects the value of the system coefficients since the steps are a height that exactly corresponds to the coefficient values.

Performing impulse testing on systems can be useful, for example, in a manufacturing environment when the functionality of a large number of systems needs verification. At the end of the assembly line, we can submit the newly manufactured system to a testing circuitry that applies impulse testing.

The waveform at the output of the system under impulse testing is called the *signature* of the system. We can set up the testing circuitry so that it automatically compares the output signature to a reference signature that comes from a system known to be functioning properly. What is remarkable here is that we can use the same testing circuitry to

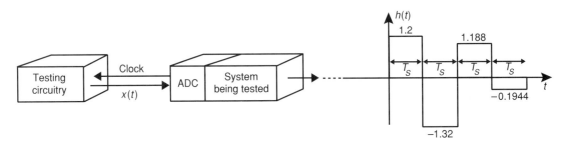

FIGURE 4–14
Example of a practical impulse test setup.

test a variety of different systems. When we test a different system, we need to change only the reference signature used in the comparison.

According to Section 3.5.3, most DSP systems limit the frequency content of the output signal by passing the DAC output through a reconstruction filter. The presence of this low-pass filter alters the shape of the signature from a staircaselike appearance to a smoother shape. In any case, even if the filter has altered the appearance of the signature, it still displays a shape whose characteristics can be used for signature matching. Figure 4–15 shows how the reconstruction filter alters the signature.

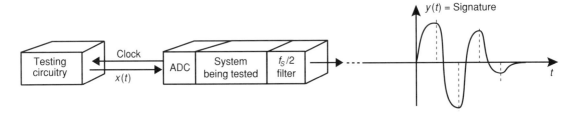

FIGURE 4–15
Signature with an output filter.

If we apply only one impulse to the system being tested, the signature appears only once at the output of the system. In most practical applications, it is preferable to apply periodic impulses so that the output signature repeats itself after periodic intervals of time. Once the signature becomes a periodic event, we can average its shape to get better precision, and we can display it on devices such as oscilloscopes or frequency analyzers.

How often should we repeat the impulse stimulus? To answer this question, we must recall that the duration of a system signature is related to the oldest input sample that the system being tested remembers. This means that we must wait until the signature pattern is complete before we apply the next impulse. According to Section 4.3.1, the oldest input remembered by a system is k_{Max} discrete-time units old; the duration of the signature therefore lasts $[k_{\text{Max}} + 1] T_S$ second.

It follows that we can apply the test at the following maximum rate of impulse testing:

$$\frac{1}{(k + 1)\, T_S \text{ second}} \text{ impulses/second}$$

Figure 4–16 illustrates a periodic impulse stimulus.

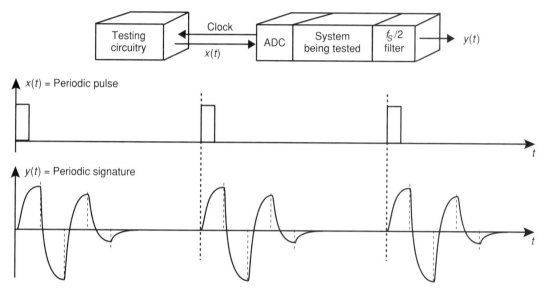

FIGURE 4–16
Periodic impulse stimulus.

The impulse function can be mathematically proven to carry a finite amount of power, which manifests itself equally over the DSP useful range of frequencies (from $\Omega = 0$ to π). This means that when we introduce an impulse function $\delta(n)$, we are simultaneously inputting a signal that contains all frequencies that this DSP system can process. If the purpose of this system is to filter the frequency contents of the input, the output of the system (the signature) shows which frequencies make it through the system and which are removed. This can be very useful to check the frequency response of systems quickly and accurately. To implement this technique, we need a *frequency analyzer* to extract the frequency response from the system signature. Figure 4–17 illustrates the frequency analysis of a system that implements a band-pass filter that allows only a band of frequencies to pass through.

4.3.3 REVERSE ENGINEERING

According to previous sections, inputting an impulse function forces a DSP system to output its signature. The shape of the signature reflects the coefficient values that define the difference equation that the system implements. Since these coefficient values completely

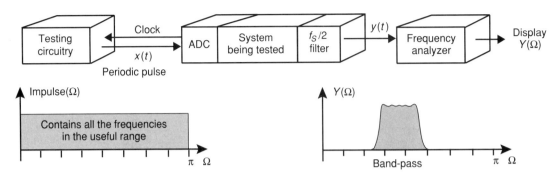

FIGURE 4–17
Analysis of the signature frequency contents.

determine the functionality of the system, we can use this information to determine how the system box works without even opening the box.

Using the same testing circuitry as that used in Section 4.3.2, we can analyze the output signature to *estimate* the coefficient values that define the system under scrutiny. Once we have done this, we can create a new system whose behavior closely resembles that of the original system. To do this, we simply insert the estimated coefficient values in the difference equation of the cloned system. The hardware structure and the processor used in the cloned system are irrelevant since the functionality of any DSP system resides only in the way that we manipulate the input numbers. If we are careful in estimating the coefficient values, the cloned system closely replicates the processing and the behavior of the original system.

4.4 SYSTEMS WITH FEEDBACK

Section 4.2.4 discussed the fact that LTI systems use a combination of present and past input values to compute the value of the output samples. All systems that we previously examined relied on memory locations to store the required input values. For convenience, the relationship between the output and the input values, developed in Section 4.2.4, is repeated here:

$$y(n) = \sum_{k=0}^{\infty} b_k \, x(n - k)$$

Expanding this series, we obtain

$$y(n) = b_0 x(n) + b_1 x(n - 1) + b_2 x(n - 2) + b_3 x(n - 3) + b_4 x(n - 4) + \ldots$$

As the processing function becomes more complex, the length of the difference equation increases. This requires more memory to store the increased number of past input values and extra computing to calculate the longer output equation.

There is an alternate way to implement complex functions without stressing the memory and computing resources of the system. This technique calls for feeding past output values back into the system difference equation. We express a system with feedback by adding an extra term to the difference equation series:

$$y(n) = \sum_{k=0}^{P} b_k x(n - k) + \sum_{k=1}^{Q} a_k y(n - k)$$

Expanding the two series for $y(n)$, we obtain the following difference equation for a system that includes output feedback:

$$y(n) = b_0 x(n) + b_1 x(n - 1) + b_2 x(n - 2) + b_3 x(n - 3) + \ldots + b_P x(n - P)$$
$$+ a_1 y(n - 1) + a_2 y(n - 2) + a_3 y(n - 3) + \ldots + b_Q x(n - Q)$$

A finite number of a_k and b_k coefficients now define the functionality of the system. In the series, each of the previous output $y(n-k)$ is weighted according to its respective weighting factor a_k. The b_k coefficients still correspond to the weighting factors applied to the input samples. Notice that a large number of *past* output values could be used to compute the *present* output value. The number of past output values that the system needs to remember depends on the oldest delayed output $y(n-Q)$ to which is applied the nonzero a_Q coefficient.

Figure 4–18 illustrates a system whose structure uses three past input values *and* three past output values to produce the present output value. Note in Figure 4–18 that the present output sample $y(n)$ is computed by combining weighted input values (on top of the adder) with previous weighted output values (under the adder).

In Figure 4–18, the past outputs are obtained by sending the present output $y(n)$ through a multistage shift register. The stages of the shift register therefore contain the delayed output values $y(n-1)$, $y(n-2)$, and $y(n-3)$. The a_1, a_2, and a_3 factors weight the past outputs before the adder considers them.

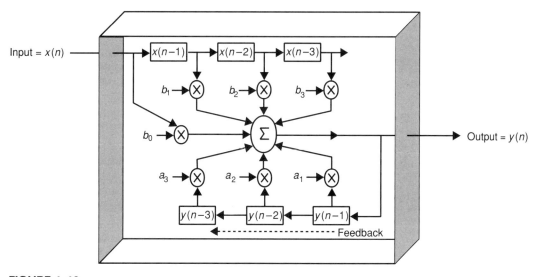

FIGURE 4–18
System with feedback.

Let's analyze a system that uses feedback to calculate the value of the output samples. To simplify the analysis, we consider the simplest possible feedback system. This minimum system must include at least one input term and one output term. The input term ensures that the output is a function of the input, and the output term ensures that there is feedback. The minimum system is therefore limited to consider only two nonzero coefficients to weight the present input and the previous output:

$$y(n) = b_0 x(n) + a_1 y(n-1)$$

Let's analyze the content of the output sample $y(n)$. We can see that it combines the weighted present input $b_0 x(n)$ and the weighted previous output $a_1 y(n-1)$. Unfortunately, this does not give a clear picture of the content of $y(n)$ since it depends on the value of the previous output sample. To analyze the content of the previous output, we substitute $n-1$ for n in the difference equation:

$$y(n-1) = b_0 x(n-1) + a_1 y(n-1-1)$$
$$= b_0 x(n-1) + a_1 y(n-2)$$

Examining the contents of $y(n-1)$, we notice that the expression contains the input $x(n-1)$. Since $y(n)$ contains $y(n-1)$, it follows that the $y(n)$ indirectly includes the previous input value $x(n-1)$. Applying the same reasoning to $y(n-2)$, we realize that $y(n)$ also indirectly includes $x(n-2)$. We can continue applying this reasoning to infinity since every output requires an even older output term that includes older inputs. It now becomes obvious that $y(n)$ indirectly corresponds to an endless series of past input terms:

$$y(n) = b_0 x(n) + a_1 y(n-1)$$
$$= b_0 x(n) + a_1 [b_0 x(n-1) + a_1 y(n-2)]$$
$$= b_0 x(n) + a_1 b_0 x(n-1) + (a_1)^2 [b_0 x(n-2) + a_1 y(n-3)]$$
$$y(n) = b_0 x(n) + a_1 b_0 x(n-1) + (a_1)^2 b_0 x(n-2) + (a_1)^3 [b_0 x(n-3) + a_1 y(n-4)]$$

and we can keep expanding $y(n)$ to include an infinite number of past input terms.

We call systems that behave this way *recursive systems*. As noted when the difference equation that describes a feedback system was expanded, every output sample is built from an infinite number of weighted past input samples. We can conclude that recursive systems behave as if they had *infinite memory* of past input events!

Systems that use output feedback allow for complex processing functions to be implemented without the need for large amounts of memory or extra computing resources.

4.4.1   INFINITE IMPULSE RESPONSE

According to Section 4.3, applying an impulse function to a system forces it to output the b_k coefficients that define the input weighting factors. Let's apply an impulse to a system that uses feedback. To simplify our analysis, we consider a simple feedback system in which $b_0 = 1.0$, $a_1 = 0.7$, and all other coefficients have a value of 0. The following difference equation describes this simple feedback system:

$$y(n) = 1.0x(n) + 0.7y(n-1)$$

Remember that the discrete impulse function $\delta(n)$ has a value of 1 at discrete-time $n = 0$ and zero at all other times. Since the discrete impulse is zero valued for $n < 0$, all system outputs are also zero valued for that interval since the input was inactive. Simply stated, the system output is at rest before the impulse is applied:

$$y(-6) = y(-5) = y(-4) = y(-3) = y(-2) = y(-1) = 0$$

Let's calculate the value of the output samples for this recursive system starting at $n = 0$ when we apply the impulse:

$$y(n) = 1.0x(n) + 0.7y(n-1)$$
$$y(0) = 1.0x(0) + 0.7y(0-1) = 1.0 \times 1 + 0.7 \times 0 = 1.0$$
$$y(1) = 1.0x(1) + 0.7y(1-1) = 1.0 \times 0 + 0.7 \times 1.0 = 0.7$$
$$y(2) = 1.0x(2) + 0.7y(2-1) = 1.0 \times 0 + 0.7 \times 0.7 = 0.49$$
$$y(3) = 1.0x(3) + 0.7y(3-1) = 1.0 \times 0 + 0.7 \times 0.49 \cong 0.34$$
$$y(4) = 1.0x(4) + 0.7y(4-1) = 1.0 \times 0 + 0.7 \times 0.343 \cong 0.24$$
$$y(5) = 1.0x(5) + 0.7y(5-1) = 1.0 \times 0 + 0.7 \times 0.2401 \cong 0.17$$
$$y(6) = 1.0x(6) + 0.7y(6-1) = 1.0 \times 0 + 0.7 \times 0.16807 \cong 0.12$$

and the series continues forever.

Figure 4–19 graphs a part of this infinite response. According to this figure, the $0.7y(n-1)$ term corresponds to an infinite number of $b_k x(n-k)$ terms. The impulse response reflects the exact value of the b_k coefficients that an equivalent FIR system would require:

$$y(n) = 1.0x(n) + 0.7y(n-1)$$

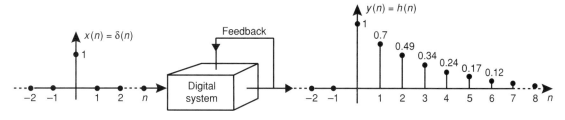

FIGURE 4–19
Impulse response of a recursive system.

is equivalent to:

$$y(n) = 1.0x(n) + 0.7y(n-1) + 0.49x(n-2) + 0.34x(n-3) +$$
$$0.24x(n-4) + 0.17x(n-5) + \ldots$$

Therefore, we call systems that use output feedback *infinite impulse response* (IIR) systems.

The impulse response in the last example decays following an exponential curve. This means that the infinite b_k values that correspond to the impulse response become smaller and smaller. The older inputs therefore carry less and less weight. This trend of the impulse response to decay is typical of *stable* feedback systems. An IIR system does behave as if it has infinite memory, but the older inputs have a tendency to carry less weight.

4.5 SYSTEM STABILITY

We must understand that practical systems are not perfect. No practical system can avoid the presence of a certain amount of noise. One typical source of noise comes from brief glitches that occur at the analog input of most systems. Another source of noise comes from the quantization of the analog input when the ADC rounds the analog amplitude into digital steps. The processor introduces additional noise when it rounds results to accommodate the binary format of the answer.

Imagine for a second that, for some unknown reason, the input becomes abnormal for a brief interval. To simplify the situation, assume that the value of a single input sample is corrupted. What effect does this have on the system processing? How is the output signal affected?

To answer these questions, we must review the difference equation that defines the system. The system may or may not have feedback; we analyze each case separately.

4.5.1 STABILITY IN FINITE IMPULSE RESPONSE SYSTEMS

An FIR system is memory limited and can remember only a certain number of past input values. If one of the input values is corrupted due to noise, the system remembers that corrupted input only for a certain amount of time.

DSP systems are time invariant, which means that the difference equation coefficients are stored as *constants* in memory. Because of this, the algorithm itself is in no way affected by the bad input. For example, consider the following FIR system whose output function requires three past input values:

$$y(n) = 1.2x(n) - 1.32x(n-1) + 1.188x(n-2) - 0.1944x(n-3)$$

When the ADC digitizes a corrupted input, the algorithm computes the output sample $y(n)$ by using the corrupted present sample $x(n)$. When the next "good" input comes in, the corrupted sample becomes one discrete-time unit older and is shifted to $x(n-1)$. As the ADC digitizes more good inputs, the corrupted sample eventually becomes four discrete-time units old, at which time it is shifted out of the system. In this particular FIR system, any disturbance at the input causes a temporary output transient that lasts for four sampling intervals. Once the bad sample is shifted out, the output signal returns to normal.

If some input value is corrupted in an FIR system, the output signal is also corrupted but only for as long as the bad sample is stored in the system memory. We can conclude that FIR systems are stable since they eventually go back to normal operation even if some abnormal condition enters them.

FIR algorithms always yield stable system operation.

4.5.2 STABILITY IN INFINITE IMPULSE RESPONSE SYSTEMS

An IIR system behaves as if it has infinite memory. If one of the inputs is abnormal, it will be remembered forever. This means that the processing algorithm uses that corrupted input in the calculation of all following output values. This information should activate an alarm bell loudly right now. IIR systems can produce unstable operation.

We must develop a technique that prevents the IIR system memory from accumulating noise levels that could render the system unstable or useless. Recall that the processing algorithm weights the past input samples before accumulating them into the output value. If the weighting factors that apply to old inputs become small enough (close to zero), the bad input samples no longer carry enough weight to significantly affect the output values. Under this condition, the output signal gradually returns to normal even if there is a bad input value.

It follows that to obtain stable operation for an IIR system, we must ensure that the weighting factors have the general tendency to become smaller and smaller as they relate to older inputs. This means that the IIR, which corresponds to the weighting factors of that system, will eventually decay to values that are close to zero.

Mathematically, for the system to be stable, the impulse response $h(n)$ must be *bounded*. This means that adding the absolute value of *all* terms of the IIR does not produce an infinite result:

$$\sum_{k=0}^{\infty} h(k) = \text{a bounded value}$$

For example, the following *infinite* series is a bounded series known to converge toward the value of 2:

$$\sum_{n=0}^{\infty} \left(\frac{1}{2}\right)^n = 1 + \frac{1}{2} + \frac{1}{4} + \frac{1}{8} + \frac{1}{16} + \ldots \Rightarrow 2$$

Let's analyze the case of a simple IIR system that uses its present input and the previous output to calculate the value of the present output sample:

$$y(n) = b_0 x(n) + a_1 y(n-1)$$

As we will see, this system could be either stable or unstable; its stability depends on the value of the coefficient a_1.

We use a fixed value of 1 for b_0 to investigate how the impulse response is affected by the value of the coefficient a_1. This helps to determine how the weighting of previous outputs influences the stability of feedback systems. Figure 4–20 illustrates the impulse response for different values of the coefficient a_1.

The first two cases illustrated in Figure 4–20 show that as a_1 approaches a value of 1, the impulse response takes longer to decay at near-zero values. The system behaves as if it contains a larger number of weighting factors that are not close to zero. In practice, if an input sample is corrupted, the resulting output transient lasts for as long as that bad sample carries a significant weight.

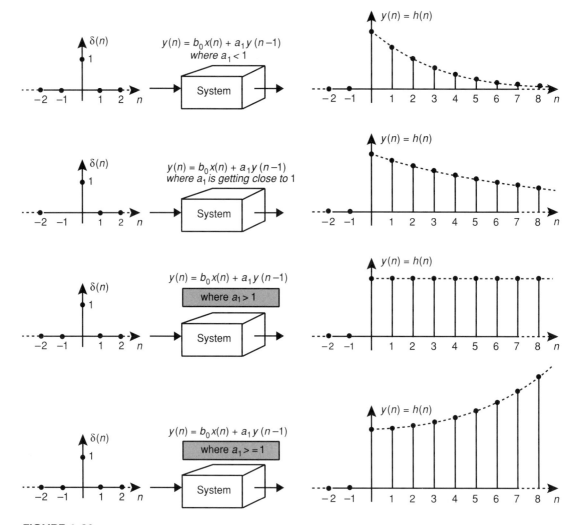

FIGURE 4–20
Impulse responses of an IIR system.

When the value of a_1 reaches 1, the impulse response no longer decays and the output transients last forever. This particular IIR system is on the edge of unstable operation.

If the value of a_1 is made greater than 1, the impulse response increases exponentially. This means that equivalent weighting factors become greater and greater. In such systems, the older the input, the more weight it carries. Such systems amplify instabilities, and the result is an unstable behavior.

To avoid unstable operation in IIR systems, we must be careful when we select the values for the a_k coefficients. Section 7.6 discusses the exact conditions that ensure stable system operation.

4.6 CONVOLUTION

DSP systems continuously receive new input samples $x(n)$ from the ADC system. For every input sample coming in, the system must compute the difference equation to generate the value of an output sample $y(n)$. Equation (4–1) is repeated here for convenience:

$$y(n) = \sum_{k=0}^{\infty} b_k\, x(n - k)$$

The calculation of each output sample $y(n)$ involves summing a number of input samples that have been scaled by their respective weighting factors. This calculation must be repeated at every sampling interval T_S. This section closely analyzes the processing steps necessary to compute each of the system output samples.

4.6.1 MECHANICS OF COMPUTING OUTPUT SAMPLES

Now we examine what is involved in computing the output of an FIR system defined by the following difference equation:

$$y(n) = 1.0x(n) - 1.1x(n - 1) + 0.99x(n - 2) - 0.16x(n - 3)$$

Figure 4–21 illustrates the impulse response of this system. Looking at the impulse response $h(n)$ in Figure 4–21, we immediately note that the impulse response values correspond exactly to the coefficient values of the system difference equation. These values are as follows:

At $n = 0$, we find $h(0) = b_0 = 1.0$, which the processor uses to weight the input sample at time n.

At $n = 1$, we find $h(1) = b_1 = -1.1$, which the processor uses to weight the input sample at time $n - 1$: $x(n-1)$.

At $n = 2$, we find $h(2) = b_2 = 0.99$, which the processor uses to weight the input sample at time $n - 2$: $x(n-2)$.

At $n = 3$ we find $h(3) = b_3 = -0.16$, which the processor uses to weight the input sample at time $n - 3$: $x(n-3)$.

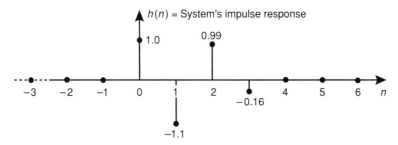

FIGURE 4–21
Impulse response of a particular FIR system.

From this we can see that each impulse response value of $h(k)$ corresponds to a coefficient b_k, which is used to weight a particular past input $x(n-k)$.

To illustrate signals, we must examine a sequence of samples. Figure 4–22 shows an arbitrarily chosen sequence of samples that illustrate the input history of a system from eight discrete-time units in the past ($n = -8$) to the present ($n = 0$).

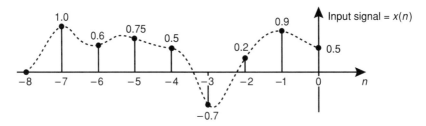

FIGURE 4–22
Sequence of input samples.

We graphically represent the mechanics necessary to compute a particular output value. Figure 4–23 illustrates the proper alignment of weighting factors and input samples. Notice that the positioning of the weighting factors in this figure corresponds to the impulse response that has been flipped horizontally. Figure 4–24 illustrates this.

Since weighting factors, coefficient values, and impulse response are the same, we can substitute the impulse response $h(k)$ for the b_k coefficients in the difference equation:

$$y(n) = h(0)x(n) + h(1)x(n-1) + h(2)x(n-2) + h(3)x(n-3)$$

This form of the difference equation is equivalent to the following general summation:

$$y(n) = \sum_{k=0}^{\infty} h(k)x(n-k) \tag{4–2}$$

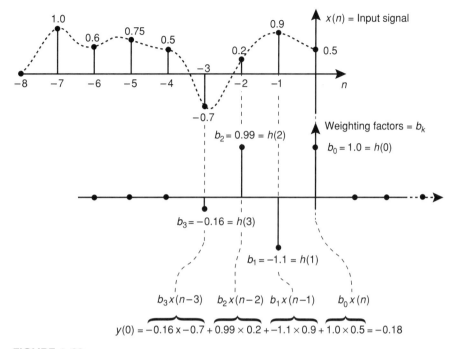

FIGURE 4–23
Aligning the input samples with the weighting factors.

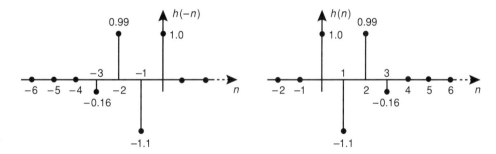

FIGURE 4–24
Flipped impulse response.

When the summation in Equation (4–2) is used repeatedly to compute a sequence of output samples, it is called the *convolution sum*. This operation is so useful that mathematicians have defined a special operator to describe it:

$$y(n) = h(n) * x(n)$$

where $*$ is the convolution operator.

The convolution sum takes the set of values that defines the system *h*(*n*) and applies it to a set of input samples *x*(*n*) to yield the value of a sequence of output samples *y*(*n*).

Repeating the convolution operation as every sample of the input signal comes in produces the sequence of samples that makes up the output signal. Notice that $h(k)$ in the summation is independent of the discrete-time variable n. This is because $h(k)$ corresponds to the coefficients of a time-invariant difference equation. The same set of $h(k)$ values is therefore used to compute *all* convolution operations for a particular system. As an example, we generate a part of the output signal by repeating the convolution operation. Figure 4–25 illustrates the process.

Notice that the flipped version of the impulse response in Figure 4–25 moves forward at every sampling interval to keep it aligned with the output sample being computed. Now that the two functions are aligned, calculating the value of each of the output samples is simply a matter of multiplying the input samples with the lined-up weighting factors and accumulating the results.

Try to verify at least one of the many outputs calculated in Figure 4–25. At this point, reflect on the amount of processing required to provide the output of a new value at every sampling interval. The example in Figure 4–25 is a rather simple system whose difference equation relies on only three past inputs. We refer to this system as a *three-tap* system. Imagine a system that relies on 100 taps (100 past inputs). These systems are common when FIR systems are implemented, and typical signal processors need to perform millions of multiplication and addition operations every second.

To simulate the operation of some DSP system, we convolve the set of weighting factors of that system with a sequence of mock input values. For most systems, calculating even just a few output values by hand proves to be a monumental task. Because calculating the convolution operation involves such a large number of multiplication and addition operations, we should use some mathematical software package (such as a spreadsheet) to simulate a convolution algorithm. Most good packages support the convolution operation.

Implementing a DSP algorithm into a practical system that samples an input signal can be done either by programming the convolution algorithm on a digital signal processor or by building a pure hardware logic circuit that implements that algorithm. Pure hardware systems are usually able to handle sampling rates that are much higher than their software counterparts. On the other hand, modifying a hardware system usually means major hardware modifications and therefore hardware solutions do not usually yield the flexibility provided by software based solutions. Figure 4–26 illustrates these two different approaches.

For example, say that we want to design a system that alternately serves in three different ways: a modem, a fax modem, and a frequency synthesizer. The software approach provides a better solution for this type of system since we can store the three different processing algorithms in the system memory. In this case, we can switch the system function by getting the processor to execute instructions belonging to a different section of memory. In this case, all three systems share the same ADC, memory, processor, and DAC hardware. Conversely, if we were to design special hardware for three different functions, we would probably need to design and build three different circuits.

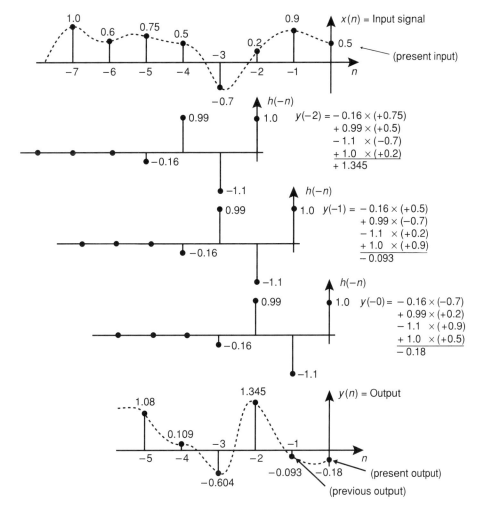

FIGURE 4–25
Convolution is used to generate the output signal.

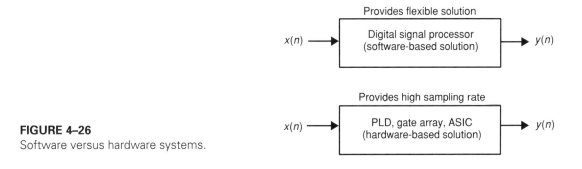

FIGURE 4–26
Software versus hardware systems.

Specialized integrated circuits such as programmable logic devices (PLDs), gate arrays, or application-specific integrated circuits (ASICs) can take us from idea to solution quickly since most manufacturers provide macrofunctions capable of implementing many standard DSP algorithms. In this case, designing hardware circuits is almost as simple as putting the set of b_k coefficients into the chip and powering it up!

4.6.2 CAUSALITY

When we developed the expression for the convolution operation, we limited the analysis to the use of present and past input samples. We call systems that never use future inputs *causal* systems. However, some systems can use future input samples to compute the output value. As soon as a system needs to refer to a future input in its calculation of the present output, it violates causality (the system is noncausal). In practice, it may seem silly even to consider that a system could look into the future; nevertheless, some applications in which the input signal is not processed in real time do exist. For example, imagine that the samples being received from a space probe are stored in some vast memory over the course of many hours. Following this data acquisition, a processor may process the stored samples to extract the data and image information. Since all signal samples are stored in memory, the processor has access to any sample. The processor may take as much time as it requires to apply very sophisticated convolution algorithms to the stored information.

To picture the operation of a noncausal system, assume that the processor is presently working on the calculation of the 1000th output sample. In this case, the processor considers the present to be discrete-time $n = 1000$. From the processor perspective, any input sample that comes after $n = 1000$ belongs to the future. However, since all the input samples are already stored in memory, the processor has access to input samples that were acquired after $n = 1000$. In this case, the difference equation that describes the processing may compute the present output by using future inputs.

In some situations, systems may purposely produce a delayed output. For example, consider that the present discrete time is $n = 25$ and that the system is presently acquiring the input sample $x(25)$. If the output is designed to be delayed by an arbitrary five discrete-time units, the processor must presently be computing the $y(20)$ output sample. The output being computed therefore lies five discrete-time units behind the input's real time reference. If we think about this system with respect to the output's point of view, this system has access to the future five input samples (since they have already come in) when it calculates the value of the $y(20)$ output.

To be able to describe both causal and noncausal systems, we must adjust the convolution sum limits to span any discrete-time interval from $-\infty$ to $+\infty$:

$$y(n) = \sum_{k=-\infty}^{\infty} h(k)x(n - k) \qquad \text{(4–3)}$$

Let's examine the impulse response for a noncausal system (see Figure 4–27). The impulse input in Figure 4–27 occurs at $n = 0$, yet a part of the impulse response $h(n)$, which results from this impulse, exists on the left side of $n = 0$ *before* the impulse input is applied.

FIGURE 4–27
Impulse response of a noncausal system.

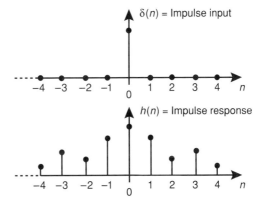

We saw in Section 4.3 that the impulse response corresponds to the weighting factors of the difference equation. A noncausal system weights future inputs, and therefore it violates the *cause-and-effect* principle; the effect happens *before* the cause.

The summation in Equation (4–3) starts at $k = -\infty$. When k is negative, the summation refers to input samples located in the future: $x(n - (-k)) = x(n + k)$. The expression $x(n + k)$ refers to an input sample that is k discrete-time units into the future.

When we calculate the convolution sum described by Equation (4–3), the variable k may access and weight input samples ranging from infinitely far into the future ($k = -\infty$) to infinitely far in the past ($k = +\infty$). For example, when the iteration reaches a value of $k = -10$, the summation generates the expression $h(-10)\, x(n + 10)$, which corresponds to weighting an input sample that lies 10 discrete-time units in the future. Later, when k iterates to a value of $k = +10$, the summation generates the expression $h(+10)\, x(n - 10)$, which corresponds to weighting an input sample that lies 10 discrete-time units into the past.

The convolution sum described in Equation (4–3) therefore completely describes how past, present, and future input samples are weighted to compute the output signal of any LTI system.

SUMMARY

- A time-invariant system always uses the same procedure to process the input signal.
- In a linear system, we can change the order in which the subsystems are sequenced.
- Most systems build their output values by linearly combining present and past input values in a time-invariant way.
- The difference equation gives the function that relates the output signal samples to the input signal samples.
- The b_k coefficient values of the difference equation completely define the system operation.

- Applying an impulse function to the input of an LTI system forces it to output all of its weighting factors in the form of the impulse response $h(n)$.
- The impulse response can be used to verify the operation of an LTI system.
- The impulse response can be used to reverse engineer an LTI system.
- If a system is limited to using a finite number of past input values to build the output, the system has a finite impulse response (FIR).
- An FIR system always provides stable system operation.
- A system may use past *output* values to calculate the present output value.
- If a system uses past output values to calculate the present output value, that system behaves as if it has infinite memory.
- A system that uses feedback always has an infinite impulse response (IIR).
- An IIR system has an element of instability, which must be monitored closely to ensure that it is controlled.
- For an IIR system to provide stable operation, its impulse response *must* decay to near-zero values.
- *Convolution* is a mathematical operation routinely implemented in a digital signal processing system to compute the values of the output signal.
- Digital systems may convolve past, present, and even *future* inputs to compute the output signal.
- Causal systems never use future inputs.

PRACTICE QUESTIONS

4-1. An 8-bit ADC has an input range of ± 10 v. This ADC samples the signal $5 \cos(1000t)$ at a rate of 1000 samples/second. Assume that the sampling starts at time $t = 0$. What is the binary value of the first five 8-bit samples produced by the ADC?

4-2. Which of the subsystems in Figure 4–28 can be sequenced to form a linear system?

Subsystem 1
$$y(n) = x(n)^2$$

Subsystem 2
$$y(n) = x(n) + 1.2\,x(n-1) + y(n-1)$$

Subsystem 3
$$y(n) = x(n) + 1.2\,x(n-1)$$

Subsystem 4
$$y(n) = x(n) + \log x(n-1)$$

Subsystem 5
$$y(n) = x(n) - 1.7\,x(n-1)$$

Subsystem 6
$$y(n) = x(n) \times x(n-1)$$

Subsystem 7
$$y(n) = 3\,x(n)$$

Subsystem 8
$$y(n) = x(n) + x(n-1)^2$$

FIGURE 4–28
Various subsystems.

4-3. A system is defined by the following difference equation:

$$y(n) = 1.0x(n) + 0.5x(n - 1) - 0.3x(n - 2)$$

    (a) What are the values of the difference equation coefficients?
    (b) How much memory is required to implement the equation?
    (c) Draw the first five terms of the impulse response of this system.
    (d) Is this system stable?

4-4. A system is defined by the following difference equation:

$$y(n) = 1.0x(n) + 0.5x(n - 1) - 0.3y(n - 1)$$

    (a) What are the values of the difference equation coefficients?
    (b) How much memory is required to implement the equation?
    (c) Draw the first five terms of the impulse response of this system.
    (d) Is this system stable?

4-5. A system implements the following difference equation:

$$y(n) = 1.2x(n) + 0.7x(n - 1) - 1.4x(n - 2)$$

Consider the input signal in Figure 4–29:
    (a) What is the flipped impulse response for this system?
    (b) What are the values of the output samples at times $n = -3$, $n = -2$, and $n = -1$?

4-6. A system implements the following difference equation:

$$y(n) = 1.0x(n) - 0.4y(n - 1)$$

    (a) How many past input values are considered when computing the output of this system?
    (b) The input signal of Figure 4–29 is applied to that system. If you know that $y(-2) = 0.9$, what is the present output of the system?

4-7. **(a)** How could you convert the IIR system of Question 4-6 into an FIR system? Your new FIR system should produce results that are very close to the original IIR system.
    (b) What do you find when you compare the amount of processing required to implement the IIR and FIR systems of part (a)?

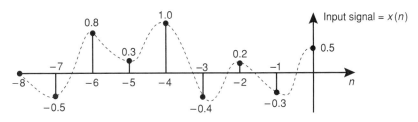

FIGURE 4–29
Input signal.

4-8. Which of the following systems are causal systems?

(a) $y(n) = x(n) + 0.3x(n-1) - 0.2x(n-2)$.

(b) $y(n) = x(n) + 0.3x(n-1) - 0.2x(n+1)$.

(c) $y(n-3) = x(n) + 0.3x(n-1) - 0.2x(n-2)$.

(d) $y(n+3) = x(n) + 0.3x(n-1) - 0.2x(n-2)$.

(e) $h(0) = 1$, $h(1) = 0.3$, $h(2) = 0.7$, all other $h(n)$ are zero valued.

(f) $h(-2) = 0.5$, $h(-1) = 0.3$, $h(0) = 0.7$, all other $h(n)$ are zero valued.

(g) $h(-1) = 0.4$, $h(0) = 0.3$, $h(1) = 0.7$, all other $h(n)$ are zero valued.

5

SPECTRAL ANALYSIS

Practical digital signal processing (DSP) often requires analyzing the frequency content of a signal. The techniques used to perform this analysis make possible important applications such as filtering, demodulation, and signal correlation. For example, one particular signal frequency band could be used to transmit a *data over voice* (DOV) signal. Another use is for *voice recognition* applications in which the processor analyzes the frequency contents to identify sounds such as vowels and consonants. Spectral analysis also could be used to compare the frequency content of two signals or two strings of numbers to determine whether they are related.

This chapter presents the concepts that lead to the development of discrete-time spectral analysis. We begin comparing continuous-time signals to discrete-time signals, which helps to identify the practical limits and problems related to discrete-time spectral analysis. We then describe the special *periodic* conditions that allow a digital processor to compute a signal spectrum.

We follow by developing a technique that generates a periodic signal from samples taken from a nonperiodic signal. We then exploit the properties of this periodic discrete-time signal to develop the Fourier analysis equation, which, when solved, yields the spectrum of a digital signal. We follow by examining different ways to solve the analysis equation. First, we use vectors because they provide an intuitive, graphical solution. Second, we consider the requirements of using a processor to compute the solution. This helps to identify some programming techniques required to handle complex numbers. Third, we analyze the amount of processing required in practice to solve the analysis equation. We then uncover a trick that reduces the amount of processing for real signals, and we identify the need to use fast algorithms to make spectral analysis possible on real time signals.

5.1 ANALYZING THE SPECTRUM OF AN INPUT SIGNAL

A common application in DSP is analysis of the spectrum of an input signal. This analysis reveals the magnitude and phase of the frequency components contained in the signal.

5.1.1 SPECTRUM OF A CONTINUOUS SIGNAL

As mentioned in Section 2.2.1, Jean Baptiste Joseph Fourier showed that a periodic signal consists of harmonically related sinusoids. This means that the spectrum of a periodic signal consists of a fundamental frequency and sinusoid harmonics that oscillate at multiples of this fundamental frequency. Figure 5–1 illustrates some of the harmonics contained in a signal.

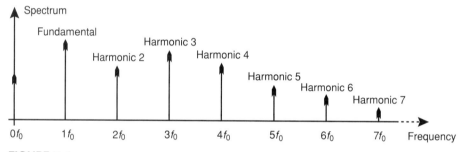

FIGURE 5–1
Some harmonics of a continuous periodic signal.

Note that the size of the frequency gaps separating the sinusoid harmonics depends on the value of the fundamental frequency f_0. The number of sinusoid harmonics that a signal contains varies, depending on the characteristics of the actual signal. Some periodic signals may contain very few sinusoid harmonics; others have a very large or even an infinite number. This creates a very special situation when an ADC samples the signal into the discrete-time domain. The sampling process generates aliased frequencies from any sinusoid that oscillates at a frequency exceeding the $f_S/2$ barrier. To avoid this (refer to Section 3.5.4), we add an antialiasing input filter to limit the signal spectrum. Figure 5–2 illustrates the frequency-limited input signal. The presence of the antialiasing filter sets limits to the range of the spectral analysis.

Digital spectral analysis is limited to frequencies ranging from zero to $f_S/2$.

5.1.2 SPECTRUM OF A DIGITIZED SIGNAL

The ADC sampling process replicates the spectrum of the frequency-limited signal around multiples of f_S as illustrated on Figure 5–3. We concentrate on the useful frequency range from zero to $f_S/2$. Assume that the fundamental frequency *decreases*. The sinusoid harmonics contained in this periodic signal then move *closer* together since they are all sepa-

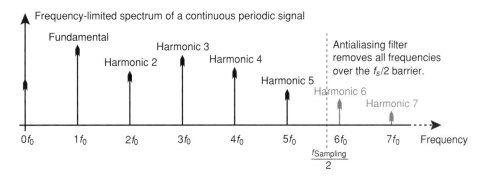

FIGURE 5–2
Frequency-limited signal.

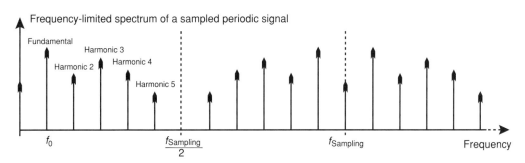

FIGURE 5–3
Spectrum of a sampled periodic signal.

rated by a range equal to the fundamental frequency. Lowering the fundamental frequency therefore makes room for more sinusoid harmonics in the $f_S/2$ range. The harmonic components of this periodic signal therefore exhibit the following characteristics:

- The frequency at which the sinusoid harmonics oscillate is an integer multiple of the fundamental frequency.
- The number of sinusoid harmonics that fit in the useful range depends on the fundamental frequency.

The lower the fundamental frequency (or equivalently, the higher the sampling rate), the more harmonics fit in the zero to $f_S/2$ range.

Since we usually do not know the value of the fundamental frequency of a system's input signal, this creates the following practical obstacles to the spectral analysis:

- Individually investigating all the frequencies values in the zero to $f_S/2$ range is practically impossible since an infinite number of values is possible. Our digital system would have to perform an infinite amount of processing.

▪ We must consequently limit the analysis to a limited set of *specific* frequencies. However, since we do not know the value of the input signal's fundamental frequency, which frequencies are we going to investigate?

Seeking a solution to these obstacles leads us to examine the digitization of signals under very special conditions, which we explore in the next section.

5.1.3 SPECTRUM OF A DIGITIZED PERIODIC SIGNAL

Digitized signals may or may not be periodic in the discrete-time domain (refer to the section in Chapter 3 on Investigation of Periodicity in the Discrete-Time Domain). In that section (see Equation [3–1]), we showed that digitized signals can become periodic with a period of N samples only if the following condition is met:

$$N = \frac{f_S}{f_{Signal}} k$$

Since N and k are integers, f_S over f_{Signal} must form a ratio of integers. Simplifying for $k = 1$, we obtain a condition that we need to make spectral analysis possible:

$$N = \frac{f_S}{f_{Signal}} \quad \text{or} \quad f_{Signal} = \frac{f_S}{N} \tag{5–1}$$

According to Equation (5–1), the sampled signal is periodic with a period of N samples if its frequency is an exact fraction of the sampling rate. For example, if the sampling rate is 8 k-samples/second and the sampled signal has a frequency of exactly 1 kHz, then

$$\text{Period} \ N = \frac{f_S}{f_{Signal}} = \frac{8000}{1000} = 8 \text{ samples}$$

In this case, there are exactly eight samples per period of the input signal. Figure 5–4 illustrates an input signal that has a period of exactly eight samples. Noticing that the signal period is an exact integer multiple of the sampling interval T_S is important. In the figure, the samples do not start where the signal crosses the horizontal axis; the phase of the sampling process is not important,

When the sampling frequency is an exact multiple of the sampled signal, the resulting sampled signal exhibits the following period and fundamental frequency:

$$\text{Signal period} = NT_S = \frac{N}{f_S} \text{ second} \qquad \text{Fundamental frequency} = \frac{f_S}{N} \text{ Hz}$$

According to Fourier, the sinusoid harmonics that make up this periodic signal must oscillate at multiples of the fundamental frequency:

$$\text{Harmonic frequency} = \text{Harmonic \#} \times \frac{f_S}{N}$$

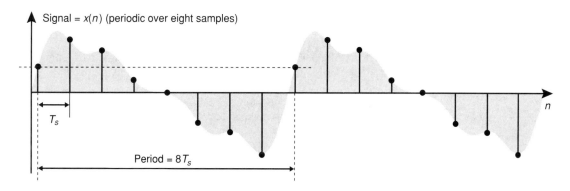

FIGURE 5-4
Input signal with a period of exactly eight samples.

Harmonic 0	Harmonic 1	Harmonic 2	Harmonic 3	Harmonic 4	$\cdots$ etc.
$0 \times \dfrac{f_S}{N}$	$1 \times \dfrac{f_S}{N}$	$2 \times \dfrac{f_S}{N}$	$3 \times \dfrac{f_S}{N}$	$4 \times \dfrac{f_S}{N}$	$\cdots$ etc.

Figure 5–5 illustrates the frequency at which these sinusoid harmonics oscillate in the frequency domain. This figure reminds us that the sampling process replicates the signal spectrum in the frequency domain. The maximum base spectrum frequency is $f_S/2$, which corresponds to the frequency of harmonic $N/2$ since:

$$\text{Harmonic frequency} = \text{Harmonic \#} \times \frac{f_S}{N}$$

$$= \frac{N}{2} \times \frac{f_S}{N} = \frac{f_S}{2}$$

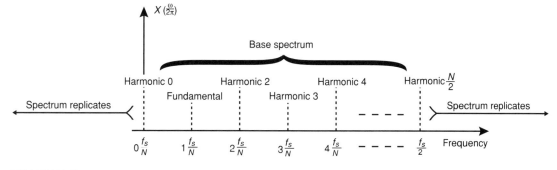

FIGURE 5-5
Frequencies contained in a signal with a period of N samples.

This immediately allows us draw two conclusions for a signal that has a period of exactly N samples:

- The base spectrum of a periodic signal contains $(N/2) + 1$ sinusoid harmonics.
- The $(N/2) + 1$ base harmonics occur at integer multiples of f_S/N.

Sampling a periodic signal at an exact multiple of its fundamental frequency provides the exact conditions that we were seeking to make spectral analysis possible:

- The sampled signal contains a *finite* number of sinusoid harmonics. Performing spectral analysis therefore requires a *finite* amount of processing.
- The sinusoid harmonics in the sampled signal occur at known frequency locations. This allows us to direct the spectral analysis to specific frequencies.

The spectrum of a digitized signal that has a period of
N samples contains $N/2 + 1$ sinusoid harmonics.
The sinusoid harmonics of this signal oscillate at multiples of f_S/N.

5.2 COMPUTING THE SPECTRUM OF A SIGNAL

In practice, a digitized signal consists of an infinite sequence of samples. The spectral analysis calculations cannot wait for all the signal samples to come in or we would wait forever. This imposes the following practical considerations:

The digital system must compute the spectral analysis
over a limited set of input samples.

Consequently, the spectral analysis applies to only a segment of the input signal. In practice, this is convenient since the frequency content of the signal typically varies over time. For example, if we consider a signal that carries music, each new musical note alters the frequency content of the signal. We should therefore be careful when selecting the size of the signal interval over which we perform the spectral analysis.

5.2.1 MAKING THE DIGITIZED SIGNAL PERIODIC

Spectral analysis becomes practically possible only if the sampled signal has a period of N samples (Section 5.1.3). Unfortunately, this happens only under the following conditions:

- The input signal contains an unchanging set of sinusoid harmonics.
- The sampling rate is an exact multiple of the input signal's fundamental frequency.

In practice, neither of these two conditions is likely to be met for the following reasons:

- The input signal is likely to contain a continuously changing spectrum.
- This spectrum most probably originates from a source that is asynchronous to the sampling circuitry.

Since the input signal is not likely to be periodic, it does not meet the essential condition that makes spectrum analysis possible. Fortunately, we can use a simple but very important trick to force a set of N input samples to become periodic. All we need to do is pretend that this set of N samples repeats itself.

To make signal processing possible, we adopt the following procedure. We select an interval of NT_S second over which to analyze the input signal. Taking samples over this interval of the input signal $x(t)$ results in a set of N samples. Figure 5–6 illustrates the digital signal $x(n)$ that results for a set of eight samples taken over a segment of an input signal $x(t)$.

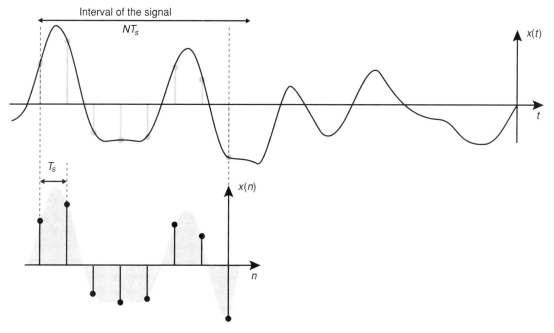

FIGURE 5–6
N samples taken from an input signal.

Since spectral analysis becomes possible in practice only for a periodic signal, we must make the N samples of $x(n)$ periodic. We do this by building a new signal $\tilde{x}(n)$ by repeating the $x(n)$ set of N samples. Figure 5–7 illustrates the new periodic signal $\tilde{x}(n)$.

Since the set of N samples repeats itself, $\tilde{x}(n)$ is a periodic signal. We have already determined that a signal that is periodic over a period of N samples contains a finite set of sinusoid harmonics.

The sinusoid harmonics contained in $\tilde{x}(n)$ represent the spectrum of the periodic signal $\tilde{x}(n)$. Since this signal contains $N/2 + 1$ sinusoid harmonics, we must now determine a way to calculate the magnitude and phase characteristics for each of them.

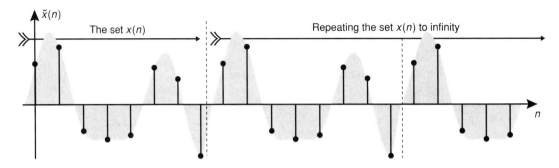

FIGURE 5–7
Periodic signal created from N samples.

5.2.2 DEFINING THE FOURIER COEFFICIENTS

We can describe each of the discrete-time sinusoid harmonics contained in a set of N samples by using Equation (2–1), which is repeated here for convenience:

$$y_k(n) = A_k \cos(2\pi[k \times f_{\text{Fundamental}}]nT_S + \phi_k)$$

where $k = 0, 1, 2, 3, 4, \ldots, N/2$.

For a periodic signal with a period of N, we know that

$$f_{\text{Fundamental}} = \frac{f_S}{N}$$

Substituting, we get

$$y_k(n) = A_k \cos\left(2\pi\left[k \times \frac{f_S}{N}\right]n T_S + \phi_k\right)$$

We know that

$$T_S = \frac{1}{f_S}$$

so we can rewrite:

$$y_k(n) = A_k \cos\left(2\pi k \frac{f_S}{N}\frac{1}{f_S} n + \phi_k\right) \tag{5–2}$$

$$k^{\text{th}} \text{ harmonic } (n) = A_k \cos\left(\frac{2\pi}{N} k n + \phi_k\right)$$

Note in Equation (5–2) that the frequency of the kth harmonic is the normalized frequency Ω_k. This can be shown by examining the following equivalencies:

$$\Omega_k = \omega_k T_S = 2\pi f_k T_S$$

where:

ω_k is the angular frequency of harmonic k and f_k is the frequency of harmonic k.

In the case of a signal, periodic with N samples, the kth harmonic oscillates at

$$f_k = \frac{f_S}{N} k$$

Substituting for f_k in the Ω_k expression, we get

$$\Omega_k = 2\pi \frac{f_S}{N} k T_S$$

Finally, since $T_S = 1/f_S$, we obtain the normalized harmonic frequency:

$$\Omega_k = 2\pi \frac{f_S}{N} k \frac{1}{f_S}$$

$$\Omega_k = \frac{2\pi}{N} k \ \text{radians/sample}$$

For example, if f_S = 16,000 samples/second and N = 8, the third harmonic has a periodic frequency of

$$f_3 = \frac{16{,}000}{8} \times 3 = 6000 \ \text{Hz}$$

and a normalized frequency of

$$\Omega_3 = \frac{2\pi}{8} \times 3 = 2.356 \ \text{radians/sample}$$

Performing spectral analysis requires that we determine values for A_k and ϕ_k in Equation (5–2). We start by expressing the sinusoid harmonics as complex exponential harmonics:

$$k^{\text{th}} \ \text{harmonic} \ (n) = \frac{A_k e^{+j\left[\frac{2\pi}{N} k n + \phi_k\right]} + A_k e^{-j\left[\frac{2\pi}{N} k n + \phi_k\right]}}{2}$$

Note that one complex exponential harmonic oscillates at a positive frequency and the other oscillates at a negative frequency. We can rewrite these complex exponentials as products of two exponentials:

$$k^{\text{th}} \ \text{harmonic} \ (n) = \frac{A_k e^{+j\phi_k}}{2} e^{+j\frac{2\pi}{N} k n} + \frac{A_k e^{-j\phi_k}}{2} e^{-j\frac{2\pi}{N} k n} \tag{5–3}$$

A single real sinusoid harmonic therefore consists of two separate complex exponential harmonics. Since the two harmonics oscillate at the negative of each other's frequency, let's call them harmonics k and $-k$. Each harmonic is multiplied by a factor that determines its amplitude and the phase:

$$\text{Complex harmonic}_k(n) = C_k e^{+j\frac{2\pi}{N}kn}$$

$$\text{where } C_k = \frac{A_k}{2} e^{+j\phi_k}$$

$$\text{Complex harmonic}_{-k}(n) = C_{-k} e^{+j\frac{2\pi}{N}(-k)n}$$

$$\text{where } C_{-k} = \frac{A_k}{2} e^{-j\phi_k}$$

(5–4)

We refer to the different C_k's as the *Fourier coefficients*. For a pure-real sinusoid harmonic, the Fourier coefficients are vectors that have

$$\text{magnitude } |C_k| = \frac{A_k}{2} = |C_{-k}| = \frac{A_k}{2}$$

$$\text{and phase } \angle C_k = \phi_k \text{ and } \angle C_{-k} = -\phi_k$$

It therefore takes a complex conjugate pair of Fourier coefficients to fully describe a real sinusoid harmonic. Figure 5–8 illustrates the C_k and C_{-k} pairs that describe real sinusoid harmonics. Determining the spectrum of the signal is therefore a question of solving for the Fourier coefficients and then grouping them into pairs to form sinusoids.

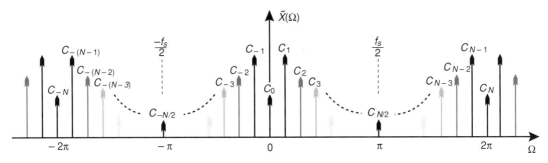

FIGURE 5–8
Fourier coefficients.

5.3 THE DISCRETE FOURIER TRANSFORM

We call the operation of finding the values of the Fourier coefficients the *discrete Fourier transform* (DFT). This operation yields a set of C_k coefficients that describe the magnitude and phase of each of the complex exponential harmonics contained in the $\tilde{x}(n)$ set of periodic samples. In other words, the DFT operation performs spectral analysis.

5.3.1   SOLVING FOR THE FOURIER COEFFICIENTS

The spectrum of a digital signal consists of an infinite number of Fourier coefficients; fortunately, the spectrum is periodic. We therefore need to solve for only one base set of Fourier coefficients to find the complete spectrum. Note that the N samples of the periodic signal $\tilde{x}(n)$ correspond to a base set of N Fourier coefficients in the frequency domain. The base set can be selected anywhere as long as it consists of N consecutive Fourier coefficients. For mathematical convenience, we choose to consider the base set starting at C_0 and ending at C_{N-1}. Figure 5–9 illustrates this choice.

Note the symmetry in Figure 5–9:

$$C_{N-1} = C_{-1} \qquad C_{N-2} = C_{-2} \qquad C_{N-3} = C_{-3} \cdots$$

The complete periodic signal $\tilde{x}(n)$ contains the sum of the N complex exponential harmonics that are scaled by their respective Fourier coefficients:

$$\tilde{x}(n) = \sum_{k=0}^{N-1} C_k e^{+j \frac{2\pi}{N} k n}$$

To solve for the value of C_k, we begin by adjusting the phase of $\tilde{x}(n)$ by multiplying both sides by a normalized complex exponential:

$$e^{-j \frac{2\pi}{N} pn} \tilde{x}(n) = e^{-j \frac{2\pi}{N} pn} \sum_{k=0}^{N-1} C_k e^{+j \frac{2\pi}{N} k n}$$

$$= \sum_{k=0}^{N-1} C_k e^{+j \frac{2\pi}{N} (k-p) n}$$

To remove the discrete-time variable n, we can take the summation over N terms on both sides:

$$\sum_{n=0}^{N-1} e^{-j \frac{2\pi}{N} pn} \tilde{x}(n) = \sum_{n=0}^{N-1} \sum_{k=0}^{N-1} C_k e^{+j \frac{2\pi}{N} (k-p) n}$$

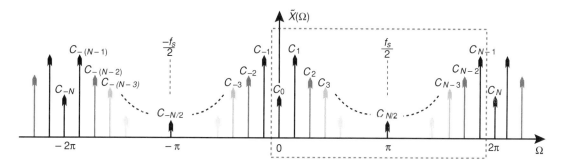

FIGURE 5–9
Base set of Fourier coefficients.

We can reverse the order of the summations on the right side of the equation:

$$\sum_{n=0}^{N-1} e^{-j\frac{2\pi}{N}pn} \tilde{x}(n) = \sum_{k=0}^{N-1} \sum_{n=0}^{N-1} C_k e^{+j\frac{2\pi}{N}(k-p)n}$$

Since C_k is independent of the summation variable n, we can take it outside the rightmost summation:

$$\sum_{n=0}^{N-1} e^{-j\frac{2\pi}{N}pn} \tilde{x}(n) = \sum_{k=0}^{N-1} C_k \sum_{n=0}^{N-1} e^{+j\frac{2\pi}{N}(k-p)n}$$

Consider the right side of the equation:

$$\sum_{k=0}^{N-1} C_k \sum_{n=0}^{N-1} e^{+j\frac{2\pi}{N}(k-p)n}$$

The complex exponentials being summed are actually vectors that have a magnitude of 1. As the inside summation iterates the value of n from 0 to $N-1$, it effectively adds a set of N different vectors. The value of $k-p$ has a key role in determining the phase angle of each vector in the set.

As the outside summation iterates through the values of k, it repeats the inside summation N times. Since the value of k affects the phase of the vectors, each new value of k sums a different set of N vectors. It is worth examining the different sets of N vectors because they exhibit some very special configurations. For example if $N = 8$, a different set of eight vectors is summed for every value of k. Figure 5–10 illustrates the vectors being summed for each of the eight values of k.

Remember that the inside summation adds the vectors in each set. Because of the symmetry, the complex vectors pair up to cancel each other out in every case except in the first case when $p = k$. In this special case, all N vectors are worth 1 since $k-p = 0$. The inside summation then yields $1 + 1 + 1 + 1 + 1 + 1 + 1 + 1 = 8 = N$.

$$\sum_{n=0}^{N-1} e^{+j\frac{2\pi}{N}(k-p)n} = \begin{cases} N, & \text{for } p = k \\ 0, & \text{otherwise} \end{cases}$$

Consequently, only the value of $k = p$ for the outside summation yields a nonzero result. Setting $p = k$ therefore allows us to simplify the outside summation to C_k. This simplifies the right side of the Fourier coefficient equation to

$$\sum_{k=0}^{N-1} C_k \sum_{n=0}^{N-1} e^{+j\frac{2\pi}{N}(k-p)n} = C_k N \qquad \text{when } p = k$$

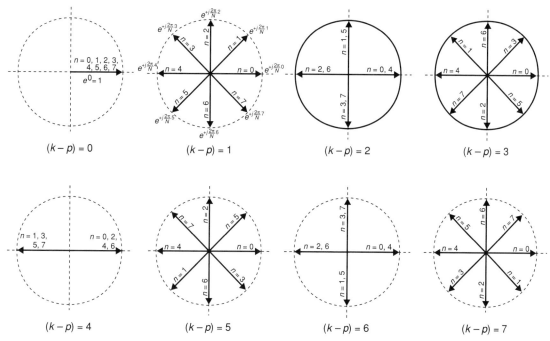

FIGURE 5–10

$\sum_{n=0}^{N-1} e^{+j\frac{2\pi}{N}(k-p)n}$ for different values of $k - p$.

Now that we have simplified the right side, let's go back to the complete Fourier coefficient equation:

$$\sum_{n=0}^{N-1} e^{-j\frac{2\pi}{N}pn} \tilde{x}(n) = \sum_{k=0}^{N-1} C_k \sum_{n=0}^{N-1} e^{+j\frac{2\pi}{N}(k-p)n}$$

$$\text{for } p = k: \sum_{n=0}^{N-1} e^{-j\frac{2\pi}{N}kn} \tilde{x}(n) = C_k N$$

Finally, we manipulate to isolate the expression for the Fourier coefficient:

$$C_k = \frac{1}{N} \sum_{n=0}^{N-1} e^{-j\frac{2\pi}{N}kn} \tilde{x}(n) \tag{5–5}$$

We call Equation (5–5) the *analysis equation* since it allows us to analyze the characteristics of each of the cosine harmonics contained in a periodic signal.

5.3.2 REPRESENTING THE FOURIER COEFFICIENTS WITH VECTORS

Computing the values of the N Fourier coefficients is a matter of solving Equation (5–5) for different values of k. For example, the following expression solves for the value of one of the two Fourier coefficients that describe the first sinusoid harmonic (the fundamental):

$$C_1 = \frac{1}{N} \sum_{n=0}^{N-1} e^{-j\frac{2\pi}{N} 1n} \tilde{x}(n)$$

Expanding this series, we get

$$C_1 = \frac{1}{N} \left[e^{-j\frac{2\pi}{N} 1\times 0} \tilde{x}(0) + e^{-j\frac{2\pi}{N} 1\times 1} \tilde{x}(1) + e^{-j\frac{2\pi}{N} 1\times 2} \tilde{x}(2) + \cdots + e^{-j\frac{2\pi}{N} 1\times(N-1)} \tilde{x}(N-1) \right]$$

The N complex exponentials correspond to vectors. For example, Figure 5–11 illustrates the eight vectors used when $N = 8$. In this case, the C_1 expression can be rewritten as

$$C_1 = \frac{1}{8} \left[\tilde{x}(0)V_0 + \tilde{x}(1)V_1 + \tilde{x}(2)V_2 + \tilde{x}(3)V_3 + \tilde{x}(4)V_4 + \tilde{x}(5)V_5 + \tilde{x}(6)V_6 + \tilde{x}(7)V_7 \right]$$

FIGURE 5–11
The eight vectors for $N = 8$.

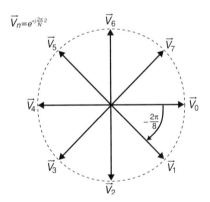

Let's consider, for example, the set of eight samples described in Figure 5–12. In this case, the expression for C_1 corresponds to

$$C_1 = \frac{1}{8} \left[0.50V_0 + 0.70V_1 - 0.30V_2 - 0.40V_3 - 0.35V_4 + 0.45V_5 + 0.32V_6 - 0.60V_7 \right]$$

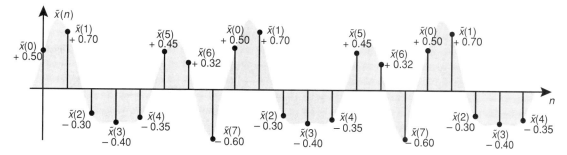

FIGURE 5–12
Periodic set of eight samples.

The values from the $\tilde{x}(n)$ set therefore determine the magnitude of each of the eight vectors. Adding vectors is a matter of sequencing them tip to tail. Subtracting vectors actually requires flipping them. This is explained mathematically by looking at a vector that rotates through π radians. Figure 5–13 illustrates this rotation. Remember that

$$e^{-j\pi} = \cos(\pi) - j\sin(\pi)$$
$$= -1$$

FIGURE 5–13
Flipping a vector.

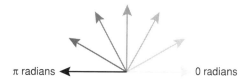

A negative vector therefore corresponds to a positive vector multiplied by the complex exponential $e^{-j\pi}$. Graphically, this is equivalent to rotating the vector through π radians (180°). Examine how we can change the sign of the vector $+Ae^{j\phi}$.

$$+Ae^{j\phi} \times e^{-j\pi} = +Ae^{j(\phi-\pi)}$$
$$= -Ae^{j\phi}$$

Figure 5–14 illustrates the scaled vectors of C_1 and their addition. In the figure, a single new vector emerges as a result of adding all the vectors making up C_1. From the value of this new vector, we can calculate the value of the C_1 coefficient:

$$C_1 = \frac{1}{8}\left[\text{Resulting vector}\right] = \frac{1}{8}\left[0.935 \angle 18.8°\right]$$
$$= 0.1169 \angle 0.328 \text{ radians}$$

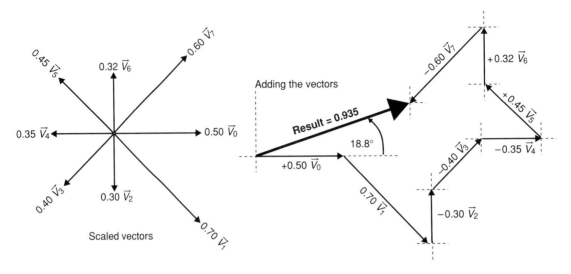

FIGURE 5–14
Adding the scaled vectors of C_1.

One of the two complex exponential harmonics making up the fundamental sinusoid of the $\tilde{x}(n)$ sequence has a normalized frequency of $2\pi/8$, an amplitude of 0.1169, and a phase shift of 0.328 radians (18.8 degrees).

To obtain the complete sinusoid harmonic, we need to compute the coefficient C_{-1} (note that we could equally use C_7), which applies to the negative frequency $-2\pi/8$. Since the N periodic samples are pure real, the sinusoid harmonics must also be pure real; consequently, the Fourier coefficient C_{-1} must be the complex conjugate of C_1. This allows us to use the already calculated value C_1 to determine the value of C_{-1}:

$$C_{-1} = \overline{C_1} = 0.1169 \angle -0.328 \text{ radians}$$

Grouping the complex conjugate harmonics described by C_1 and C_{-1} yields a pure-real sinusoid harmonic. The fundamental frequency (harmonic 1) of the periodic signal being analyzed is as follows:

$$\text{Fundamental}(n) = C_1 \, e^{+j\frac{2\pi}{8}n} + C_{-1} \, e^{-j\frac{2\pi}{8}n}$$

$$= 0.1169 \, e^{0.328j} \, e^{+j\frac{2\pi}{8}n} + 0.1169 \, e^{-0.328j} \, e^{-j\frac{2\pi}{8}n}$$

$$= 0.1169 \, e^{+j\left[\frac{2\pi}{8}n + 0.328\right]} + 0.1169 \, e^{-j\left[\frac{2\pi}{8}n + 0.328\right]}$$

$$= 2 \times 0.1169 \cos\left(\frac{2\pi}{8}n + 0.328\right)$$

$$= 0.2338 \cos\left(\frac{2\pi}{8}n + 0.328\right)$$

Figure 5–15 illustrates the first eight samples of the fundamental sinusoid. Similarly, we could use vectorial analysis to determine the amplitude and phase characteristics of the other sinusoid components of $\tilde{x}(n)$.

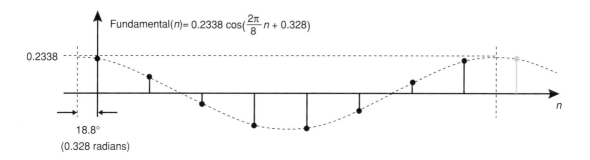

FIGURE 5–15
Fundamental with an amplitude of 0.2338 and a phase shift of 18.8 degrees.

5.3.3 COMPUTING THE DISCRETE FOURIER TRANSFORM

In practice, we want a digital system to compute the values of the Fourier coefficients. DSP systems use the coefficients of a difference equation to weight the present and past values of the input (refer to Section 4.2.4). Calculating a DFT is slightly different because the answer consists of a set of output values.

Assume that a DAC is sampling an input signal. The samples are stored in memory locations; eventually, the digital system will have stored a set of N samples. These samples are labeled from $x(n-N+1)$ to the present input $x(n)$. We then use this set to build a periodic signal $\tilde{x}(n)$. Figure 5–16 illustrates the relationship between the sampled signal $x(n)$ and its periodic counterpart when $N = 8$. The figure establishes the following relationship between the labeling of $x(n)$ and $\tilde{x}(n)$:

$$x(n-7) = \tilde{x}(0) \qquad x(n-6) = \tilde{x}(1) \qquad x(n-5) = \tilde{x}(2) \qquad x(n-4) = \tilde{x}(3)$$
$$x(n-3) = \tilde{x}(4) \qquad x(n-2) = \tilde{x}(5) \qquad x(n-1) = \tilde{x}(6) \qquad x(n) = \tilde{x}(7)$$

Note that $\tilde{x}(0) = x(n-N+1)$

To put it simply, the input samples are numbered in reversed order. Using this discovery, we can translate the $\tilde{x}(n)$ of the eight-point DFT into $x(n)$ terms. For example, take the expanded series for an eight-point Fourier coefficient:

$$C_k = \frac{1}{N}\left[e^{-j\frac{2\pi}{N}k \times 0}\tilde{x}(0) + e^{-j\frac{2\pi}{N}k \times 1}\tilde{x}(1) + e^{-j\frac{2\pi}{N}k \times 2}\tilde{x}(2) + \cdots \right.$$
$$\left. \cdots + e^{-j\frac{2\pi}{N}k \times 7}\tilde{x}(7) \right]$$

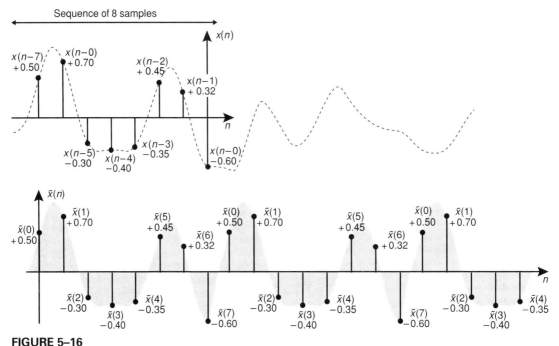

FIGURE 5–16
Relating $x(n)$ to its periodic counterpart.

Translating the $\tilde{x}(n)$ terms into $x(n)$ input samples yields

$$C_k = \frac{1}{N}\left[\begin{array}{c} e^{-j\frac{2\pi}{N}k\times 0}x(n-7) + e^{-j\frac{2\pi}{N}k\times 1}x(n-6) + \\ e^{-j\frac{2\pi}{N}k\times 2}x(n-5) + \cdots + e^{-j\frac{2\pi}{N}k\times 7}x(n) \end{array}\right]$$

We can compare the expression of the C_k Fourier coefficients to an eight-term difference equation:

$$C_k = y_k(n) = b_0 x(n) + b_1 x(n-1) + b_2 x(n-2) + b_3 x(n-3) + \ldots + b_7 x(n-7)$$

where

$$k = 0, 1, 2, 3, 4, 5, 6, 7$$

The weighting factors of the difference equation correspond to the complex exponential of the DFT series:

$$b_0 = \frac{e^{-j\frac{2\pi}{N}k\times(N-1)}}{N}, \quad b_1 = \frac{e^{-j\frac{2\pi}{N}k\times(N-2)}}{N}, \quad \ldots \quad,$$

$$b_{N-2} = \frac{e^{-j\frac{2\pi}{N}k\times 1}}{N}, \quad b_{N-1} = \frac{e^{-j\frac{2\pi}{N}k\times 0}}{N}$$

The difference equation coefficients are therefore complex vectors. This discovery provides the key to program the DFT on the digital system. Translating the N Fourier coefficients into N difference equations yields

$$C_k = y_k(n) \quad \text{where} \quad k = 0, 1, 2, \cdots (N-1)$$

$$y_k(n) = \frac{e^{-j\frac{2\pi}{N}k \times (N-1)}}{N} x(n) + \frac{e^{-j\frac{2\pi}{N}k \times (N-2)}}{N} x(n-1) + \cdots$$

$$\cdots + \frac{e^{-j\frac{2\pi}{N}k \times 0}}{N} x(n-N+1)$$

The complex coefficients of the difference equation may be translated into real and imaginary parts using Euler's identity:

$$e^{-j\phi} = \cos(\phi) - j\sin(\phi)$$

For example, consider the difference equation coefficients used to calculate the Fourier coefficient $C_1 = y_1(n)$ for a set of N samples:

$$b_0 = \frac{1}{8} e^{-j\frac{2\pi}{8} 1 \times 7} \cong +0.088 + 0.088j$$

$$b_1 = \frac{1}{8} e^{-j\frac{2\pi}{8} 1 \times 6} \cong +0.000 + 0.125j$$

$$b_2 = \frac{1}{8} e^{-j\frac{2\pi}{8} 1 \times 5} \cong -0.088 + 0.088j$$

$$b_3 = \frac{1}{8} e^{-j\frac{2\pi}{8} 1 \times 4} \cong -0.125 + 0.000j$$

$$b_4 = \frac{1}{8} e^{-j\frac{2\pi}{8} 1 \times 3} \cong -0.088 - 0.088j$$

$$b_5 = \frac{1}{8} e^{-j\frac{2\pi}{8} 1 \times 2} \cong +0.000 - 0.125j$$

$$b_6 = \frac{1}{8} e^{-j\frac{2\pi}{8} 1 \times 1} \cong +0.088 - 0.088j$$

$$b_7 = \frac{1}{8} e^{-j\frac{2\pi}{8} 1 \times 0} \cong +0.125 + 0.000j$$

Calculating $y_k(n)$, which is the kth Fourier coefficient, therefore involves using complex numbers to weight the input samples. In practice, performing a complex multiplication requires separate operations for the real and for the imaginary parts. For example, when $N = 8$, the difference equation for C_1 is $y_1(n)$:

$$y_1(n) = \begin{bmatrix} 0.088\ x(n) + 0.000\ x(n-1) - 0.088\ x(n-2) - \\ 0.125\ x(n-3) - 0.088\ x(n-4) + 0.000\ x(n-5) + \\ 0.088\ x(n-6) + 0.125\ x(n-7) \end{bmatrix}$$

and

$$+\ j \begin{bmatrix} 0.088\ x(n) + 0.125\ x(n-1) + 0.088\ x(n-2) + \\ 0.000\ x(n-3) - 0.088\ x(n-4) - 0.125\ x(n-5) - \\ 0.088 x(n-6) + 0.000\ x(n-7) \end{bmatrix}$$

When implementing this calculation, we must compute the contents of the two square brackets separately. We interpret one result as being real and the other as being imaginary. We use the values in Figure 5–16 to calculate C_1:

$$C_1 = y_1(n) = \begin{bmatrix} 0.088\ (-0.60) + 0.000\ (0.32) - 0.088\ (0.45) - 0.125\ (-0.35) - \\ 0.088\ (-0.40) + 0.000\ (-0.30) + 0.088\ (0.70) + 0.125\ x(0.50) \end{bmatrix}$$

$$+\ j \begin{bmatrix} 0.088\ (-0.60) + 0.125\ (0.32) + 0.088\ (0.45) + 0.000\ (-0.35) - \\ 0.088\ (-0.40) - 0.125\ x(-0.30) - 0.088(0.70) + 0.000\ (0.50) \end{bmatrix}$$

$$= 0.1107 + 0.0377 j$$

Translating this answer into polar coordinates yields the characteristics of the C_1 Fourier coefficient:

$$C_1 = 0.1169 \angle 18.8°$$

Compare this result with the one obtained when we used vectorial analysis in Section 5.3.2. To obtain the full spectrum for $\tilde{x}(n)$, the processor must repeat the procedure for each of the other Fourier coefficients. Table 5–1 lists the results of these computations. Notice that the complex conjugate pairs C_1 and C_7, C_2 and C_6, C_3 and C_5 each control the amplitude and phase of a real cosine. Note in the table that C_4 can stand by itself since it contains no imaginary part because it applies to $f_S/2$, which is the normalized frequency π:

$$C_4 e^{+j\pi n} = |C_4| \cos(\pi n) - j |C_4| \sin(\pi n)$$

TABLE 5–1
Fourier coefficients.

$$C_0 = +0.0400 + 0.0000j \xrightarrow{Polar} |C_0| = 0.0400 \text{ and } \angle C_0 = 0°$$

$$C_1 = +0.1107 + 0.3477j \xrightarrow{Polar} |C_1| = 0.1169 \text{ and } \angle C_1 = +18.8°$$

$$C_2 = +0.0163 - 0.2688j \xrightarrow{Polar} |C_2| = 0.2692 \text{ and } \angle C_2 = -86.5°$$

$$C_3 = +0.1018 - 0.1173j \xrightarrow{Polar} |C_3| = 0.1553 \text{ and } \angle C_0 = -49.0°$$

$$C_4 = +0.0025 + 0.0000j \xrightarrow{Polar} |C_4| = 0.0025 \text{ and } \angle C_4 = 0°$$

$$C_5 = +0.1018 + 0.1173j \xrightarrow{Polar} |C_5| = 0.1553 \text{ and } \angle C_5 = +49.0°$$

$$C_6 = +0.0163 + 0.2688j \xrightarrow{Polar} |C_6| = 0.2692 \text{ and } \angle C_6 = +86.5°$$

$$C_7 = +0.1107 - 0.3477j \xrightarrow{Polar} |C_7| = 0.1169 \text{ and } \angle C_7 = -18.8°$$

We know that $\sin(\pi n) = 0$ and therefore

$$C_4 \, e^{+j\pi n} = |C_4| \cos(\pi n)$$

Applying the Fourier coefficients to their respective sinusoid harmonics describes the base spectrum contained in $\tilde{x}(n)$:

$$\tilde{x}(n) = + \ 0.0400 + 2 \times 0.1169 \cos\left(1 \frac{2\pi}{8} n + 18.8°\right)$$

$$+ \ 2 \times 0.2692 \cos\left(2 \frac{2\pi}{8} n - 86.5°\right)$$

$$+ \ 2 \times 0.1553 \cos\left(3 \frac{2\pi}{8} n - 49.0°\right)$$

$$+ \ 0.0025 \cos\left(4 \frac{2\pi}{8} n + 0°\right)$$

5.3.4 PROGRAMMING THE DISCRETE FOURIER TRANSFORM

A digital system that performs continuous spectral analysis on an input signal must solve for all N Fourier coefficients. Each C_k equation corresponds to a distinct difference equation described by its own set of complex weighting factors. For example, Table 5–2 lists the coefficients for the N difference equations when $\tilde{x}(n)$ consists of a set of eight samples. A quick examination of the table reveals that the coefficients can adopt one of eight different values:

$$\frac{V_0}{8},\quad \frac{V_1}{8},\quad \frac{V_2}{8},\quad \frac{V_3}{8},\quad \frac{V_4}{8},\quad \frac{V_5}{8},\quad \frac{V_6}{8},\quad \frac{V_7}{8}$$

Each of the N difference equations assigns a different sequence of these values to the coefficients. Since these values are time-invariant weighting factors, we can store them permanently in the digital system memory. Actually, because the values are complex numbers, it is necessary to store the real and imaginary parts in two separate tables:

Real Table   *Imaginary Table*

$$\text{Real}\left\{\frac{V_0}{8}\right\} \qquad \text{Im}\left\{\frac{V_0}{8}\right\}$$

$$\text{Real}\left\{\frac{V_1}{8}\right\} \qquad \text{Im}\left\{\frac{V_1}{8}\right\}$$

$$\ldots \qquad\qquad \ldots$$

$$\text{Real}\left\{\frac{V_{N-1}}{8}\right\} \qquad \text{Im}\left\{\frac{V_{N-1}}{8}\right\}$$

TABLE 5–2
Difference equation coefficients for a DFT with $N = 8$.

$$y_0(n): b_0 = \frac{V_0}{8},\ b_1 = \frac{V_0}{8},\ b_2 = \frac{V_0}{8},\ b_3 = \frac{V_0}{8},\ b_4 = \frac{V_0}{8},\ b_5 = \frac{V_0}{8},\ b_6 = \frac{V_0}{8},\ b_7 = \frac{V_0}{8}$$

$$y_1(n): b_0 = \frac{V_7}{8},\ b_1 = \frac{V_6}{8},\ b_2 = \frac{V_5}{8},\ b_3 = \frac{V_4}{8},\ b_4 = \frac{V_3}{8},\ b_5 = \frac{V_2}{8},\ b_6 = \frac{V_1}{8},\ b_7 = \frac{V_0}{8}$$

$$y_2(n): b_0 = \frac{V_6}{8},\ b_1 = \frac{V_4}{8},\ b_2 = \frac{V_2}{8},\ b_3 = \frac{V_0}{8},\ b_4 = \frac{V_6}{8},\ b_5 = \frac{V_4}{8},\ b_6 = \frac{V_2}{8},\ b_7 = \frac{V_0}{8}$$

$$y_3(n): b_0 = \frac{V_5}{8},\ b_1 = \frac{V_2}{8},\ b_2 = \frac{V_7}{8},\ b_3 = \frac{V_4}{8},\ b_4 = \frac{V_1}{8},\ b_5 = \frac{V_6}{8},\ b_6 = \frac{V_3}{8},\ b_7 = \frac{V_0}{8}$$

$$y_4(n): b_0 = \frac{V_4}{8},\ b_1 = \frac{V_0}{8},\ b_2 = \frac{V_4}{8},\ b_3 = \frac{V_0}{8},\ b_4 = \frac{V_4}{8},\ b_5 = \frac{V_0}{8},\ b_6 = \frac{V_4}{8},\ b_7 = \frac{V_0}{8}$$

$$y_5(n): b_0 = \frac{V_3}{8},\ b_1 = \frac{V_6}{8},\ b_2 = \frac{V_1}{8},\ b_3 = \frac{V_4}{8},\ b_4 = \frac{V_7}{8},\ b_5 = \frac{V_2}{8},\ b_6 = \frac{V_5}{8},\ b_7 = \frac{V_0}{8}$$

$$y_6(n): b_0 = \frac{V_2}{8},\ b_1 = \frac{V_4}{8},\ b_2 = \frac{V_6}{8},\ b_3 = \frac{V_0}{8},\ b_4 = \frac{V_2}{8},\ b_5 = \frac{V_4}{8},\ b_6 = \frac{V_6}{8},\ b_7 = \frac{V_0}{8}$$

$$y_7(n): b_0 = \frac{V_1}{8},\ b_1 = \frac{V_2}{8},\ b_2 = \frac{V_3}{8},\ b_3 = \frac{V_4}{8},\ b_4 = \frac{V_5}{8},\ b_5 = \frac{V_6}{8},\ b_6 = \frac{V_7}{8},\ b_7 = \frac{V_0}{8}$$

When the digital system solves a particular equation, it uses a pointer to access these values from the tables. To compute the value of a particular Fourier coefficient, the processor multiplies the real and imaginary weighting factors by the appropriate input value (Im represents *imaginary*):

$$
\left.
\begin{array}{l}
\begin{array}{l}
\text{Real}\left\{b_0\right\} x(n) \\
+\ \text{Real}\left\{b_1\right\} x(n-1) \\
+\ \text{Real}\left\{b_2\right\} x(n-2) \\
\qquad \cdots \\
+\ \text{Real}\left\{b_{N-1}\right\} x(n-N+1) \\
\hline
\text{Real}\left\{y_k(n)\right\}
\end{array} \\[1em]
\begin{array}{l}
\text{Im}\left\{b_0\right\} x(n) \\
+\ \text{Im}\left\{b_1\right\} x(n-1) \\
+\ \text{Im}\left\{b_1\right\} x(n-2) \\
\qquad \cdots \\
+\ \text{Im}\left\{b_{N-1}\right\} x(n-N+1) \\
\hline
\text{Im}\left\{y_k(n)\right\}
\end{array}
\end{array}
\right\}
\begin{array}{l}
\left[\text{Required for each of the}\right. \\
\left. N \text{ difference equations}\right]
\end{array}
$$

Processing any of these terms requires a multiplication followed by an addition (called a *MAC operation*). The total volume of processing required to solve all N equations is therefore determined as follows:

$$
\frac{2 \times (N \ \text{MAC operations})}{\text{Equation}} \times (N \ \text{Equations}) = 2\, N^2 \ \text{MAC operations}
$$

The number of samples in a period of $\tilde{x}(n)$ determines the precision of the spectral analysis. If N is too small, we can extract only a small set of harmonics, which yields a rough spectrum. If N is too large, the dynamic signal spectrum may have time to vary too much over the NT_S interval and although we extract a large number of harmonics, these may not correspond to reality. Making N larger also escalates the amount of processing. For example, consider the following cases:

$$
\begin{aligned}
N &= \ \ 8 \text{ requires: } 2 \times 8^2 \ = \ \ \ 128 \text{ MACs} \\
N &= 16 \text{ requires: } 2 \times 16^2 = \ \ \ 512 \text{ MACs} \\
N &= 32 \text{ requires: } 2 \times 32^2 = 2048 \text{ MACs} \\
N &= 64 \text{ requires: } 2 \times 64^2 = 8192 \text{ MACs}
\end{aligned}
$$

Clearly, as N increases, the volume of processing increases exponentially. Practical applications often require spectral analysis that covers thousands of harmonics. In such cases, the amount of computing can overwhelm most digital systems.

5.3.5 SPEEDING UP THE DISCRETE FOURIER TRANSFORM CALCULATION

The DFT calculations are quite time consuming in practice, and consequently the processing time required to obtain a spectrum may limit some applications. Fortunately, we can use many techniques to drastically minimize the number of operations required to solve the N difference equations of a DFT. The specific technique or combination of techniques that can be used depends on the characteristics of the actual application.

Taking Advantage of Real Samples

Most practical applications acquire the $\tilde{x}(n)$ set from an ADC. In this case, the set consists of *pure-real samples,* which results in a *pure-real spectrum* that regroups *pure-real cosines.* A real signal $\tilde{x}(n)$ consists of a combination of $N/2 + 1$ sinusoid harmonics. For example, when $N = 8$, the eight complex exponential harmonics may be regrouped as follows:

$$\tilde{x}(n) = C_0 \, e^{j\Omega_0 n} + \left[C_1 \, e^{j\Omega_1 n} + C_{-1} \, e^{-j\Omega_1 n} \right] + \left[C_2 \, e^{j\Omega_2 n} + C_{-2} \, e^{-j\Omega_2 n} \right] +$$

$$\left[C_3 \, e^{j\Omega_3 n} + C_{-3} \, e^{-j\Omega_3 n} \right] + C_4 \, e^{j\Omega_4 n}$$

where

$$\Omega_k = \frac{2\pi}{N} k$$

Since these are pure-real input samples, we should be able to regroup the complex exponential harmonics into pure-real cosines.

    Examine the spectrum of $\tilde{x}(n)$ in Figure 5–17. In the figure, since $\tilde{x}(n)$ is a *real* discrete-time signal, the frequency components must be mirrored about the origin. This means that the images are conjugates of each other. They have the same magnitude and conjugate phases:

$$C_{-1} = \overline{C_1} \qquad C_{-2} = \overline{C_2} \qquad C_{-3} = \overline{C_3}$$

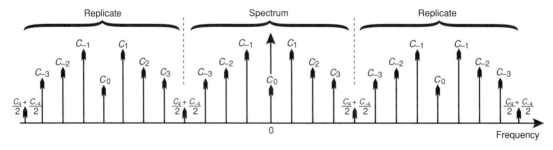

FIGURE 5–17
DFT spectrum of a set of eight real samples.

We can therefore substitute for the values of C_{-1}, C_{-2}, and C_{-3} as shown in Figure 5–18. Another consequence of discrete-time signals is that their spectrum is periodic at intervals of $\Omega_S = 2\pi$ (the normalized sampling rate). The complete signal spectrum must therefore include all of these replicates. Figure 5–18 illustrates the pure-real spectrum and some of its replicates.

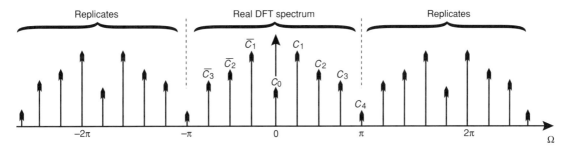

FIGURE 5–18
Periodic DFT spectrum of a set of eight real samples.

Figure 5–19 illustrates pairs of complex conjugate frequency components being grouped into cosines. Note that C_4 in the figure stands by itself but that its corresponding cosine is not multiplied by a factor of 2. This is so because it is located at $f_S/2$, which is the normalized frequency π:

$$e^{+j\pi n} = e^{-j\pi n}$$

$$C_4 e^{+j\pi n} = \frac{C_4 e^{+j\pi n} + C_4 e^{-j\pi n}}{2}$$

$$= \left| C_4 \right| \cos(\pi n)$$

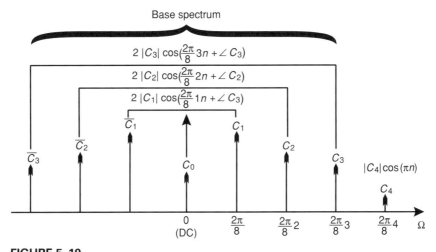

FIGURE 5–19
Matching components to create real cosines.

Also note in Figure 5–19 that harmonic 0 oscillates at 0 Hz. It therefore represents the direct current (DC) component of the signal. The expression for $\tilde{x}(n)$ therefore reduces to a DC term and four cosine terms:

$$\tilde{x}(n) = C_0 + 2|C_1|\cos\left(\frac{2\pi}{8} 1n + \angle C_1\right) + 2|C_2|\cos\left(\frac{2\pi}{8} 2n + \angle C_2\right)$$
$$+ 2|C_3|\cos\left(\frac{2\pi}{8} 3n + \angle C_3\right) + |C_4|\cos\left(\frac{2\pi}{8} 4n\right)$$

This means that we do not need to compute the Fourier coefficients C_{-1}, C_{-2}, and C_{-3}.

**If the *N* samples are pure real, we need
to compute only *N*/2 + 1 Fourier coefficients.**

For example, to perform a DFT on 512 samples (a 512-point DFT), we need to compute only

$$\frac{512}{2} + 1 = 257 \text{ Fourier coefficients}$$

Taking advantage of the fact that $\tilde{x}(n)$ is real cuts the computing time almost by half.

Using the Fast Fourier Transform

A number of astute techniques allow a substantial reduction of the number of operations required to solve an N-point DFT. The previous sections independently solved each of the Fourier coefficients as a separate difference equation.

If we examine the eight-point DFT coefficients listed in Table 5–2, we immediately notice that every equation uses a different combination of the *same* eight coefficients. This suggests immediately that if we solve all equations simultaneously, we can reduce the number of operations. For example, the b_7 coefficient is the same for all equations so that the term $b_7 x(n-7)$ needs to be calculated only once. Actually, additional opportunities for simplification exist when we start exploiting the symmetry of the vectors:

$$\text{If: } V_k = e^{j\frac{2\pi}{N}k} \quad \text{then: } V_k = -V_{k + N/2}$$

Examine Figure 5–20 in which the symmetry is demonstrated for $N = 8$. More opportunities emerge when we take advantage of the periodicity property:

$$\text{If: } V_k = e^{j\frac{2\pi}{N}k} \quad \text{then: } V_k = V_{k+N}$$

The many techniques that exploit these properties to reduce the amount of processing are known as *fast Fourier transforms* (FFT). These techniques work best when N is chosen as a number that is the square of a power of 2:

$$N = \left(2^m\right)^2$$

FIGURE 5–20
Symmetry of vectors when $N = 8$.

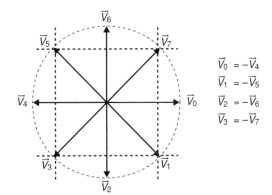

$$\vec{V}_0 = -\vec{V}_4$$
$$\vec{V}_1 = -\vec{V}_5$$
$$\vec{V}_2 = -\vec{V}_6$$
$$\vec{V}_3 = -\vec{V}_7$$

As an example of their computational efficiency, most FFT algorithms reduce the number of calculations required for a popular 256-point transform ($N = 256 = [2^4]^2$) by a factor of 64!

The development of FFT algorithms exceeds the introductory nature of this book, but this should not prevent you from taking advantage of them. An enormous selection of public domain software implements these techniques for almost any processor. In practice, there is no point in reinventing the wheel. Most designers do not bother coding these algorithms into their programs; they simply acquire one of these programs and incorporate it into their application.

What is really exciting, from an engineering point of view, is that even the slowest digital processors available today can perform these transforms at rates that allow the real-time processing of many signals, including voice.

Using Hardware Circuits to Compute the Discrete Fourier Transforms

A number of digital circuits perform DFTs at incredible rates. These are pure hardware circuits that have been optimized to yield extraordinary speed. Such devices are available as

- Specialized integrated circuits.
- DFT circuit templates that can be dropped into PLDs. Some of these PLDs offer enough density to include the control circuitry necessary to interface ADC, memory, and so on.
- Templates for ASICs.

SUMMARY

- Continuous-time signals can contain harmonics at any frequency, but discrete-time signals are limited to the zero to $f_S/2$ range.
- If a continuous-time signal contains harmonics above the $f_S/2$ limit, they should be removed with an anti-aliasing filter before the digitization process.
- All harmonics of a signal occur at frequency intervals equal to the fundamental frequency. As the fundamental frequency decreases, the signal harmonics are packed closer together.

- Spectral analysis is possible in practice only if

  The signal contains a finite number of harmonics.

  We know the position of these harmonics.

- Digital signals are periodic with a period of N samples when they are sampled at a rate of exactly N times their fundamental frequency.
- A periodic digital signal contains a base spectrum consisting of N harmonics.
- The harmonics of a periodic digital signal occur at multiples of f_S/N.
- Spectral analysis is possible only for a periodic digital signal.
- The digital system computes the spectral analysis over a limited set of input samples.
- The frequency content of a signal is likely to change over time; therefore, we must carefully select the signal interval over which to perform the spectral analysis.
- A set of N samples $x(n)$ can be turned into a periodic signal $\tilde{x}(n)$ simply by repeating the set forever in time.
- The Fourier coefficients C_k are vectors whose magnitude and phase angle respectively describe the amplitude and the phase shift of the complex harmonics contained in $\tilde{x}(n)$.
- Spectral analysis is performed by determining the value of the Fourier coefficients. This operation is called a *discrete Fourier transform* (DFT).
- The Fourier coefficients are found by solving the *analysis equation:*

$$C_k = \frac{1}{N} \sum_{n=0}^{N-1} e^{-j\frac{2\pi}{N}kn} \tilde{x}(n)$$

- The analysis equation generates a sum of complex vectors.
- Complex multiplication is performed by computing the real and imaginary parts separately.
- Selecting the value of N too small results in a rough spectrum.

  Selecting the value of N too large requires too much computing and may result in poor results if the harmonic content of the signal changed over the sampling interval.

- If $\tilde{x}(n)$ is real, the DFT can be computed by using approximately half as many operations.
- Fast Fourier transforms (FFTs) are algorithms that exploit the symmetry of periodic discrete-time vectors to drastically reduce the number of operations required to perform the Fourier transform.
- DFTs can be performed incredibly fast using pure hardware digital circuits.

PRACTICE QUESTIONS

5-1. An analog signal has a fundamental frequency of 300 Hz and an infinite number of harmonics. A DSP system applies an antialiasing filter and an ADC samples the resulting signal at a rate of 10 k-samples/second.

  (a) How many harmonics of that signal can this system analyze?

  (b) Assume that the analog fundamental frequency is reduced to 100 Hz. Repeat part (a).

5-2. Consider the analog sinusoid $x(t) = \cos(2\pi \times 2000t)$.
  (a) What is the frequency of this signal in hertz?
  (b) What is the frequency of this signal in radians/second?
  (c) Assume that we start sampling this signal at time $t = 0$ and at a rate of 10 k-samples/second. What are the values of the first 10 samples?

5-3. An ADC samples a 1000 Hz analog *sinusoid* at a synchronous rate of 10 k-samples/second.
  (a) Is the resulting $x(n)$ signal periodic? If so, what is the period?
  (b) Assume that the analog sinusoid frequency is reduced to 300 Hz. Repeat part (a).
  (c) Assume that the analog sinusoid angular frequency is 1000 *radians*/second. Repeat part (a).

5-4. Following the use of an antialiasing filter, an ADC samples a 500 Hz periodic analog signal at a rate of 8 k-samples/second.
  (a) At what frequencies oscillate the sinusoid harmonics contained in the resulting digital signal $x(n)$?
  (b) If 10 samples of $x(n)$ are converted to a periodic $\tilde{x}(n)$, at what frequencies oscillate the harmonics of the resulting $\tilde{x}(n)$?
  (c) If 25 samples of $x(n)$ are converted to a periodic $\tilde{x}(n)$, at what frequencies oscillate the sinusoid harmonics of the resulting $\tilde{x}(n)$?

5-5. The fifth Fourier coefficient of a 16-sample real $\tilde{x}(n)$ signal is represented by $C_5 = 0.34e^{j1.4}$.
  (a) Assume that $f_S = 8$ kHz. What is the frequency of the complex exponential harmonic being controlled by C_5?
  (b) What is the *amplitude* of the fifth sinusoid harmonic described by C_5 and $\overline{C_5}$?
  (c) What is the *phase* of the fifth sinusoid harmonic?
  (d) How many *samples* are there on one period of the fifth harmonic?

5-6. An ADC acquires eight samples at a rate of 10 k-samples/second. These samples are used to define one period of the periodic signal $\tilde{x}(n)$:

$$\tilde{x}(0) = 0.5, \quad \tilde{x}(1) = 0.8, \quad \tilde{x}(2) = -0.4, \quad \tilde{x}(3) = 0.3,$$
$$\tilde{x}(4) = -0.6, \quad \tilde{x}(5) = -0.9, \quad \tilde{x}(6) = -0.7, \quad \tilde{x}(7) = 0.4$$

  (a) Find the value of C_2 by adding the individual vectors of the analysis equation on graph paper.
  (b) Compute the value of C_2 by performing the complex multiplication operations of the analysis equation.
  (c) Use a software package that supports the DFT or FFT operations to determine the complex value of all Fourier coefficients.
  (d) Applying the results of part (c), express $\tilde{x}(n)$ as the sum of a DC component and a number of cosines (give the exact amplitude, phase, and angular frequency of each cosine).

5-7. Consider that a period of $\tilde{x}(n)$ consists of 4096 *real* samples.
  (a) Including the DC component, how many sinusoid harmonics can we extract in practice?
  (b) How many Fourier coefficients do we need to compute to analyze the spectrum of these samples?

6

FREQUENCY RESPONSE OF DIGITAL SIGNAL PROCESSING SYSTEMS

Chapter 4 discussed the fact that linear/time invariant (LTI) systems compute their output signal by weighting the past, present, and future input samples according to a set of weighting factors. The entire operation of the system depends on the value of these weighting factors. This chapter analyzes the effect of the weighting factor values on the response of the system.

Chapters 2 and 5 showed that signals consist of a mix of sinusoids. For this reason, we specifically investigate how systems react to sinusoid inputs. We determine that sinusoids traveling through an LTI system have their amplitude and phase modified. The amount by which the amplitude and/or the phase changes depends on the sinusoid frequency for any particular system. We can trace curves that describe the gain and the phase shift that a particular system applies over a range of frequencies. Such curves completely describe the frequency response of systems; we use these curves to represent the overall behavior of systems. What is fascinating is that the shape of these system response curves depends entirely on the value of the weighting factors that define the system.

All systems have a frequency response; the digital-to-analog (DAC) system used to generate the output is no exception. Since DACs are hybrid systems that interface the digital world to the analog world, we develop the theory that links the operation of analog and digital systems. This helps us to understand the process DACs use to translate a digital signal into an analog one. We find that the DAC has a frequency response that actually changes the amplitude of the sinusoids it is reproducing. We follow this by developing the concept of signal equalization to compensate for the frequency response of the DAC. Finally, we briefly explain the necessity for a reconstruction filter to improve the quality of the system output.

6.1 MECHANICS OF LINEAR/TIME-INVARIANT SYSTEMS

Chapter 4, particularly in Section 4.6, discussed the operations used by linear/time-invariant (LTI) systems when processing a sequence of input samples into an output signal. At that time, we developed the *convolution operation,* which describes the interaction between the system impulse response and the input samples. When the processor computes the convolution operation, the difference equation is repeatedly used to weight the past, present, and future input samples $x(n - k)$ to produce a sequence of output samples $y(n)$. The difference equation expression is repeated here for convenience:

$$y(n) = \sum_{k=-\infty}^{\infty} h(k)\, x(n - k)$$

We know that the input signal consists of a mix of sinusoidal components. Processing the input samples therefore alters the characteristics of these components. Each particular system alters the content of the input signal according to its own special characteristics, which we call the *system response.*

We use the *superposition approach* to break the response of the system to an input signal into smaller parts that are easier to manage. This allows us to separately calculate the system response to every sinusoid component contained in the input signal. Once we know how the system responds to each of the individual sinusoid component, we superpose the results to draw the complete system response.

6.2 SYSTEM RESPONSE TO SINUSOIDAL INPUTS

Because we can achieve the overall system response by methodically analyzing the response to individual sinusoidal components, we investigate LTI systems that have a *single* sinusoid applied at the system input.

A digitized sinusoid consists of a pair of discrete-time conjugate complex exponentials as Euler showed:

$$A\cos(\omega T_S n) = \frac{A e^{+j\omega T_S n} + A e^{-j\omega T_S n}}{2} \tag{6-1}$$

$$A\sin(\omega T_S n) = \frac{A e^{+j\omega T_S n} - A e^{-j\omega T_S n}}{2j} \tag{6-2}$$

The system processor uses the difference equation to compute the value of the output samples. Figure 6–1 illustrates the input and output functions of an LTI system.

FIGURE 6–1

Input and output of an LTI system.

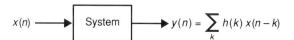

6.2.1 DISCRETE-TIME FOURIER TRANSFORM

To simplify the analysis, we break the input sinusoid function into two scaled complex conjugate components $x(n)$ and $\bar{x}(n)$:

$$A \cos(\omega T_S n) = \frac{Ae^{+j\omega T_S n} + Ae^{-j\omega T_S n}}{2}$$

$$= \frac{A}{2} \times \left[x(n) + \bar{x}(n) \right]$$

$$A \sin(\omega T_S n) = \frac{Ae^{+j\omega T_S n} - Ae^{-j\omega T_S n}}{2j}$$

$$= \frac{A}{2j} \times \left[x(n) - \bar{x}(n) \right]$$

where

$$x(n) = e^{+j\omega T_S n} \quad \text{and} \quad \bar{x}(n) = e^{-j\omega T_S n}$$

Applying a complex exponential to the input of an LTI digital system is no different than applying any other discrete-time signal. Since we want to analyze the system response to input sinusoids, we apply the two complex exponential components as separate inputs. Figure 6–2 illustrates the two system outputs that result. As the figure indicates, we can build the system output $y(n)$ for an input sinusoid by superposing the two individual outputs $y_1(n)$ and $y_2(n)$.

Let's have a closer look at the individual output expressions:

$$y_1(n) = \sum_{k=-\infty}^{\infty} h(k)e^{+j\omega T_S (n-k)} \qquad y_2(n) = \sum_{k=-\infty}^{\infty} h(k)e^{-j\omega T_S (n-k)}$$

$$y_1(n) = \sum_{k=-\infty}^{\infty} h(k)e^{\left[+j\omega T_S n - j\omega T_S k\right]} \qquad y_2(n) = \sum_{k=-\infty}^{\infty} h(k)e^{\left[-j\omega T_S n + j\omega T_S k\right]}$$

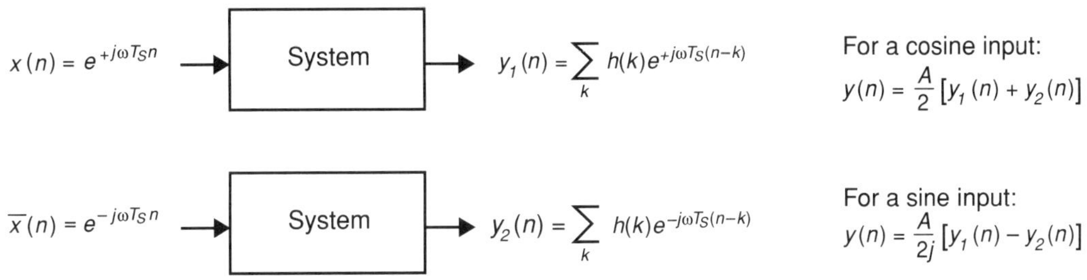

$x(n) = e^{+j\omega T_S n} \rightarrow$ System $\rightarrow y_1(n) = \sum_k h(k)e^{+j\omega T_S(n-k)}$

For a cosine input:
$$y(n) = \frac{A}{2}\left[y_1(n) + y_2(n)\right]$$

$\bar{x}(n) = e^{-j\omega T_S n} \rightarrow$ System $\rightarrow y_2(n) = \sum_k h(k)e^{-j\omega T_S(n-k)}$

For a sine input:
$$y(n) = \frac{A}{2j}\left[y_1(n) - y_2(n)\right]$$

FIGURE 6–2
System output for conjugate complex exponential inputs.

Since the exponent of the complex exponential contains the sum of two terms, we can expand it into a product of two complex exponentials:

$$y_1(n) = \sum_{k=-\infty}^{\infty} h(k) e^{+j\omega T_S n} e^{-j\omega T_S k}$$

$$y_2(n) = \sum_{k=-\infty}^{\infty} h(k) e^{-j\omega T_S n} e^{+j\omega T_S k}$$

You will notice in each of the two separate cases that one of the two complex exponentials is completely independent of the summation variable k. This allows this complex exponential to be moved out of the summation without changing the validity of the output relation:

$$y_1(n) = e^{+j\omega T_S n} \sum_{k=-\infty}^{\infty} h(k) e^{-j\omega T_S k} \qquad y_2(n) = e^{-j\omega T_S n} \sum_{k=-\infty}^{\infty} h(k) e^{+j\omega T_S k}$$

The complex exponentials that we just moved out of the summation are actually the system input functions $x(n)$ and $\bar{x}(n)$. Rewriting the output relationship in terms of the input function helps to analyze this result:

$$y_1(n) = x(n) \sum_{k=-\infty}^{\infty} h(k) e^{-j\omega T_S k} \qquad y_2(n) = \bar{x}(n) \sum_{k=-\infty}^{\infty} h(k) e^{+j\omega T_S k}$$

In each of the two cases, the output of the system consists of the input signal, which is *multiplied* by the result of a summation. Note that these summations depend only on the impulse response that defines the system. We conclude that these summations represent what the system does to the input signal to produce the output signal. Each of the summation expressions represents how the system responds to one of the two conjugate discrete-time complex exponentials. Figure 6–3 illustrates this system response.

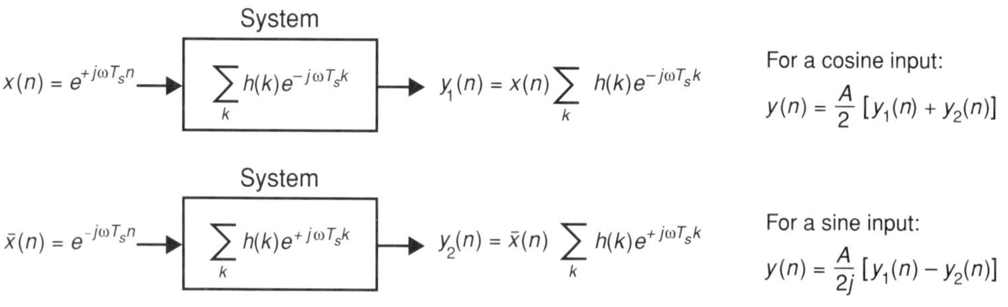

FIGURE 6–3
System response to conjugate complex exponential inputs.

To clarify the nature of the system response, we interpret the content of the system response summations. In each of the two cases in Figure 6–3, the summation adds a series of complex exponentials. Since complex exponentials are vectors, the system response therefore corresponds to a sum of vectors.

What are the practical implications of this result? Having a detailed look at the vectors being added will help answer this question. The vectors have a length $h(k)$ and an angle of $\omega T_S k$, which is negative for the $x(n)$ input and positive for the conjugate input $\bar{x}(n)$. Let's review the significance of the terms used by the system response summation:

- k is the iteration variable. The iteration limits of k must cover all nonzero $h(k)$.
- $h(k)$ corresponds to the constant values of the difference equation coefficients that we use to design the operation of the system.
- ω is the angular frequency of the input sinusoid.
- T_S is the time interval between samples.

Each of the k vectors has a magnitude of $h(k)$ and an angle of multiples of ωT_S. Figure 6–4 gives a graphical representation of one of the vectors used by the system response summation.

FIGURE 6–4
One of the many vectors making up the system response.

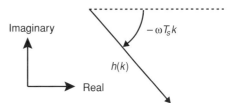

As Section 5.3.2 discussed, adding vectors is a simple question of sequencing them tail to tip. When the value of k is iterated, the expression $\omega T_S k$ generates multiples of the angle ωT_S, which is the normalized frequency Ω of the input sinusoid. For example, if the ADC acquires samples at a rate of 8000 samples/second and if the input sinusoid has a frequency of 1555 Hz, then:

$$\Omega = \omega T_S = 2\pi \times 1555 \times \frac{1}{8000} = 1.221 \text{ radians/sample}$$

We can convert this normalized frequency into degrees as follows:

$$1.221 \text{ radians/sample} \Leftrightarrow 1.221 \times \frac{180°}{\pi} \cong 70 \text{ degrees/sample}$$

In this case, the system response vectors occur at multiples of

$$1.221 \text{ radians/sample} \Leftrightarrow 70 \text{ degrees/sample}$$

Let's apply this normalized input $\omega T_S = 70°$/sample to the system defined by the impulse response described in Figure 6–5. The system described in Figure 6–5 has a six-term impulse response. The system response summation therefore iterates k from 0 to 5. If we apply a 1555 Hz input sinusoid, the response summation adds six vectors whose angles are multiples of 70 degrees. Figure 6–6 illustrates the addition of the six vectors that shape the response of this system for each of the two conjugate exponential input components.

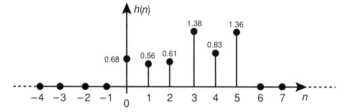

$$y(n) = 0.68\, x(n) + 0.56\, x(n-1) + 0.61\, x(n-2) + 1.38\, x(n-3) + 0.83\, x(n-4) + 1.36\, x(n-5)$$

FIGURE 6–5
Impulse response of a particular LTI system.

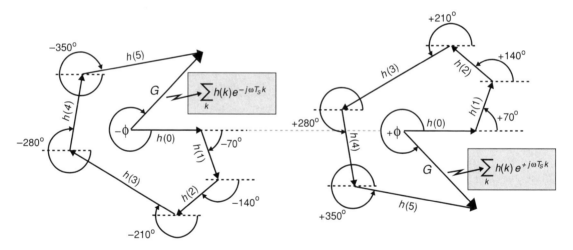

FIGURE 6–6
Vectorial components of the system response.

In the two cases illustrated in Figure 6–6, we can see that the six vectors add up to yield an overall system response vector of length G and of angle $-\phi$ and $+\phi$. The values of G and ϕ depend on

- The impulse response $h(k)$, which consists of a constant set of values since the system is time invariant.

- The normalized frequency $\Omega = \omega T_S$, which depends on the frequency of the input sinusoid and on the sampling interval.

When we plan the system, we select the ADC sampling rate and design a set of values for the difference equation coefficients. Once this is done, the impulse response terms $h(k)$ have *constant* values. The only *variable* that affects the system response is the normalized frequency of the input signal. As ω or T_S vary, the normalized frequency $\Omega = \omega T_S$ changes, and the system responds with a new value for G and ϕ. Because of this, the gain and phase responses are considered functions of the normalized frequency $G(\Omega)$ and $\phi(\Omega)$.

A sinusoid traveling through an LTI system of fixed sampling rate has its magnitude and phase altered by an amount that is a function of its normalized *frequency*.

Figure 6–7 illustrates that the input signal, the system response, and the output signal are all functions of Ω. Let's expand the $y_1(n)$ and $y_2(n)$ outputs illustrated in Figure 6–7:

$$y_1(n) = e^{+j\Omega n}G(\Omega)e^{-j\phi(\Omega)} \qquad y_2(n) = e^{-j\Omega n}G(\Omega)e^{+j\phi(\Omega)}$$
$$= G(\Omega)e^{+j(\Omega n - \phi(\Omega))} \qquad\qquad = G(\Omega)e^{-j(\Omega n - \phi(\Omega))}$$

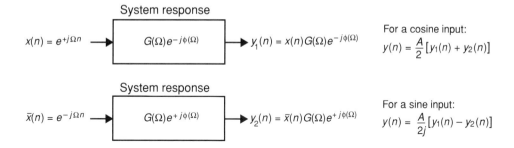

FIGURE 6–7
System parameters are functions of Ω.

When the input signal is $x(n) = A\cos(\Omega n)$, we can determine the output signal by superposing as follows:

$$y(n) = \frac{A}{2} \times \left[y_1(n) + y_2(n) \right]$$
$$= \frac{G(\Omega)\,Ae^{+j(\Omega n - \phi(\Omega))} + G(\Omega)\,Ae^{-j(\Omega n - \phi(\Omega))}}{2}$$
$$= G(\Omega)\,A\cos\left(\Omega n - \phi(\Omega)\right)$$

We can equally build the output signal in the case of a sine input of the form $x(n) = A \sin(\Omega n)$ by superposing as follows:

$$
\begin{aligned}
y(n) &= \frac{A}{2j} \times \left[y_1(n) - y_2(n) \right] \\
&= \frac{G(\Omega) A e^{+j(\Omega n - \phi(\Omega))} - G(\Omega) A e^{-j(\Omega n - \phi(\Omega))}}{2j} \\
&= G(\Omega) A \sin\left(\Omega n - \phi(\Omega)\right)
\end{aligned}
$$

Figure 6–8 illustrates the system output for input sinusoids.

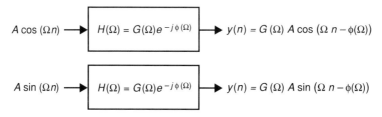

FIGURE 6–8
System alters the amplitude and the phase of input sinusoids.

Comparing Figure 6–7 and Figure 6–8, we note that the system gain $G(\Omega)$ and the phase delay $-\phi(\Omega)$ applied to the sinusoid inputs correspond exactly to the system response for the complex exponential input $x(n) = e^{+j\Omega n}$.

To determine the system response to a real sinusoid input, we need to analyze only what happens to its $e^{+j\Omega n}$ component.

The result illustrated in Figure 6–8 is very important since it tells how LTI systems respond to any sinusoidal input. The amount by which the system scales the amplitude and delays the phase depends on the normalized frequency of the input sinusoid. For this reason, we call this response the *system frequency response $H(\Omega)$*.

An LTI system responds to a sinusoidal input by scaling its amplitude and delaying its phase.

$$
\begin{aligned}
H(\Omega) &= \sum_{k=-\infty}^{+\infty} h(k) e^{-j\Omega k} \\
&= G(\Omega) e^{-j\phi(\Omega)}
\end{aligned}
\tag{6--3}
$$

We call Equation (6–3) the discrete-time *Fourier transform;* take care not to confuse it with the DFT. Remember from Chapter 5 that we use the DFT to analyze the characteristics of the individual harmonics contained in one period of a periodic digitized signal. The discrete-time Fourier transform is quite different since we use it to determine how the system processing alters the amplitude and phase characteristics of the sinusoids that enter the system.

Notice that computing the discrete-time Fourier transform completely eliminates the discrete-time variable n. The operation described by Equation (6–3) takes the impulse response $h(n)$, which completely describes the system in the discrete-time domain, and translates it into $H(\Omega)$, which completely describes the response of the system in the frequency domain. This confirms that the set of difference equation coefficients effectively controls the amount of change applied to the amplitude and the phase of input sinusoids.

The next section shows how to calculate the discrete-time Fourier transform and how to draw the frequency response curves that describe a system in the frequency domain.

6.3 PRACTICAL USE OF THE DISCRETE-TIME FOURIER TRANSFORM

Calculating the discrete-time Fourier transform reveals what happens to a particular sinusoid waveform when it travels through a system. Since signals consist of a mix of sinusoid components, calculating the transform at each of these sinusoid frequencies identifies the change in amplitude and phase that the system applies to each frequency component.

6.3.1 CALCULATION OF THE DISCRETE-TIME FOURIER TRANSFORM

To calculate the discrete-time Fourier transform, we need to know the frequency of the input sinusoid, the system sampling rate, and the impulse response that defines the system under scrutiny. For example, we input a 1555 Hz sinusoid to a system that samples the input at a rate of 8000 samples/second. We use the system described on Figure 6–5 in this example. The impulse response of this system is limited to the following six nonzero terms:

$$h(0) = 0.68 \quad h(1) = 0.56 \quad h(2) = 0.61 \quad h(3) = 1.38 \quad h(4) = 0.83 \quad h(5) = 1.36$$

The normalized frequency of the input is

$$\Omega = \omega T_S = 2\pi \times 1555 \times \frac{1}{8000} \cong 1.221 \text{ radians/sample}$$

In this case, we resolve the discrete-time Fourier transform by computing the following series:

$$H(\Omega) = \sum_{k=-\infty}^{+\infty} h(k)e^{-j\Omega k}$$

$$= 0.68e^{-j\,1.221 \times 0} + 0.56e^{-j\,1.221 \times 1} + 0.61e^{-j\,1.221 \times 2} + 1.38e^{-j\,1.221 \times 3} +$$
$$0.83e^{-j\,1.221 \times 4} + 1.36e^{-j\,1.221 \times 5}$$

Translating the terms into complex numbers yields

$$H(\Omega) = (0.68) + (0.192 - 0.526j) + (-0.467 - 0.393j) + (-1.20 + 0.688j) +$$
$$(0.143 + 0.818j) + (1.34 + 0.239j)$$
$$= 0.690 + 0.827j$$

The result is a complex number that corresponds to a vector that has the following magnitude and angle:

$$G(\Omega) = |H(\Omega)| \qquad\qquad \phi(\Omega) = \angle H(\Omega)$$

$$= \sqrt{(0.690)^2 + (0.827)^2} \qquad = \arctan\!\left(\frac{0.827}{0.690}\right)$$

$$= 1.08 \qquad\qquad = 0.87 \text{ radians}$$

The magnitude and the angle of the discrete-time Fourier transform represent the gain and phase shift, respectively, that this system applies to a 1555 Hz input sinusoid. Figure 6–9 illustrates these changes. Note in the figure that the phase shift is +0.87 radians. The system seems to be moving the phase forward when it should be delaying it. This can be explained by representing the 2π radians of a circle as extending from 0 to $-\pi$ and from 0 to $+\pi$. In this case, we can translate any delay that extends beyond $-\pi$ to an equivalent positive phase shift. A phase delay of $-2\pi + 0.87 = -5.4$ radians produces the same result as a phase shift of +0.87 radians. Figure 6–10 illustrates this change of perspective.

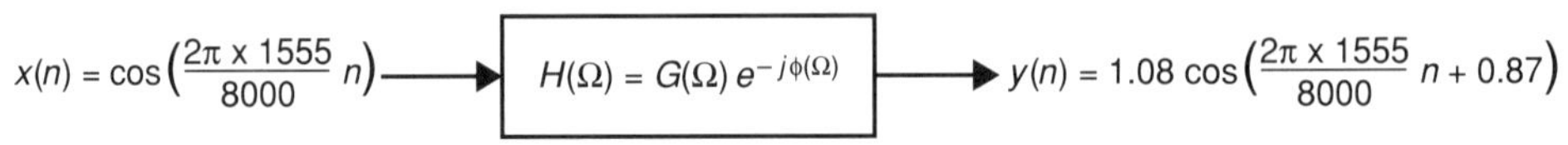

$$x(n) = \cos\!\left(\frac{2\pi \times 1555}{8000}\,n\right) \longrightarrow \boxed{H(\Omega) = G(\Omega)\,e^{-j\phi(\Omega)}} \longrightarrow y(n) = 1.08 \cos\!\left(\frac{2\pi \times 1555}{8000}\,n + 0.87\right)$$

FIGURE 6–9
System changes the amplitude and shifts the phase.

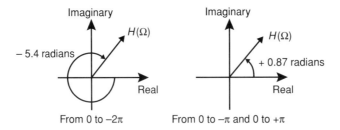

FIGURE 6–10
Representation of the phase shift.

The calculation of the discrete-time Fourier transform for a specific value of the normalized input frequency Ω requires the addition of the vectors described by the Fourier transform summation. As the previous example indicates, evaluating this summation is tedious since it requires a considerable number of rather repetitive operations.

Many software packages are available to perform these calculations. In practice, any mathematical software package capable of adding a set of vectors can be used to calculate the discrete-time Fourier transform. For example, all spreadsheets can be set up easily to calculate the discrete-time Fourier transform series. Many spreadsheets even include the discrete-time Fourier transform as a standard preprogrammed function.

6.3.2 CALCULATING THE FREQUENCY RESPONSE OF A SYSTEM

Solving the discrete-time Fourier transform yields the system gain and phase response for *one* input sinusoid that oscillates at a particular input frequency. The amounts of gain and phase shift that a system applies are functions of the frequency of the input sinusoid. In a digital system, the harmonic components of a signal are limited to oscillating within a finite range of frequencies. The sampling theorem fixes this range to extend from zero to $f_S/2$. Expressed in terms of normalized frequency, a digital system can process signals that contain harmonic frequencies ranging from zero to π.

Once we select the limits of the frequency range of interest, we can calculate the gain and the phase shift that the system applies at a number of frequency positions along this range. Plotting the values of $G(\Omega)$ and $\phi(\Omega)$ along the selected range of frequencies yields curves that respectively describe the system gain and phase responses. A curve provides an overall view for the system frequency response and allows the behavior of different systems to be compared.

The value of the discrete-time Fourier transform at any particular frequency entirely depends on the system impulse response. Since $h(n)$ is an image of the values of the difference equation coefficients, the design of the coefficient values entirely determines the shape of the system response curves. Chapter 7 discusses how to control the system response by designing appropriate values for these coefficients.

We can calculate the discrete-time Fourier transform quickly by using software applications. These repeat the discrete-time Fourier transform calculations over a range of frequencies to generate system response curves. Once the software is set up, all you need to input are the terms of the system impulse response and the range of frequencies that you want analyzed.

The software then produces the following output:

- The gain response curve is drawn by plotting the *magnitude* of the discrete-time Fourier transform over a range of normalized frequencies. The magnitude of the discrete-time Fourier transform is

$$|H(\Omega)| = G(\Omega)$$

This *curve* describes how the system alters the amplitude of the different input sinusoids.

- The phase response curve is drawn by plotting the *argument* of the discrete-time Fourier transform over a range of normalized frequencies. The argument of the discrete-time Fourier transform is

$$\angle H(\Omega) = \phi(\Omega)$$

This *curve* describes how the system shifts the phase of the different input sinusoids.

Plotting the system response over a range of frequencies poses a practical problem since any *continuous* range contains an infinite number of different frequency values. The software cannot calculate the discrete-time Fourier transform at an infinite number of locations along a range. In practice, the only way to solve this problem is to limit the calculations to a number of specific frequency values located along the range. To do this, we subdivide the range into a reasonable number of uniform frequency segments. We then let the software application compute the system response at the beginning of each of these frequency segments.

When setting up the software to compute the system response, we must define the following items:

- A *matrix* that contains the impulse response terms $h(n)$.
- An expression that describes the Fourier transform:

$$H(\Omega) = \sum_{k=-\infty}^{+\infty} h(k)e^{-j\Omega k}$$

- A graph that plots the value of the magnitude $|H(\Omega)|$ and/or the phase $\angle H(\Omega)$ at specific locations along the required range of normalized frequencies.

The frequency response graphs are only an *approximation* of the system response since the calculations occur at a limited number of locations along the range of frequencies. If the points that make up the graph are spaced too far apart, the system response could vary significantly between the points. We must therefore ensure an adequate size for the frequency segments that separate graph points. When we increase the number of graph points, we effectively use segments that cover smaller frequency ranges. This reduces the risk that the system response will vary significantly inside the smaller frequency segments. If we define enough segments, the software will plot so many points that the graph will appear to be a continuous curve over the range of frequencies that we want to analyze.

For example, consider plotting a system gain curve using different sizes for the frequency segments. Figure 6–11 shows that choosing smaller segment sizes results in a better approximation of the system gain response curve.

At each point along the graph, the software must evaluate the system response by calculating the discrete-time Fourier transform. Using more points therefore implies more computing, which means that we must wait longer for the software to display the result. Although we may be tempted to reduce the number of points to obtain faster results, we must take care not to miss important details of a system response curve. For example, the graph with the large segments in Figure 6–11 misses the fact that the response curve drops

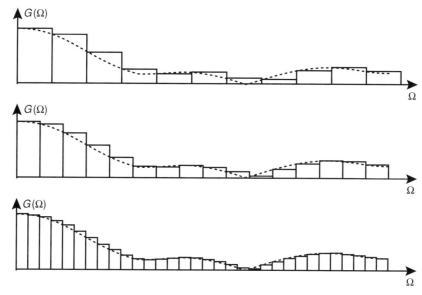

FIGURE 6–11
Computing $G(\Omega)$ over different segment sizes.

to zero at a specific frequency. Making the segments smaller increases the accuracy of the approximation and consequently reveals more detail of the system response curve.

The complete system response requires plotting distinct gain and phase response graphs. Figure 6–12 combines the amplitude and the phase response graphs to display a complete representation of the system frequency response. Note that the frequency segments are now small enough to produce what looks like a continuous response curve.

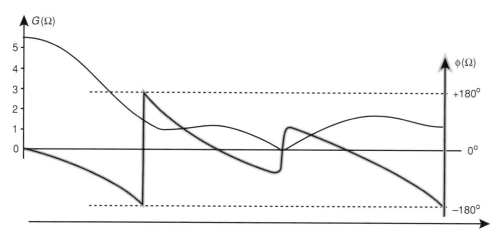

FIGURE 6–12
Gain and phase response curves combined.

Notice that the phase curve in Figure 6–12 wraps to the opposite side when the phase exceeds the ±180° limits. This permits keeping the phase plot within ±180° boundaries. This behavior is a direct result of the periodicity of sinusoids. For a periodic waveform, delaying the phase by 180° produces that same result as advancing it by 180°. For example, instead of saying that the phase of the sinusoid is shifted by −200°(−180° − 20°), we can equally say that it is shifted by +160° (+180° − 20°). Figure 6–13 illustrates this property of periodic signals.

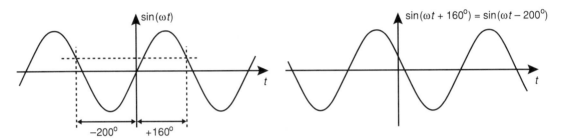

FIGURE 6–13
Wrapping around 180°.

6.3.3 PERIODICITY OF THE FREQUENCY RESPONSE

Our investigation of periodicity in the frequency domain, in Chapter 3, revealed on page 78 that discrete signals are periodic in the frequency domain. The discrete-time Fourier transform operation sums periodic harmonically related complex exponentials and, because of this, the result must also be a periodic function. This means that the curves that describe the frequency response of all discrete-time systems are periodic. Let's illustrate this with an example.

Consider a very simple system whose weighting factors consist of the following two b_k coefficients:

$$b_0 = 1 \quad \text{and} \quad b_1 = 0.3$$

The difference equation of this system uses two input samples:

$$y(n) = 1x(n) + 0.3x(n-1)$$

This system therefore computes each of its output samples by weighting the previous input sample $x(n-1)$ by a factor of 0.3 and by adding the result to the present input sample $x(n)$.

When a discrete impulse function is applied to this system, the system responds by outputting an impulse response that reflects the value of the weighting factors:

$$h(0) = 1 \quad \text{and} \quad h(1) = 0.3$$

We calculate the discrete-time Fourier transform for this system as follows:

$$H(\Omega) = \sum_{k=0}^{1} h(k)e^{-j\Omega k}$$
$$= 1e^{-j\Omega \times 0} + 0.3e^{-j\Omega \times 1}$$
$$= 1 + 0.3e^{-j\Omega}$$

Clearly, the discrete-time Fourier transform for this system consists of two periodic complex exponentials, which are vectors. The first vector has a modulus (length) of 1 and an angle of 0, which means that it is always horizontal. The second vector has a modulus of 0.3 and an angle that depends on the normalized input frequency Ω. Figure 6–14 illustrates the addition of these two vectors.

FIGURE 6–14

$H(\Omega)$ for a system defined by $h(0) = 1$ and $h(1) = 0.3$.

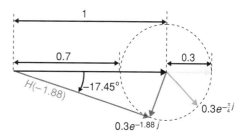

Referring to Figure 6–14, the value of the transform $H(\Omega)$ depends on the normalized input frequency Ω. As Ω increases from zero to π, the $0.3e^{-j\Omega}$ vector rotates clockwise to trace the lower half circle. As the input frequency increases, the response vector goes through the following points of interest:

- At a DC input of $\Omega = 0$, the resulting response vector reaches its longest elongation (the largest gain): $|H(0)| = G(0) = 1.3$; the system phase shift is 0 degrees.
- As the normalized frequency of the input sinusoid moves from zero to π, the tip of the response vector follows the lower half circle. Along the way, its length gets smaller and its angle reaches a largest negative value of –17.45 degrees when $\Omega = 1.88$ as shown in Figure 6–15.
- When the normalized frequency reaches a value of π, the $0.3e^{-j\Omega}$ vector points exactly left and the response vector reaches its shortest elongation (the smallest gain): $|H(\pi)| = G(\pi) = 0.7$. At that input frequency, the system phase shift returns to 0 degrees.

A normalized frequency of π corresponds to an input frequency of $f_S/2$, which is normally the highest frequency processed by a digital system. Nevertheless, if there is no antialiasing filter, nothing prevents inputting frequencies above this limit. Returning to Figure 6–14, the $0.3e^{-j\Omega}$ vector continues its rotation if Ω increases beyond a value of π. The tip of the response vector continues tracking by following the upper half circle. This results in the

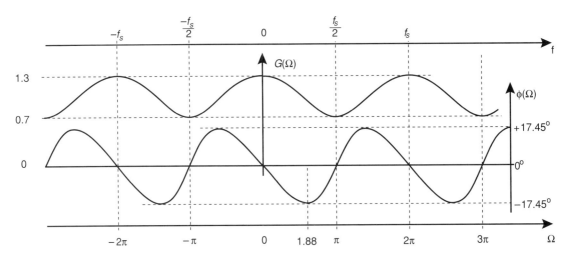

FIGURE 6–15
Fourier transform is periodic.

$H(\Omega)$ vector adopting values that are exact conjugates of the corresponding lower half-circle positions. If Ω increases beyond a value of 2π, the response vector repeats the same pattern.

We can determine the system response to negative input frequencies in the same way. Again, the response vector repeats the response pattern. Figure 6–15 illustrates the periodicity of the frequency response by plotting the length (gain) and the angle (phase shift) of $H(\Omega)$ for ranges of negative and positive input frequencies.

Note that the Fourier transform in the Figure 6–15 curves is periodic with a period of 2π. This means that any plot covering a range of 2π provides enough information to determine the system frequency response at any frequency.

Further simplification of the calculation of the frequency response graph is possible when the system impulse response consists of pure-real terms. In the previous example, the system used pure-real numbers to weight the input samples; this resulted in pure-real output samples. Effectively, a system that has a pure-real response that processes a pure-real input signal will produce a pure-real output waveform. Remember that pure-real numbers may be expressed as containing two parts that are complex conjugates of each other. For this reason, a real system response requires that the negative frequency range be a complex conjugate reflection of the positive frequencies range.

A pure-real system response requires that (1) the gain response of the system display even symmetry about 0 Hz and (2) the phase response of the system display odd symmetry about 0 Hz.

Figure 6–16 illustrates the symmetries of a real system response by plotting a 2π period that spans a range of $-\pi$ to $+\pi$. When the system has a pure-real response, we need to plot only the positive frequency side of Figure 6–16. This side shows the system gain and the phase shift applied to the $e^{+j\Omega n}$ component of an input sinusoid. According to Section 6.2.1, this response exactly reflects the changes that the system applies to real sinusoids.

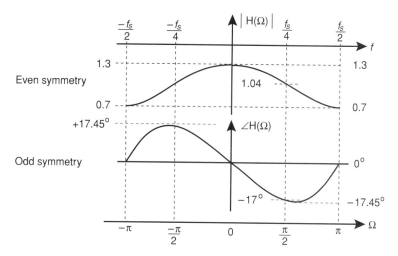

FIGURE 6–16
Symmetries of a system with a real impulse response.

For example, we can use the positive frequency side of Figure 6–16 to determine the system gain and phase shift for any frequency in the normalized range. Assuming that the ADC sampling rate is 8000 samples/second, a 2000 Hz input cosine oscillates at the following normalized frequency:

$$\Omega = \omega T_S = 2\pi \times 2000 \times \frac{1}{8000} = \frac{\pi}{2} \text{ radians/sample}$$

For the system illustrated in Figure 6–16, a sinusoid oscillating at a normalized input frequency of $\Omega = \pi/2$ has its amplitude modified by a factor of 1.04 and its phase shifts by –17 degrees (–0.297 radians) as it travels through the system:

$$\cos(2\pi \times 2000n) \Rightarrow \text{SYSTEM} \Rightarrow 1.04 \cos(2\pi \times 2000n - 0.297)$$

6.4 FREQUENCY RESPONSE OF A DIGITAL-TO-ANALOG CONVERTER

Section 2.1.3 discussed the use of a DAC to translate the processor output samples to the analog world. We know that a DAC converts the value of a digital sample to a specific voltage, which it maintains over a sampling interval to produce a voltage step. Figure 6–17 illustrates the input and output functions of a DAC in both time and frequency domains.

We know that the output function of a system must be related to its input function. We compare the discrete-time input to the continuous-time output of the DAC in the upper part of Figure 6–17. A relationship definitely exists between the waveforms since their components have the same height; however, they exhibit fundamental differences in continuity. If we examine the frequency contents of the input signal in the lower part of Figure 6–17, we

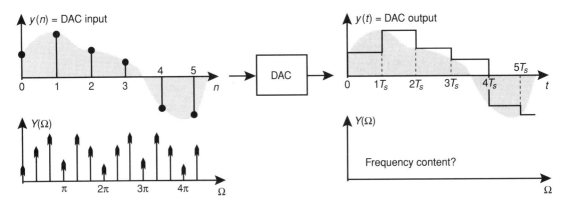

FIGURE 6–17
Input and output of a DAC.

see that the discrete-time input consists of periodic harmonically related sinusoids. The frequency domain representation of the output signal must be related to these harmonics, yet somehow it must also be different. What are the relationships and the differences that establish the link between the input and output signals in the frequency domain?

The way that the DAC system modifies the frequency content of the input signal is the subject of the following subsections in which we investigate the frequency response of DAC systems.

6.4.1 OPERATION OF CONTINUOUS-TIME SYSTEMS

According to Section 4.2.4, discrete-time systems compute their outputs by combining weighted inputs according to an LTI process, defined by a difference equation. Once we know the constant weighting factors that define the system, we can use the convolution operation to compute the sequence of output values.

Imagine that a particular system must use the last five seconds of past input history to compute the output value. If the sampling rate is 1 sample/second, this system requires a difference equation that uses five weighting factors. If we increase the sampling rate to 2 samples/second, the system must use 10 weighting factors to look back 5 seconds. We can picture the operation of a continuous-time system by imagining a discrete-time system whose sampling rate increases toward infinity. If the system always considers a constant segment of past history, the system requires more weighting factors as the sampling rate increases. It follows that a system that has an infinite sampling rate must be defined by an infinite number of weighting factors. Such a system uses a *continuous* weighting function to operate continuously on the input signal. An analog system *continuously* weights its input signal using a continuous weighting function. Figure 6–18 compares the impulse response and the output function of a discrete-time system to its counterpart in the continuous-time domain.

Note in Figure 6–18 that a continuous impulse response defines the continuous-time system. This function reflects the continuous weighting function that the system uses to compute its analog output. Since the weighting function is continuous, the system output

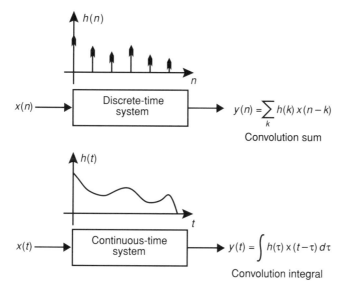

FIGURE 6–18
Discrete-time and continuous-time systems compared.

function uses a continuous convolution *integral* to weight the input signal over the range of time defined by the nonzero values of $h(t)$:

$$y(t) = \int h(\tau)\, x(t - \tau)\, d\tau \tag{6–4}$$

Compare this integral to the convolution *sum* (Equation 4–3), which we use to compute the output of discrete-time systems.

6.4.2 MECHANICS OF A DAC SYSTEM

The DAC is a continuous-time system that outputs an analog signal. Since it provides a continuous output, its operation must be defined by a continuous weighting function. To determine the weighting function of a DAC system, we can apply an impulse function to its input. Submitting this continuous-time impulse to the DAC is equivalent to clocking the DAC latch with a new input that has a value of 1 at time $t = 0$. The DAC latch records the impulse value and produces the corresponding voltage step. Since the impulse function has a value of 0 at all other times, the latch is clocked with a value of 0 at all other times. Consequently, at time $t = T_S$, the latch receives the next clock pulse, records a value of zero, and loses memory of the last impulse value. We can therefore conclude that a DAC system has memory of the past T_S interval of time. Its impulse response therefore consists of a constant voltage step lasting over an interval of T_S seconds. The amplitude of the step is determined by the weighted contents of the latched value. Figure 6–19 illustrates the DAC impulse response.

As Figure 6–19 indicates, the impulse response is continuous, lasts for T_S seconds, and has a constant amplitude that corresponds to the DAC weighting factor. If we know the value of the digital input, we can use the convolution integral of Equation (6–4) to determine the

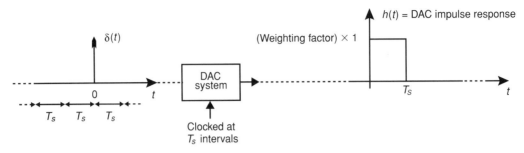

FIGURE 6–19
Duration of a DAC impulse response.

value of the output voltage. For example, if the DAC has a *unity gain,* we expect an impulse value of 0.5 (a numerical input of 0.5) to yield an output of 0.5 volt and an impulse value of 0.3 to yield an output of 0.3 volt.

We next determine the value of the weighting factor that defines a unity gain DAC. When we apply an impulse (a numerical value of 1), we can compute the output at time T_S by using the convolution integral as follows:

$$y(t) = \int_0^{T_S} h(\tau) \, x(t - \tau) d\tau$$

Note that we need to evaluate the integral only from zero to T_S because the impulse response function is zero valued at all other times. The weighting factor $h(\tau)$ is constant over the T_S interval, and the input $x(T_S - \tau)$ is maintained at a constant value of 1 by the latch. We can therefore write

$$y(t) = \int_0^{T_S} (\text{Constant weighting factor}) \times 1 \ d\tau$$

$$= (\text{Constant weighting factor}) \times \int_0^{T_S} 1 \ d\tau$$

$$= (\text{Constant weighting factor}) \times 1 \, [\tau]_0^{T_S}$$

$$= (\text{Constant weighting factor}) \times 1 \, [T_S - 0]$$

$$= (\text{Constant weighting factor}) \times 1 \, T_S$$

For the DAC under study to have a unity gain, a numerical input of 1 must yield an output of 1 volt. We can therefore calculate the weighting factor as follows:

$$y(t) = 1 = (\text{Constant weighting factor}) \times 1 \, T_S$$

$$\text{Constant weighting factor} = \frac{1}{T_S}$$

Figure 6–20 illustrates the impulse response of a unity gain DAC.

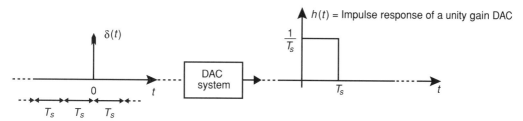

FIGURE 6–20
Impulse response of a unity gain DAC.

6.4.3 FREQUENCY RESPONSE OF CONTINUOUS-TIME SYSTEMS

Applying the discrete-time Fourier transform (Equation [6–3]) to a digital system impulse response yields the frequency response of the system (refer to Section 6.3.1). Drawing the frequency response curves allows us to determine the gain and phase changes applied to input sinusoids oscillating at any frequency. To obtain the frequency response for a DAC system, we need to develop a continuous-time equivalent to the discrete-time Fourier transform.

We can develop the continuous-time Fourier transform using the same approach as the one we adopted to develop the discrete-time Fourier transform. To do this, we use the convolution integral to examine the output of a system when we apply a continuous periodic complex exponential to the analog system input:

$$y(t) = \int_{\infty}^{-\infty} h(\tau)\, x(t - \tau) d\tau$$

For the periodic input $x(t) = e^{j\omega t}$, we get the following output:

$$y(t) = \int_{\infty}^{-\infty} h(\tau)\, e^{j\omega(t-\tau)} d\tau = \int_{\infty}^{-\infty} h(\tau)\, e^{j\omega t} e^{-j\omega \tau} d\tau$$

$$y(t) = e^{j\omega t} \int_{\infty}^{-\infty} h(\tau)\, e^{-j\omega \tau} d\tau$$

The input signal is multiplied by an integral that must correspond to the system response for which we are looking. Notice that the integral depends only on ω, which means that the system response is a function of the input frequency:

$$H(\omega) = \int_{\infty}^{-\infty} h(\tau)\, e^{-j\omega \tau} d\tau \qquad \text{(6–5)}$$

We call Equation (6–5) the *continuous-time Fourier transform*. Compare it to Equation (6–3), the discrete-time equivalent. Notice that the summation of its discrete-time equivalent is simply replaced by a continuous integral. This transform provides the tool to analyze the frequency response of a DAC system.

6.4.4 FREQUENCY RESPONSE OF A DIGITAL-TO-ANALOG CONVERTER

We determine the frequency response of a DAC system by applying the continuous-time Fourier transform (Equation [6–5]) to the DAC impulse response. We determined in Section 6.4.2 that the impulse response of a unity gain DAC is a step function with a height of $1/T_S$ and a duration of T_S second as depicted on Figure 6–20.

The continuous-time Fourier transform integral is defined in Equation (6–5) but, since the DAC impulse response $h(t)$ is zero valued outside the zero to T_S range, we can redefine the integration limits as follows:

$$H(\omega) = \int_0^{T_S} h(\tau)\, e^{-j\omega\tau} d\tau$$

The value of the impulse response $h(t)$ is constant at a value of $1/T_S$ over the integration limits. Substituting this into the Fourier transform yields

$$H(\omega) = \int_0^{T_S} \frac{1}{T_S} e^{-j\omega\tau}\, d\tau = \frac{1}{T_S} \int_0^{T_S} e^{-j\omega\tau}\, d\tau$$

Integrating, we get

$$H(\omega) = \frac{1}{T_S} \left[\frac{e^{-j\omega\tau}}{-j\omega} \right]_0^{T_S} = \frac{1}{T_S} \left[\frac{e^{-j\omega T_S} - 1}{-j\omega} \right]$$

Recognizing that

$$-1 = -\left(e^{-j\frac{\omega T_S}{2}} \times e^{+j\frac{\omega T_S}{2}} \right)$$

we can substitute and then factor

$$H(\omega) = \frac{1}{T_S} \left[\frac{e^{-j\omega T_S} - \left(e^{-j\frac{\omega T_S}{2}} \times e^{+j\frac{\omega T_S}{2}} \right)}{-j\omega} \right]$$

$$= \frac{1}{T_S} e^{-j\frac{\omega T_S}{2}} \left[\frac{+e^{-j\frac{\omega T_S}{2}} - e^{+j\frac{\omega T_S}{2}}}{-j\omega} \right]$$

Moving the denominator ω out of the brackets yields

$$H(\omega) = \frac{1}{\omega T_S} e^{-j\frac{\omega T_S}{2}} \left[\frac{+e^{-j\frac{\omega T_S}{2}} - e^{+j\frac{\omega T_S}{2}}}{-j} \right]$$

$$= \frac{1}{\omega T_S} e^{-j\frac{\omega T_S}{2}} \left[\frac{+ e^{+j\frac{\omega T_S}{2}} - e^{-j\frac{\omega T_S}{2}}}{j} \right]$$

According to Euler, the expression

$$\left[\frac{+ e^{+j\frac{\omega T_S}{2}} - e^{-j\frac{\omega T_S}{2}}}{j} \right]$$

corresponds to

$$2 \sin\left(\frac{\omega T_S}{2} \right)$$

We can substitute this value in the Fourier transform expression:

$$H(\omega) = \frac{1}{\omega T_S} e^{-j\frac{\omega T_S}{2}} \times 2 \sin\left(\frac{\omega T_S}{2} \right)$$

We rearrange the Fourier transform vector to isolate the magnitude and the angle of the resulting vector:

$$H(\omega) = \frac{2 \sin\left(\dfrac{\omega T_S}{2} \right)}{\omega T_S} e^{-j\frac{\omega T_S}{2}}$$

$$= \frac{\sin\left(\dfrac{\omega T_S}{2} \right)}{\dfrac{\omega T_S}{2}} e^{-j\frac{\omega T_S}{2}}$$

As we have established, the magnitude and the phase of $H(\omega)$ determine the gain and the phase shift, respectively, applied to an input sinusoid of frequency ω.

$$\text{DAC gain: } |H(\omega)| = G(\omega) = \frac{\sin\left(\dfrac{\omega T_S}{2}\right)}{\dfrac{\omega T_S}{2}} \tag{6–6}$$

$$\text{DAC phase shift: } \angle H(\omega) = \phi(\omega) = -\frac{\omega T_S}{2} \tag{6–7}$$

If we now define practical values for ω and T_S, we can calculate values for the gain and phase shift that a DAC system applies when it translates digitized sinusoids into continuous signals. As an example, we assume that a DAC converts digital input values at the rate of 10 k-samples/second and that these samples carry a full amplitude (± 1) sinusoid waveform oscillating at 2 kHz. In this particular case, the sampling period is

$$T_S = \frac{1}{10\,k}$$

The angular frequency of the input sinusoid is

$$\omega = 2\pi \text{ radians/cycle} \times 2\,k\text{-cycles/second} = 4\pi\,k\text{-radians/second}$$

We use Equation (6–6) to calculate the gain that the DAC applies to the sinusoid:

$$G(4\pi\ k) = \frac{\sin\left(\dfrac{4\pi\ k \times \dfrac{1}{10\,k}}{2}\right)}{\dfrac{4\pi\ k \times \dfrac{1}{10\,k}}{2}} = \frac{\sin\left(\dfrac{\pi}{5}\right)}{\dfrac{\pi}{5}} = 0.9355$$

The DAC gain response is less than 1 (an attenuation); therefore, the sinusoid has its amplitude lowered from ± 1 to ± 0.9355 by the DAC. In this particular example, the DAC has a gain of 5 producing a full-scale output amplitude of ± 5 volt. The 2 kHz sinusoid contained in the analog output oscillates with an amplitude of

$$\text{Output sinusoid amplitude} = \pm 5v \times 0.9355 = \pm 4.6775v$$

We use Equation (6–7) to calculate the phase shift that the DAC system applies to the output sinusoid:

$$\phi(4\pi k) = -\frac{4\pi\ k T_S}{2} = -\frac{4\pi k \times \dfrac{1}{10\,k}}{2} = -\frac{\pi}{5}$$

The 2 kHz sinusoid component at the output of the DAC therefore has the following characteristics:

$$\cos\left(2\pi \times 2k \times \frac{1}{10k}n\right) \xrightarrow{\text{DAC}} 4.6775\cos\left(2\pi \times 2k\left(2\pi \times 2kt - \frac{\pi}{5}\right)t - \frac{\pi}{5}\right)$$

To describe the DAC response over a range of frequencies, we draw gain and phase curves. This means that we must calculate $G(\omega)$ and $\phi(\omega)$ for a range of ω values. In digital systems, the useful range of frequency spans from zero to $f_S/2$. This means that ω should cover the following range:

$$\omega = 0 \text{ to } \frac{2\pi f_S}{2} = 0 \text{ to } \pi f_S$$

This corresponds to the following normalized frequency range:

$$\Omega = \omega T_S = 0 \text{ to } \pi$$

We draw the phase delay response curve by plotting Equation (6–7), the value of $\phi(\omega)$ over that range. Figure 6–21 illustrates the phase response of a DAC. Note that the phase response is a straight line in this example. Such a response does not create any signal distortion. Section 8.1.2 discusses the advantages of such linear-phase response systems.

To simplify Equation (6–6) and Equation (6–7), which describe the gain and phase response of a DAC, we define the following variable:

$$x = \frac{\omega T_S}{2}$$

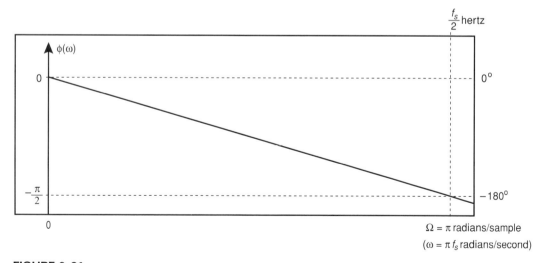

FIGURE 6–21
Phase response of a DAC.

Since the maximum frequency in a digital system is $\omega_{Max} = \pi f_S$, the value of x that corresponds to this maximum is

$$x_{Max} = \frac{\omega_{Max} T_S}{2} = \frac{\pi f_S T_S}{2}$$

$$= \frac{\pi f_S \times \dfrac{1}{f_S}}{2} = \frac{\pi}{2}$$

Defining the variable x in this way is quite convenient when we analyze the output of a DAC since we can use it to simplify its frequency response expression:

$$H(\omega) = \frac{\sin\left(\dfrac{\omega T_S}{2}\right)}{\dfrac{\omega T_S}{2}} e^{-j\left(\frac{\omega T_S}{2}\right)} = G(\omega)\, e^{-\phi(\omega)}$$

$$= \frac{\sin x}{x}\, e^{-jx}$$

The $\sin x/x$ expression corresponds to the gain response of the DAC, and it occurs routinely in many frequency analysis applications. Because of this, mathematicians have a special name for it:

$$\text{sinc } x = \frac{\sin x}{x}$$

Figure 6–22 illustrates the appearance of the sinc x function.

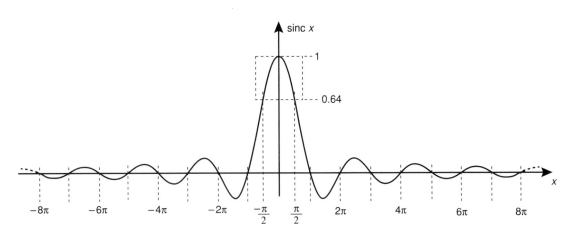

FIGURE 6–22
Sinc x function.

The range from zero to $\pi/2$ is of particular interest since it corresponds to the useful range of frequencies that apply to DSP systems. Over this range, as Figure 6–22 indicates, the sinc x function starts at a value of 1 and decays as the value of x increases. Figure 6–23 magnifies the useful range of the sinc x curve and shows the relationship between $x, f,$ and Ω.

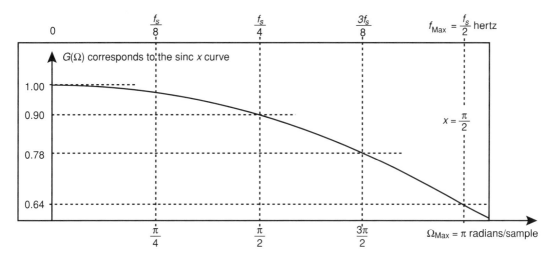

FIGURE 6–23
Magnified sinc x curve.

The portion of the sinc x curve displayed in Figure 6–23 corresponds to the gain response of a DAC system when it tries to reproduce the useful range of frequencies handled by DSP systems. Notice that the frequency axis calibration depends on the value of the sampling rate. For example, if the sampling rate is 8 kHz, the DAC gain response curve is interpreted as follows:

- The maximum input frequency is $f_S/2 = 4$ kHz where the DAC gain response is 0.64.
- An input frequency of $3/4 \times f_S/2 = 3f_S/8 = 3$ kHz where the DAC gain response is 0.78.
- An input frequency of $1/2 \times f_S/2 = f_S/4 = 2$ kHz where the DAC gain response is 0.90.

Clearly, the gain response droops significantly as the DAC reproduces frequencies that approach the limits of the practical range. If the frequency response of the DAC is not compensated for, the higher frequencies contained in the output signal suffer serious gain distortions.

6.4.5 COMPENSATION WITH AN EQUALIZER

The previous section indicated that the gain response of a DAC system produces a significant attenuation of the higher frequencies contained in the output signal. To ensure that the DSP system reproduces all the frequencies equally, we must ensure that the gain response curve is as "flat" as possible. A perfectly flat gain-response curve means that the system re-produces all the frequencies with exactly the same gain. In practice, it is possible to com-pensate partially for the DAC droop, but the system will never achieve a perfectly flat gain response.

The DAC frequency response is inherent to the way DACs operate. The gain droop is impossible to avoid unless a different method is used to convert the digital signal to the analog world. Building such an alternative would be complicated and expensive. A better way to deal with this problem is to use some form of compensation circuit that restores the frequencies to their proper respective amplitudes. We call such circuit an *equalizer*.

An equalizer works by canceling the effect of the DAC frequency response. To do this, the frequency response of the equalizer is engineered to invert that of the DAC. Math-ematically, the equalizer gain response must match the $1/(\text{sinc } x)$ curve. Figure 6–24 illus-trates the ideal equalizer response.

A system that includes an equalizer ideally cancels the DAC droop if the equalizer response is the exact inverse of the DAC gain curve. Figure 6–25 illustrates the addition of an equalizer in the system.

From a practical point of view, building the ideal equalizer is not an easy task. Get-ting a close match to the $1/(\text{sinc } x)$ curve requires either expensive analog components or a fair amount of signal processing. Nevertheless, it is still required, and the designer should

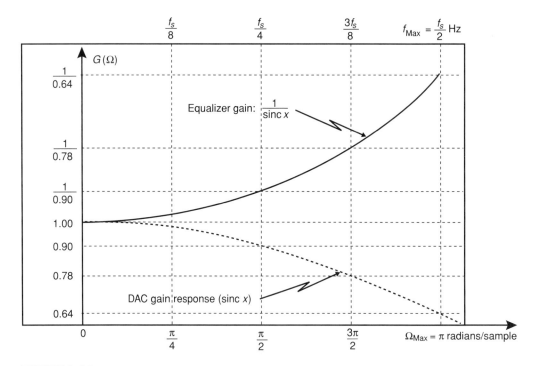

FIGURE 6–24
Ideal equalizer response.

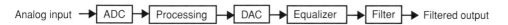

FIGURE 6–25
Use of a system equalizer.

try to put the equalizer in a part of the system where its cost will have the least impact on the final cost of the product.

Fortunately, in an LTI system, we can arrange the sequence of subsystems in any order. This means that we can place the equalizer *before* the DAC. In this case, the equalizer preemphasizes the frequencies, and the DAC droop restores the amplitude to a flat response.

Let's look at two cases in which it makes sense to position the equalizer before the DAC. In the first case, consider the recording of digital multitrack music. The studio engineers use many high-quality microphones to generate low-noise signals that are sent through an antialiasing filter. A number of ADCs digitize the filtered microphone signals, and the samples are stored on a tape or on a hard drive in digital format. Following the recording session, the production engineer mixes the digitized signals to produce a stereo master. A manufacturing process then reproduces the content of the master onto commercial CDs.

The consumer buys the CD and inserts it into a player. A laser reads the digital information and sends the samples to a pair of DACs, which produce staircaselike stereo outputs. These outputs are then sent through a pair of reconstruction filters before being amplified to drive the speakers. Figure 6–26 illustrates the entire system.

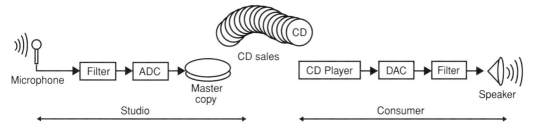

FIGURE 6–26
Production of a music CD.

The natural frequency response of DACs attenuate the high frequencies (refer to Section 6.4.4). To obtain an accurate reproduction of the recorded music, the system must compensate the frequency response of the DACs. We can build expensive equalizers into every CD player commercially available to the consumer, or we can preemphasize the signal *once* on the master before transferring it to the CDs. Where would you put the equalizer?

In the second case, consider the telephone system. Many businesses already use digital telephones, and private homes will soon benefit from this technology. A digital telephone contains an ADC that digitizes the voice at a rate of 8000 samples/second. The digital telephone transfers the voice samples to a central office so that they can be transmitted to the destination. The telephone network routes the voice samples to the destination telephone where a DAC translates them back to the analog world. A reconstruction filter cleans up the staircaselike DAC output, and an amplifier drives the resulting signal to a speaker, which produces the sound waves that reach the ear. Figure 6–27 illustrates the main components of a pure digital telephone network.

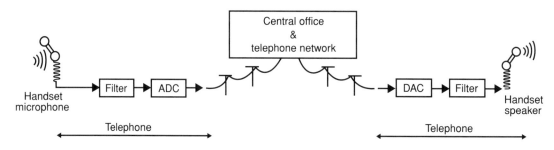

FIGURE 6–27
Digital telephone network.

To obtain good sound quality, the signal must be equalized somewhere along the digital telephone system. We can build equalizers into every commercially available digital telephone receiver, or we can preemphasize the signal *once* in the central office before transferring it to the destination telephone. Where would you put the equalizer?

Figure 6–28 illustrates what happens to a sinusoid as it travels through a digital system. As the figure indicates,

- An antialiasing filter has already ensured that the analog input contains only frequencies inside the $f_S/2$ barrier. Any analog sinusoid inside the useful range of frequencies corresponds to two conjugate complex exponentials.
- The ADC digitizes an analog cosine $\omega_0 = 2\pi f_0$; this has the effect of replicating the original spectrum components around multiples of f_S.
- The DAC receives the numerical samples coming from the ADC and stretches them over the sampling interval T_S. This operation results in a frequency response, which attenuates the frequency components of the signal according to a sinc x curve.
- The reconstruction filter removes all the frequencies outside the $f_S/2$ limit so that only frequencies in the useful range remain. Figure 6–28 assumes an ideal reconstruction filter but, in practice, it may not do such a perfect job. We examine the characteristics of this output filter in the next section.
- Note that the amplitude of the output sinusoid is lower than it was at the digital input of the ADC. Also note that the frequency response of the DAC has also delayed the sinusoid phase.

If the system requires a flat frequency response, the effect that the DAC has on the amplitude creates a practical problem. To compensate for the gain response of the DAC, we must add an equalizer at some point along this linear digital system.

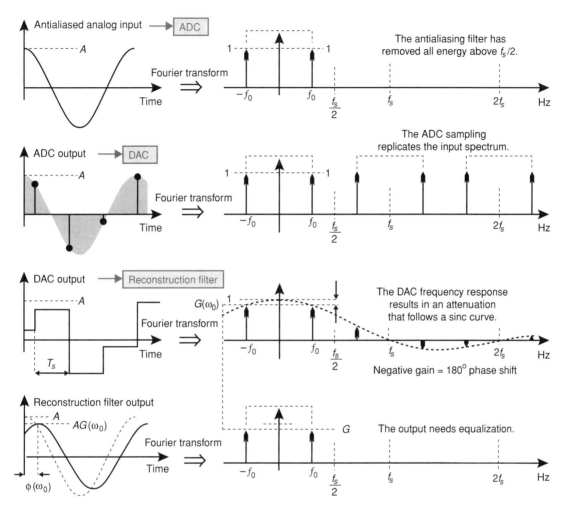

FIGURE 6–28
Sinusoid traveling through a digital system.

6.4.6 RECONSTRUCTION FILTERS

Translating the samples of a discrete-time signal to the continuous-time domain is simple in theory. The replicated frequency components that lie outside the $f_S/2$ barrier must be removed. To do this, we need to design a filter that will ideally pass all frequencies that reside inside the $f_S/2$ barrier and completely eliminate those that are located outside this limit. Figure 6–29 illustrates the ideal gain characteristics of a reconstruction filter.

The unity gain DAC that we used in the previous section is called a *zero-order hold circuit* because its output value is related to a single input sample through the following equation for t ranging from nT_S to $(n + 1)T_S$:

$$y(t) = x(n)t^0$$

FIGURE 6–29
Ideal reconstruction filter.

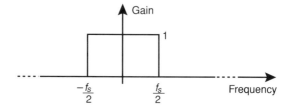

The zero-order equation of a DAC results in a poor reconstruction filter that follows the shape of a sinc *x* curve. Figure 6–30 compares the ideal shape of a reconstruction filter to the poor reconstruction filter that results when using a zero-order DAC.

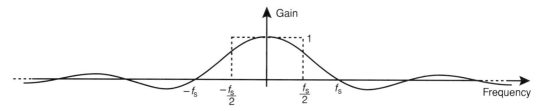

FIGURE 6–30
Comparison of the sinc *x* curve to an ideal reconstruction filter.

The reason for the staircaselike appearance of the DAC output is that the sinc *x* response is not able to completely remove the replicated spectrum frequencies located outside the $f_S/2$ barrier. The DAC has managed to translate the discrete-time domain samples into the continuous-time domain, but the output signal still contains relics of the sampling process. This type of DAC is by far the most common type of converter commercially available. In a zero-order hold circuit, the present output voltage totally depends on the last value latched by the DAC circuit.

If we wanted, we could try to improve the zero-order hold circuitry by designing a new type of DAC. For example, using the last two sample values to produce the output voltage would allow linear interpolation between each set of two samples. Such a converter uses *first-order hold circuitry* because its output value is related to two input samples through a first-order equation for t ranging from nT_S to $(n + 1)T_S$:

$$y(t) = \left[x(n + 1) - x(n)\right]\frac{t^1 - nT_S}{T_S} + x(n)$$

The output of such converter would no longer be flat but would consist of straight lines that join the digital sample values. Figure 6–31 compares the output of such converter to a zero-order hold converter.

Comparing the output of the first-order hold circuit depicted in Figure 6–31 to the zero-order hold converter, you can certainly appreciate that the first-order hold circuit produces

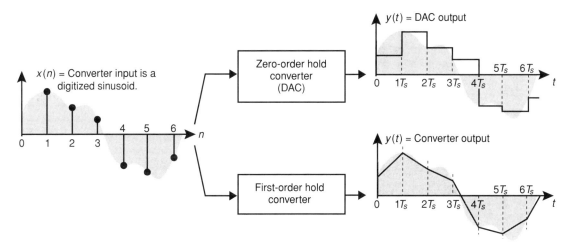

FIGURE 6–31
Converters of different order.

results that are closer to the ideal shaded waveform. This is so because the frequency response of a first-order hold circuit produces a superior approximation to the ideal reconstruction filter. Although a first-order hold converter improves the appearance of the output, it is still unable to completely eliminate the relics of the sampling process. We could try designing second- or third-order converters, but this would be too complicated and expensive. For this reason, commercially available DACs are of the zero-order hold variety.

We must therefore add extra reconstruction filtering at the output of the DAC to get closer to the ideal reconstruction filter characteristics. Many different types of filters are commercially available to do this job; these are discussed in Chapter 8. Special attention should be given to the Bessel filter.

SUMMARY

- In an LTI system, we can separately analyze the system response to each sinusoid component contained in the input signal.
- Sinusoids consist of complex exponentials; sending these through an LTI system reveals how the system responds to such inputs.
- An LTI system modifies the amplitude and the phase of sinusoid inputs.
- In an LTI system of fixed sampling rate, the amount of amplitude change and the phase delay that the system applies depend on the *frequency* of the input sinusoid.
- The Fourier transform translates the system impulse response into an expression that completely defines the system in the frequency domain.
- Calculating the system Fourier transform at a specific input frequency yields a vector whose magnitude and phase describe the system gain and phase shift, respectively, at that frequency.
- Practical calculation of the Fourier transform requires tedious calculations that are best handled by specialized software tools.

- We can obtain the system response over a range of frequencies by calculating and plotting values of the Fourier transform along that range of frequencies. This graphical representation requires separate gain and phase curves.
- The Fourier transform of discrete-time systems is a periodic function.
- Systems that are defined using pure-real weighting factors yield a pure-real system response. In such systems, the negative frequency side of the system response is the complex conjugate image of the positive frequency side.
- DACs have a frequency response, which modifies the amplitude and the phase of the sinusoid components of a signal.
- The gain response of the DAC must usually be corrected by adding an equalizer.
- The equalizer can be added anywhere along the LTI system, but positioning it *before* the DAC often yields substantial simplifications and cost savings.
- The staircaselike appearance at the output of a DAC exists because a zero-order hold DAC is incapable of completely removing the frequency components lying outside the $f_S/2$ limit. An additional reconstruction filter is required to help remove these frequencies.

PRACTICE QUESTIONS

6-1. Consider the following signal:

$$x(n) = 0.6e^{-j(2n+0.5)} + 2.4je^{-jln} + 0.7e^{+j0n} + 2.4je^{+jln} + 0.6e^{+j(2n+0.5)}$$

(a) Is the signal $x(n)$ pure real?

(b) Find the amplitude and the phase of the sinusoids contained in $x(n)$.

(c) Is $x(n)$ periodic? If so, what is the duration of the period when the sampling rate is 8 k-samples/second?

6-2. A system has the following impulse response:

$$h(0) = 1, h(1) = 0.6, h(2) = 0.4$$

(a) What are the three vectors that describe the Fourier transform of this system?

(b) On graph paper draw the three vectors that describe the system response to an input sinusoid oscillating at a normalized frequency of $\pi/10$.

(c) Use your drawing of part (b) to estimate the gain and the phase shift that the system applies to an input sinusoid oscillating at a normalized frequency of $\pi/10$.

6-3. A system has the following impulse response:

$$h(0) = 1, h(1) = 0.6, h(2) = 0.4$$

Use a software package of your choice to plot the system gain over the range of 0 to $f_S/2$.

6-4. A system has a pure-real frequency response. For an input sinusoid oscillating at a normalized frequency of 1.2 radians/sample, this system applies a gain of 2.8 and a phase shift of 30 degrees.

(a) If the input sinusoid is $3.2 \sin(1.2n + \pi/10)$, what expression describes the output sinusoid?

(b) What is the system gain at a normalized frequency of –1.2 radians/sample?

(c) What is the system phase shift at a normalized frequency of –1.2 radians/sample?

6-5. The frequency response of a system is described by the curves in Figure 6–32:

(a) If the sampling rate is 8 k-samples/second, use the response curves to estimate the gain and the phase shift that the system applies to an input sinusoid oscillating at 1 kHz.

(b) If the sampling rate is reduced to 4 k-samples/second, estimate the gain and the phase shift that the system applies to the 1 kHz input sinusoid.

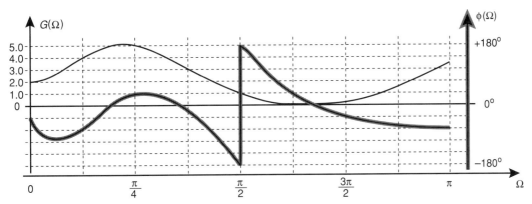

FIGURE 6–32
Frequency response of a system.

6-6. A DSP system uses a sampling rate of 8 k-samples/second and has no output reconstruction filter. Consider that the unity gain DAC of that system receives samples of a synthesized cosine waveform that has a frequency of 10 k-radians/second and an amplitude of 0.8.

(a) What are four of the many frequency components contained in the synthesized cosine?

(b) What is the amplitude of the 10 k-radians/second component at the output of the DAC?

(c) Determine the amount of phase shift that the DAC applies to the 10 k-radians/second component at the output of the DAC?

(d) Consider that the frequency of the synthesized cosine being outputted is changed to 20 k-radians/second. Determine the gain and phase shift that the DAC applies to the 20 k-radians/second component.

(e) Consider that the phase of the synthesized cosine being synthesized is shifted by 1 radian. Does this have an effect on the amplitude and the phase of the 10 k-radians/second component at the output of the DAC?

6-7. A digital signal processing system uses a sampling rate of 8 k-samples/second. What is the necessary equalizer gain at a frequency of 2 kHz?

7

DESIGNING THE SYSTEM RESPONSE

Linear/time-invariant (LTI) systems are described by their impulse response. This response completely determines how the system responds to various types of input signals. Designing the system so that it responds in a certain way therefore implies that we must engineer distinct values for the terms of the impulse response.

Chapter 7 develops and explores an approach to calculate the value of the impulse response terms. We do this by breaking terms of the impulse response into a number of individual *factors*, each controlling a part of the total system response. We call these factors the *poles* and the *zeros* of the system; by plotting them on a special diagram, we can visualize the effect they have on the overall system response. Shaping the response of the system becomes a question of arranging a pattern of poles and zeros on this diagram.

We investigate the way that the position of the system poles and zeros influences the overall system response. The discoveries we make during this investigation provide the key to control the frequency response of LTI systems. We then experiment with designing the gain and phase shift that the system applies to the different signal harmonics as they travel from input to output.

We establish that certain pole positions are dangerous since they produce instabilities in the system. We close by positioning poles along a special set of locations to create very precise oscillators.

7.1 LOOKING AT THE SYSTEM IN A NEW WAY

The difference equation that we developed in Section 4.2.4 *completely* describes the operations that a digital processor needs to perform to calculate the value of each output sample:

$$y(n) = b_0 x(n) + b_1 x(n-1) + b_2 x(n-2) + b_3 x(n-3) + \ldots$$

The b_k coefficients are the characteristic elements that describe the measure of past, present, or future inputs that are included in the system output. The difference equation coefficients therefore represent weighting factors that the system applies to the input samples as it computes output samples.

When we apply an impulse to the input of an LTI system, it responds by outputting a characteristic response $h(n)$, which is *unique* to the system. This impulse response exactly reflects the value of the b_k coefficients and therefore corresponds to the value of the system weighting factors. From the impulse response, we developed the convolution operation, which provides a general description of the operation of an LTI system. The general form for discrete-time convolution (Equation [4–3]) is repeated here for convenience:

$$y(n) = \sum_{k=-\infty}^{\infty} h(k)x(n-k)$$

The convolution operation completely describes the operations that an LTI system performs on the input signal $x(n)$ to produce the output signal $y(n)$. All systems are characterized by their particular impulse response $h(n)$; in other words, we can alter what the system does by changing the value of the system weighting factors.

In Chapter 6, we inputted periodic complex exponentials, and this led to the Fourier transform, which describes the gain and phase response of systems when we input a frequency. Equation (6–3) is repeated here for convenience:

$$H(\Omega) = \sum_{k=-\infty}^{\infty} h(k)e^{-j\Omega k}$$

The difference equation, the convolution operation, and the Fourier transform provide a way to describe, program, and analyze the operation of the system given its impulse response. Unfortunately, none of these tells us how to design the terms of the impulse response to achieve a particular system response.

Next we develop a new way to look at an LTI system. This new description shows how to relate the impulse response terms to the system response. Using this, we will be able to engineer the system response so that is meets particular functional specifications.

7.1.1 DEVELOPING THE Z-TRANSFORM

When we developed the Fourier transform in Chapter 6, we applied the periodic complex exponential $e^{j\Omega n}$ to the input of the system. In this section, we want to determine the response of the system for a broad class of discrete-time domain input signals that we define generally as $x(n) = Z^n$. We define Z as a vector adopting any complex value of the form

$$Z = Real + j Imaginary$$

The Z-domain therefore covers any real, complex, or imaginary value. We can change the value of Z to make the input vary over a broad range of different *types* of input signal. For example, if $Z = e^{j\Omega}$, the input signal becomes $x(n) = e^{j\Omega n}$, which corresponds to the periodic complex exponential that we used to develop the Fourier transform. Note that

$Z = e^{j\Omega}$ is only one of the many types of input signals that we can create by altering the value of Z.

We can represent the complex value of Z on a Cartesian coordinate system the same way that we plot any complex number on an Argand diagram. We use the x axis to represent the real part of Z and the y axis to represent the imaginary part of Z. In this case, the Argand diagram represents a two-dimensional surface, which we refer to as the *Z-plane*. Figure 7–1 plots an arbitrarily chosen value of Z with a magnitude $R = 1$ and an angle $\Omega = \pi/5$. Note in the figure that we can use Cartesian, polar, or complex exponential notations to define the location of Z. Also note that we use a small crosshair to represent the position of Z on the Z-plane.

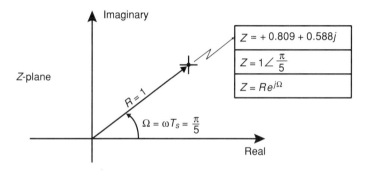

FIGURE 7–1
Z plotted on the Z-plane.

The idea here is to analyze how the behavior of an LTI system changes when we change the *type* of input. We start the analysis by using the time-domain input-to-output relationship described by the convolution equation. Figure 7–2 illustrates this input-to-output relationship. As the figure indicates, the relationship between the input and the output of a time-invariant system is described by convolving the impulse response $h(n)$ with the input signal $x(n)$:

$$y(n) = \sum_{k=-\infty}^{\infty} h(k)x(n-k)$$

FIGURE 7–2
Time-domain input-to-output relationship.

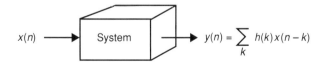

If we now define the discrete-time input as being equal to the signal $x(n) = Z^n$, the output of the LTI system becomes

$$y(n) = \sum_{k=-\infty}^{\infty} h(k)Z^{n-k}$$

The exponent of Z in the summation corresponds to a product of two exponentiated Z terms:

$$y(n) = \sum_{k=-\infty}^{\infty} h(k) Z^n Z^{-k}$$

Notice that the term Z^n is completely independent of the summation variable k. This allows us to take it out of the summation:

$$y(n) = Z^n \sum_{k=-\infty}^{\infty} h(k) Z^{-k} \qquad (7\text{–}1)$$

We notice that the output in Equation (7–1) consists of the input signal Z^n, which multiplies a summation expression. Since the summation expression multiplies the input signal, we must conclude that the summation expression represents *how* the system manipulates the input signal to produce the output. The summation therefore describes the system response to a discrete-time input of the form $x(n) = Z^n$.

Considering that the input signal $x(n) = Z^n$ is a discrete-time domain expression, the value of Z is a constant complex number and n is the discrete-time variable. This interpretation does not allow us to vary the value of Z; therefore, we need to redefine the way that we interpret our input signal.

Since we want to analyze how the system responds to different types of inputs, we need to vary the value of Z and freeze the value of n. When we do this, the input signal becomes a Z-domain function since it depends only on the value of the variable Z:

$$X(Z) = Z^n$$

Now that the discrete-time variable n is no longer the variable, we can analyze the way a system responds to different types of input signals. What is important here is the fact that we can assign different values to Z to analyze the system response to different types of input signals. Each time that we apply a different input $X(Z)$, the system responds in a different way to produce a new output function. In this context, the output is labeled $Y(Z)$ since it also becomes a function of Z.

Once we make Z the variable, we can rewrite Equation (7–1) to reflect this change of perspective:

$$Y(Z) = X(Z) \sum_{k=-\infty}^{\infty} h(k) Z^{-k} \qquad (7\text{–}2)$$

The summation represents the Z-domain response of the system to the input $X(Z)$. Consequently, the system response is a function of the value of Z, which we can express as follows:

$$H(Z) = \sum_{k=-\infty}^{\infty} h(k) Z^{-k} \qquad (7\text{–}3)$$

We call Equation (7–3) the *Z-transform* of the system. Note the important fact that the summation that describes the Z-transform is completely independent of the discrete-time variable n. The value of the Z-transform depends only on the impulse response terms—in other words, on the value of the weighting factors that define the system.

The *Z*-transform relies on the value of the weighting factors to describe the system response as a function of the *type* of input signal.

How a system responds therefore depends on the value of the weighting factors and on the type of input signal (the value that we assign to Z). The Z-transform consequently carries the discrete-time domain impulse response $h(n)$ into the Z-domain, where it becomes $H(Z)$. Figure 7–3 combines the results of Equation (7–2) and Equation (7–3) to illustrate the system in the Z-domain.

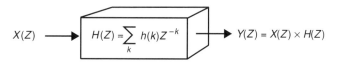

$X(Z) \longrightarrow \boxed{H(Z) = \sum_k h(k)Z^{-k}} \longrightarrow Y(Z) = X(Z) \times H(Z)$

FIGURE 7–3
System in the *Z*-domain.

Figure 7–3 indicates that the output of the system in the Z-domain is found by multiplying the Z-domain input by the system Z-transform. Working in the Z-domain therefore reduces the mathematical complexity of computing the output to a single product operation. This is quite a simplification from the discrete-time domain, where we had to compute the difference equation repeatedly to produce a new output value at every sampling interval.

Manipulating the output expression of Figure 7–3, we can express the Z-transform as a ratio of the output signal over the input signal:

$$Y(Z) = X(Z) \times H(Z)$$

$$H(Z) = \frac{Y(Z)}{X(Z)} = \sum_{k=-\infty}^{\infty} h(k)Z^{-k} \qquad \textbf{(7–4)}$$

7.1.2 INTERPRETING THE TERMS OF THE Z-TRANSFORM SUMMATION

One simple way to interpret the significance of the terms contained in the Z-transform is to compare the output of the system in the discrete-time domain to the output obtained in the Z-domain:

Discrete-Time Domain Output

$$y(n) = \sum_{k=-\infty}^{\infty} h(k)x(n-k)$$

Z-Domain Output

$$Y(Z) = X(Z) \sum_{k=-\infty}^{\infty} h(k)Z^{-k}$$

Both expressions that describe the output of the system use the system weighting factors in the form of the impulse response $h(k)$. Since both $y(n)$ and $Y(Z)$ represent the output of exactly the same system, they *must* represent the same signal. We expand a few terms for both the discrete-time domain and the Z-domain output summations:

$$y(n) = \sum_{k=-\infty}^{\infty} h(k) x(n-k) \quad = \quad \ldots + h(0) x(n-0) +$$

$$h(1) x(n-1) + h(2) x(n-2) + \ldots$$

$$Y(Z) = X(Z) \times \sum_{k=-\infty}^{\infty} h(k) Z^{-k} = \ldots + h(0) X(Z) Z^{-0} +$$

$$h(1) X(Z) Z^{-1} + h(2) X(Z) Z^{-2} + \ldots$$

Since $y(n)$ and $Y(Z)$ represent exactly the same output signal, we can establish the following direct relationship between the elements of the two expressions:

$$
\begin{array}{lll}
h(0)x(n-0) & \text{corresponds to} & h(0)X(Z)Z^{-0} \\
h(1)x(n-1) & \text{corresponds to} & h(1)X(Z)Z^{-1} \\
h(2)x(n-2) & \text{corresponds to} & h(2)X(Z)Z^{-2} \\
\ldots & \ldots & \ldots \\
h(k)x(n-k) & \text{corresponds to} & h(k)X(Z)Z^{-k}
\end{array}
$$

Since $h(k)$ is common to both the discrete-time domain and the Z-domain expressions, clearly the following relationship exists:

$$
\begin{array}{lll}
x(n-0) & \text{corresponds to} & X(Z)Z^{-0} \\
x(n-1) & \text{corresponds to} & X(Z)Z^{-1} \\
x(n-2) & \text{corresponds to} & X(Z)Z^{-2} \\
\ldots & \ldots & \ldots \\
x(n-k) & \text{corresponds to} & X(Z)Z^{-k}
\end{array}
$$

The expression $x(n-k)$ represents the value of the input sample that was acquired by the A to D converter k discrete-time units ago. Expressed in other terms, $x(n-k)$ corresponds to the input signal delayed by k discrete-time units. Since there is a direct relationship between the discrete-time domain and the Z-domain, the expression $X(Z)Z^{-k}$ must also represent the input signal delayed by k time units. The input signal corresponds to $X(Z)$, and, therefore, the expression Z^{-k} must be interpreted as a delay of k discrete-time units since it multiplies the input signal $X(Z)$.

We can generalize and say that the exponent of Z in the Z-domain corresponds to a time shift in the discrete-time domain:

Z-Domain   *Discrete-Time Domain*

$$X(Z) Z^{\text{shift}} \longleftrightarrow x(n + \text{shift})$$

A *Z*-domain signal multiplied by *Z*shift suffers a discrete-time shift, which is equal to the value of the exponent of *Z*.

This knowledge provides a basic rule to convert the system difference equation to and from the Z-domain. Consider the following examples:

$$X(Z)Z^{-4} \xleftarrow{\text{Z-domain} \qquad \text{Discrete-time domain}} x(n-4)$$

$$Y(Z)Z^{+2} \xleftarrow{\text{Z-domain} \qquad \text{Discrete-time domain}} y(n+2)$$

$$H(Z)Z^{+3} \xleftarrow{\text{Z-domain} \qquad \text{Discrete-time domain}} h(n+3)$$

7.1.3   CALCULATING THE *Z*-TRANSFORM

Before we can make practical use of the Z-transform to design the system response, we must first learn to express systems in the Z-domain. If we start from the discrete-time domain, a system is entirely defined by its impulse response $h(n)$. To transfer a system to the Z-domain, we must apply the Z-transform, as described by Equation (7–4), to the impulse response of the system.

To make life interesting, two different kinds of systems exist:

1. Those that have a finite impulse response (FIR).
2. Those that have an infinite impulse response (IIR).

Calculating the Z-transform of an FIR system is straightforward since the summation uses a *finite* number of $h(n)$ terms. Calculating the Z-transform of an IIR system is rather tricky since these systems have an *infinite* impulse response. Fortunately, a trick allows us to describe an IIR system using a finite Z-transform expression. The next two sections deal with the Z-transforms of FIR and IIR systems, respectively.

Z-Transform of Finite Impulse Response Systems

We can best explain the procedure to find the Z-transforms of FIR systems using a practical example. We arbitrarily define an FIR system with the impulse response illustrated in Figure 7-4. The example FIR system in Figure 7–4 is defined by three $h(n)$ terms, which indicates that this system uses three weighting factors:

$$h(0) = 1 \qquad h(1) = -0.3 \qquad h(2) = -0.4$$

Since this is an FIR system, the number of nonzero $h(k)$ values is limited. The Z-transform calculation requires using only values of the iteration variable k that points to nonzero $h(k)$ values. In the example system, since there are only three nonzero values for $h(k)$, we need to iterate only through $k = 0$, 1, and 2. Calculating the Z-transform generates the following three terms:

$$H(Z) = \frac{Y(Z)}{X(Z)} = \sum_{k=-\infty}^{\infty} h(k)Z^{-k}$$

$$\frac{Y(Z)}{X(Z)} = 1Z^{-0} - 0.3Z^{-1} - 0.4Z^{-2}$$

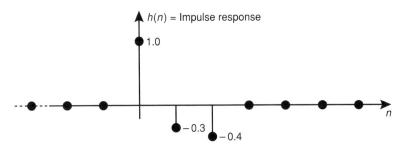

FIGURE 7–4
Arbitrarily chosen FIR response.

Isolating the Z-domain output $Y(Z)$, we can see that it consists of the input $X(Z)$ multiplied by the Z-transform:

$$Y(Z) = X(Z) \times H(Z)$$
$$= X(Z) [1Z^{-0} - 0.3Z^{-1} - 0.4Z^{-2}]$$
$$= X(Z) Z^{-0} - 0.3X(Z) Z^{-1} - 0.4X(Z)Z^{-2}$$

The Z-domain output expression for this system consists of three terms that all contain negative exponents. We know that negative exponents of Z correspond to a delay in the discrete-time domain (refer to Section 7.1.2). Using this knowledge, we can convert the output expression back to a discrete-time domain expression. To do this, we separately interpret each of the three terms in the Z-domain output expression:

$X(Z)Z^{-0}$ represents the present input (the delay is 0): $\qquad x(n) = x(n-0)$

$X(Z)Z^{-1}$ represents the input, delayed by one discrete-time unit: $\qquad x(n-1)$

$X(Z)Z^{-2}$ represents the input, delayed by two discrete-time units: $\qquad x(n-2)$

Using this interpretation, we can translate the Z-domain terms into the following discrete-time output expression:

$$y(n) = x(n) - 0.3x(n-1) - 0.4x(n-2)$$

Note that this expression is the difference equation that describes the output of the system.

Z-Transform of Infinite Impulse Response Systems

A system that has an IIR is defined by an infinite number of weighting factors. Such a system needs to weight an infinite number of past inputs to provide every single output.

Because we cannot compute an infinite number of terms (refer to Section 4.4), we use a subterfuge that achieves the equivalent of looking infinitely into the past. We include past *outputs* in the difference equation that defines the system. As an example, we examine an IIR system that we arbitrarily define using the following *recursive* difference equation:

$$y(n) = x(n) + 0.2y(n-1) + 0.15y(n-2)$$

To translate this system into the Z-domain, we use the fact that the negative exponents of Z represent discrete-time delays. This provides the following equivalencies:

$$y(n) \xrightarrow{\quad Z-\text{domain} \quad} Y(Z)\,Z^{-0}$$

$$y(n-1) \xrightarrow{\quad Z-\text{domain} \quad} Y(Z)\,Z^{-1}$$

$$y(n-2) \xrightarrow{\quad Z-\text{domain} \quad} Y(Z)\,Z^{-2}$$

Using these equivalencies, we can *translate* the recursive difference equation to the Z-domain:

$$y(n) = x(n) + 0.2y(n-1) + 0.15y(n-2)$$

becomes

$$Y(Z)Z^{-0} = X(Z) + 0.2Y(Z)Z^{-1} + 0.15Y(Z)Z^{-2}$$

We can regroup and factor the $Y(Z)$ terms on the left side of the equation:

$$Y(Z)[Z^{-0} - 0.2Z^{-1} - 0.15Z^{-2}] = X(Z)$$

We can now isolate the output expression $Y(Z)$ on the left side:

$$Y(Z) = \frac{X(Z)}{Z^{-0} - 0.2\,Z^{-1} - 0.15\,Z^{-2}}$$

Finally, since $Z^{-0} = 1$, and since the Z-transform is the ratio of the output signal over the input signal, we can write

$$H(Z) = \frac{Y(Z)}{X(Z)} = \frac{1}{1 - 0.2\,Z^{-1} - 0.15\,Z^{-2}}$$

To avoid using an infinite number of past inputs, we used past *outputs* in the difference equation that describes this IIR system. The recursive part of the system therefore corresponds to the *denominator* part of the Z-transform.

7.2 MANIPULATING THE *Z*-TRANSFORM

The intent of this chapter is to develop a technique for writing the system Z-transform from a target frequency response. Once we have the Z-transform, we can extract the impulse response and use its terms to compute the output samples using the convolution operation.

7.2.1 FINDING THE ROOTS OF THE *Z*-TRANSFORM

This section establishes the link between the value of the system weighting factors and the system response. To visualize the system response in terms of the values of the system weighting factors, it is best to rewrite $H(Z)$ to eliminate all negative exponents of Z. We can remove the negative exponents of Z by multiplying the Z-transform by an appropriate unity factor. This operation converts $H(Z)$ to a new form, which is mathematically equivalent to the original form.

To illustrate this, we combine the examples in the previous sections on the Z-Transform of Finite Impulse Response Systems and the Z-Transform of Infinite Impulse Systems to create a Z-transform that contains both numerator and denominator expressions:

$$H(Z) = \frac{Y(Z)}{X(Z)} = \frac{1Z^{-0} - 0.3Z^{-1} - 0.4Z^{-2}}{1Z^{-0} - 0.2Z^{-1} - 0.15Z^{-2}}$$

Since the highest negative exponent is Z^{-2}, we must multiply the Z-transform by the following second-order unity factor:

$$\frac{Z^{+2}}{Z^{+2}}$$

Applying this factor to the Z-transform, we obtain

$$H(Z) = \frac{1Z^{-0} - 0.3Z^{-1} - 0.4Z^{-2}}{1Z^{-0} - 0.2Z^{-1} - 0.15Z^{-2}} \times \frac{Z^2}{Z^2}$$

$$= \frac{Z^{+2} - 0.3Z^{+1} - 0.4}{Z^{+2} - 0.2Z^{+1} - 0.15}$$

Any function can be represented by its roots; the $H(Z)$ function is no exception. The roots of a function specify the value(s) of the variable (Z in this case) that will make the function assume a value equal to 0. We consider the numerator and the denominator of the $H(Z)$ function separately. Both numerator and denominator are of the second order and therefore both are quadratic expressions.

We can find the roots of any quadratic expression by using the following well-known formula for $f(x) = ax^2 + bx + c$:

$$\text{Root 1} = \frac{-b + \sqrt{b^2 - 4ac}}{2a} \qquad \text{Root 2} = \frac{-b - \sqrt{b^2 - 4ac}}{2a}$$

In the case of the numerator of the example Z-transform function

$$a = 1 \qquad b = -0.3 \qquad c = -0.4$$

Solving yields the following values for the roots of the Z-transform numerator N:

$$N_{\text{Root 1}} = \frac{-(-0.3) + \sqrt{(-0.3)^2 - 4 \times 1 \times (-0.4)}}{2 \times 1}$$

$$N_{\text{Root 1}} = +0.8$$

$$N_{\text{Root 2}} = \frac{-(-0.3) - \sqrt{(-0.3)^2 - 4 \times 1 \times (-0.4)}}{2 \times 1}$$

$$N_{\text{Root 2}} = -0.5$$

In the case of the denominator of the example Z-transform function,

$$a = 1 \qquad b = -0.2 \qquad c = -0.15$$

Solving, we obtain the following two denominator (D) roots:

$$D_{\text{Root 1}} = \frac{-(-0.2) + \sqrt{(-0.2)^2 - 4 \times 1 \times (-0.15)}}{2 \times 1}$$

$$D_{\text{Root 1}} = +0.5$$

$$D_{\text{Root 2}} = \frac{-(-0.2) - \sqrt{(-0.2)^2 - 4 \times 1 \times (-0.15)}}{2 \times 1}$$

$$D_{\text{Root 2}} = -0.3$$

We can now express the numerator and the denominator of the Z-transform as the product of factors based on the position of the roots:

$$H(Z) = \frac{Z^{+2} - 0.3Z^{+1} - 0.4}{Z^{+2} - 1.1Z^{+1} + 0.3} = \frac{\left(Z - N_{\text{Root 1}}\right)\left(Z - N_{\text{Root 2}}\right)}{\left(Z - D_{\text{Root 1}}\right)\left(Z - D_{\text{Root 2}}\right)}$$

$$= \frac{(Z - 0.8)(Z + 0.5)}{(Z - 0.5)(Z + 0.3)}$$

You can verify that the numerator of $H(Z)$ is equal to zero when

$$Z = \text{NRoot1} = +0.8 \text{ or when } Z = \text{NRoot2} = -0.5$$

You can verify that the denominator is equal to zero when

$$Z = \text{DRoot1} = +0.5 \text{ or when } Z = \text{DRoot2} = -0.3$$

Since the Z-transform $H(Z) = 0$ when Z assumes a value of either N_{Root1} or N_{Root2}, these two special values are referred to as the *zeros* of this system:

$$\text{Zero1} = N_{\text{Root1}} \text{ and Zero2} = N_{\text{Root2}}$$

A *zero* is defined as the position of *Z* in the Z-plane that makes *H*(*Z*) = 0.

When Z assumes the value of D_{Root1} or of D_{Root2}, the *denominator of H(Z)* is equal to zero, and this makes the Z-transform assume an infinite value: $H(Z) = \infty$. We call the value of the variable Z that makes a function reach an infinite value a *pole*. In our example, D_{Root1} and D_{Root2} are the poles of the system:

$$\text{Pole1} = D_{Root1} \text{ and Pole2} = D_{Root2}$$

A *pole* is defined as the position of *Z* in the *Z*-plane that makes *H(Z)* = ∞.

Remember that Z is a complex number that we can illustrate on the Z-plane. Figure 7–5 positions the poles and the zeros that define the denominator and numerator, respectively, of the example system along with an arbitrary value for the input variable Z. Notice that Figure 7–5 illustrates the input variable Z using a crosshair, the zeros using Os, and the poles using Xs. Plotting the position of the poles and the zeros creates a *pole-zero diagram*. This diagram completely defines a particular system since we can use the pole and the zero positions to completely rebuild the equation that defines a system. We can therefore design a complete system by positioning poles and zeros on the Z-plane.

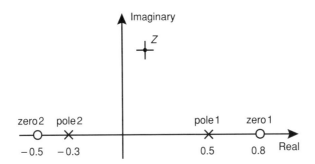

FIGURE 7–5
Input variable Z, the poles, and the zeros on the Z-plane.

7.2.2 EXPRESSING THE *Z*-TRANSFORM USING VECTORS

We can express the numerator of a system Z-transform using factored expressions that use the system *zeros*. We now examine such an expression:

$$H(Z) = \frac{(Z - 0.8)(Z + 0.5)}{\text{Denominator}} = \frac{(Z - \text{Zero1})(Z - \text{Zero2})}{\text{Denominator}}$$

where

$$\left\{ \begin{array}{l} \text{Zero1} = +0.8 \\ \text{Zero2} = -0.5 \end{array} \right\}$$

Since the Z-transform represents the way that the entire system responds to the input signal, the numerator factors directly contribute to the system response. The numerator factors use complex values; consequently, we can represent these values as vectors:

$$H(Z) = \frac{(Z - \text{Zero}1)(Z - \text{Zero}2)}{\text{Denominator}} \qquad \overrightarrow{H(Z)} = \frac{\left(\overrightarrow{V_Z} - \overrightarrow{V_{\text{Zero}1}}\right)\left(\overrightarrow{V_Z} - \overrightarrow{V_{\text{Zero}2}}\right)}{\text{Denominator}}$$

Plotting these vectors allows the visualization of the numerator part of the system response. Figure 7–6 illustrates the three individual vectors that are used in the numerator of the example Z-transform. However, the numerator of the Z-transform combines these vectors to create two factored expressions: $\left(\overrightarrow{V_Z} - \overrightarrow{V_{\text{Zero}1}}\right)$ and $\left(\overrightarrow{V_Z} - \overrightarrow{V_{\text{Zero}2}}\right)$. Each of these expressions represents the difference between the input vector Z and one of the zero vectors. Figure 7–7 illustrates the bracketed factors as the difference between two vectors.

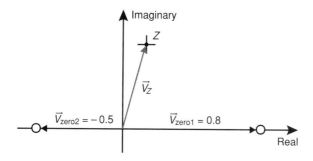

FIGURE 7–6
Vector components of the numerator.

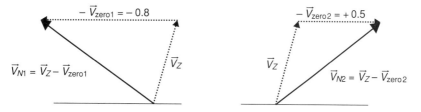

FIGURE 7–7
Vectors that correspond to the zero factors.

Examining Figure 7–8, we can see that the numerator factors correspond to the resultant vectors $\overrightarrow{V_{N1}}$ and $\overrightarrow{V_{N2}}$. We can fit these resultant vectors on the Z-plane to get the overall picture. Figure 7–8 illustrates the response vectors along with the position of the system zeros. The figure represents the response vectors that correspond to the numerator part of the system Z-transform. Note that a response vector exists for each of the numerator

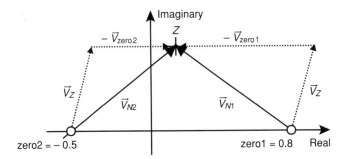

FIGURE 7–8
Response vectors that correspond to the system zeros.

root factors. Each response vector starts at the position of its respective zero and extends to the location of the Z input variable.

We now shift attention to the denominator of the system Z-transform. In this case, we are dealing with the poles of the system:

$$H(Z) = \frac{\text{Numerator factors}}{(Z - 0.5)(Z + 0.3)} = \frac{\text{Numerator factors}}{(Z - \text{Pole 1})(Z - \text{Pole 2})}$$

where

$$\text{Pole 1} = +0.5 \quad \text{and} \quad \text{Pole 2} = -0.3$$

Following the reasoning that we used for the numerator part, each of the denominator root factors is associated with a pole response vector that starts at the position of its respective pole and extends to the input variable location Z:

$$\overrightarrow{H(Z)} = \frac{\text{Numerator vectors}}{\left(\overrightarrow{V_Z} - \overrightarrow{V_{\text{Pole 1}}}\right)\left(\overrightarrow{V_Z} - \overrightarrow{V_{\text{Pole 2}}}\right)}$$

Figure 7–9 illustrates the resultant pole response vectors, $\overrightarrow{V_{D1}}$ and $\overrightarrow{V_{D2}}$, of the example system. The Z-transform therefore consists of a number of numerator and denominator response vectors. In the example case, the Z-transform simplifies to the following four response vectors:

$$\overrightarrow{H(Z)} = \frac{\left(\overrightarrow{V_Z} - \overrightarrow{V_{\text{Zero 1}}}\right)\left(\overrightarrow{V_Z} - \overrightarrow{V_{\text{Zero 2}}}\right)}{\left(\overrightarrow{V_Z} - \overrightarrow{V_{\text{Pole 1}}}\right)\left(\overrightarrow{V_Z} - \overrightarrow{V_{\text{Pole 2}}}\right)} = \frac{\overrightarrow{V_{N1}} \times \overrightarrow{V_{N2}}}{\overrightarrow{V_{D1}} \times \overrightarrow{V_{D2}}}$$

The next section analyzes the significance of the response vectors that correspond to the Z-transform factors. This analysis provides us an intuitive tool that will allow us to visualize, understand, and design the system response.

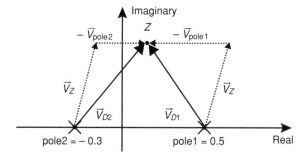

FIGURE 7–9
Response vectors that correspond to the system poles.

7.3 INTERPRETING VECTORS IN THE *Z*-PLANE

The Z-transform describes a system's response to different types of input signals. Finding the roots of a system Z-transform allows us rewrite its numerator and denominator in a form that consists of factored expressions:

$$H(Z) = \frac{(Z - \text{Zero1})(Z - \text{Zero2})(Z - \text{Zero3})(\dots}{(Z - \text{Pole1})(Z - \text{Pole2})(Z - \text{Pole3})(\dots} \qquad (7\text{–}5)$$

where

Zero1, Zero2, Zero3, ... are the roots of the numerator.

Pole1, Pole2, Pole3, ... are the roots of the denominator.

Equation (7–5) therefore describes the complete system response in terms of the position of the system poles and zeros. Each of the root factors of the Z-transform contributes to the overall system response. Since the roots used in Equation (7–5) completely describe a time-invariant system, the roots must occupy time-invariant *positions* on the Z-plane. Consequently, the system response changes only when we modify the type of input signal. This corresponds to changing the value of the variable Z.

7.3.1 CHANGING THE VALUE OF THE INPUT VARIABLE Z

If we have the Z-transform of a system, changing the value of input variable Z yields the system response for that particular input signal. Let's change the value of the input signal Z in the case of the system that we used as an example throughout Section 7.2. This system has the following Z-transform:

$$H(Z) = \frac{Z^{+2} - 0.3Z^{+1} - 0.4}{Z^{+2} - 0.2Z^{+1} - 0.15} = \frac{(Z - 0.8)(Z + 0.5)}{(Z - 0.5)(Z + 0.3)}$$

According to Section 7.2.2, we can rewrite this Z-transform as vectors, showing that each factored expression corresponds to a distinct resultant vector.

$$\overrightarrow{H(Z)} = \frac{\left(\overrightarrow{V_Z} - \overrightarrow{V_{Zero1}}\right)\left(\overrightarrow{V_Z} - \overrightarrow{V_{Zero2}}\right)}{\left(\overrightarrow{V_Z} - \overrightarrow{V_{Pole1}}\right)\left(\overrightarrow{V_Z} - \overrightarrow{V_{Pole2}}\right)} = \frac{\overrightarrow{V_{N1}} \times \overrightarrow{V_{N2}}}{\overrightarrow{V_{D1}} \times \overrightarrow{V_{D2}}}$$

Each resultant numerator and denominator vector links one of the root positions to the position of the input signal Z. Figure 7–10 illustrates what happens to the response vectors when the input signal Z changes location. As the figure indicates, moving the input variable Z from location Z_A to Z_B in the Z-plane alters the length (magnitude) and angle (phase) of each response vector.

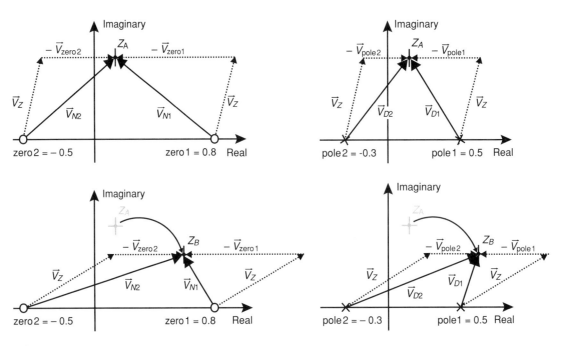

FIGURE 7–10
Response vectors of the system.

7.3.2 COMBINING THE RESPONSE VECTORS

Remember that each response vector contributes to the overall system response. To find the numerator response, we multiply the numerator response vectors. Repeating the operation for the denominator vectors yields the denominator response:

$$\overrightarrow{H(Z)} = \frac{\overrightarrow{V_{N1}} \times \overrightarrow{V_{N2}} \times \overrightarrow{V_{N3}} \times \overrightarrow{V_{N4}} \times \ldots}{\overrightarrow{V_{D1}} \times \overrightarrow{V_{D2}} \times \overrightarrow{V_{D3}} \times \overrightarrow{V_{D4}} \times \ldots}$$

Multiplying vectors is done by multiplying their magnitudes and adding their phases. If we define

$$G_x = \left| \overrightarrow{V_X} \right| \quad \text{and} \quad \phi_x = \angle \overrightarrow{V_X}$$

we can rewrite the Z-transform as

$$H(Z) = \frac{G_{N1} \angle \phi_{N1} \times G_{N2} \angle \phi_{N2} \times G_{N3} \angle \phi_{N3} \times G_{N4} \angle \phi_{N4} \times \ldots}{G_{D1} \angle \phi_{D1} \times G_{D2} \angle \phi_{D2} \times G_{D3} \angle \phi_{D3} \times G_{D4} \angle \phi_{D4} \times \ldots}$$

$$= \frac{\left[G_{N1} \times G_{N2} \times G_{N3} \times G_{N4} \times \ldots \right] \angle \left[\phi_{N1} + \phi_{N2} + \phi_{N3} + \phi_{N4} + \ldots \right]}{\left[G_{D1} \times G_{D2} \times G_{D3} \times G_{D4} \times \ldots \right] \angle \left[\phi_{D1} + \phi_{D2} + \phi_{D3} + \phi_{D4} + \ldots \right]}$$

The product of any number of vectors yields a *single* new vector that has a new magnitude and angle. We can therefore rewrite the Z-transform as the ratio of an equivalent numerator vector over an equivalent denominator vector:

$$H(Z) = \frac{V_{\text{Numerator}}}{V_{\text{Denominator}}} = \frac{G_N \angle \phi_N}{G_D \angle \phi_D}$$

G_N and ϕ_N represent the portion of the system gain and phase shift that is attributable to the zeros that define the system.

The denominator vector corresponds to the portion of the system response that is attributable to the poles that define the system. To simplify the pole response vector, we can move it to the numerator by inverting its magnitude and by reversing the sign of its phase angle:

$$H(Z) = \left(G_N \angle \phi_N \right) \times \left(\frac{1}{G_D} \angle - \phi_D \right)$$

The poles of the system therefore contribute a gain of $1/G_D$ and a phase delay of $-\phi_D$.

We find the complete system gain and phase shift by combining the numerator and denominator responses into a single vector:

$$H(Z) = \frac{G_N \angle \phi_N}{G_D \angle \phi_D} = \frac{G_N}{G_D} \angle \{ \phi_N - \phi_D \}$$

$$= V_{\text{System}} \quad = G \angle \phi$$

Solving the *Z*-transform for a specific value of the input variable *Z* yields the gain *G* and the phase shift ϕ that the system applies to that input signal.

7.3.3 CALCULATING THE SYSTEM RESPONSE TO A SINUSOID INPUT

According to Equation (7–4), calculating a system output in the Z-domain is done by multiplying the input signal by the system Z-transform:

$$Y(Z) = X(Z) \times H(Z)$$

Since the Z-transform reduces itself to a single resultant vector $G \angle \phi$, the system modifies the input signal $X(Z)$ according to the magnitude and the phase of the Z-transform resultant vector:

$$Y(Z) = X(Z) \times G \angle \phi$$

The input $X(Z)$ can assume any complex value. We use the Z-transform to determine the system response to an input sinusoid of the form $x(n) = A\cos(\Omega n)$. Euler showed that this basic waveform consists of the following two complex exponentials:

$$x(n) = \frac{A\,e^{+j\Omega n} + A\,e^{-j\Omega n}}{2}$$

Since $x(n)$ contains two complex values, we can break it into two separate inputs before we translate it to the Z-domain:

$$X_1(Z) = \frac{AZ^n}{2}$$

where

$$Z = e^{+j\Omega}$$

$$X_2(\overline{Z}) = \frac{A\overline{Z}^n}{2}$$

where

$$\overline{Z} = e^{-j\Omega}$$

To find the Z-domain output, we can superpose these two Z-domain inputs:

$$Y(Z) = [X_1(Z) \times H(Z)] + [X_2(\overline{Z}) \times H(\overline{Z})]$$

We must therefore find $H(Z)$ and $H(\overline{Z})$. Note that these two values are complex conjugates of each other. Figure 7–11 illustrates the complement Z values on the Z-plane.

We now apply the sinusoid input to the example system used in Section 7.2.2. This system has zeros at locations –0.5 and +0.8 and poles at locations –0.3 and +0.5. Figure 7–12 illustrates the resulting system response vectors.

Each of the response vectors in Figure 7–12 has a magnitude and a phase that contribute to the overall system response. Since this system is defined using two zeros and two

FIGURE 7–11
Input sinusoid contains conjugate values of Z.

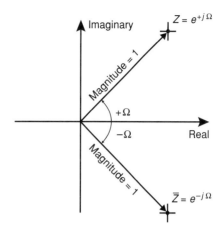

FIGURE 7–12
Response vectors resulting from a sinusoid input.

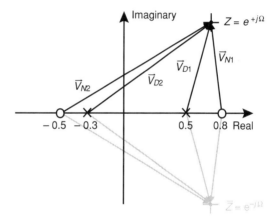

poles, there are four vectors leading to each of the conjugate locations Z and $\overline{Z}$. These sets are mirror images of each other for systems that have a pure-real response. The symmetry found in pure-real systems allows us to simplify many calculations.

Since the two sets of response vectors are conjugates of each other, we can evaluate the system Z-transform for each of the two conjugate locations of Z as follows:

$$\overrightarrow{H(Z)} = \frac{\overrightarrow{V_{N1}} \times \overrightarrow{V_{N2}}}{\overrightarrow{V_{D1}} \times \overrightarrow{V_{D2}}} \qquad \overrightarrow{H(\overline{Z})} = \frac{\overrightarrow{\overline{V}_{N1}} \times \overrightarrow{\overline{V}_{N2}}}{\overrightarrow{\overline{V}_{D1}} \times \overrightarrow{\overline{V}_{D2}}}$$

$$H(Z) = G \angle \phi \qquad H(\overline{Z}) = \overline{H(Z)} = G \angle -\phi$$

Note that the two system responses differ only by the sign of the response angle. This is so because complex conjugate vectors differ only by the sign of their angle.

Using these Z-transform results, we can write the following output expression when the input is a sinusoid:

$$Y(Z) = \left[X_1(Z) \times H(Z) \right] + \left[X_2(\overline{Z}) \times H(\overline{Z}) \right]$$

$$= \left[\frac{AZ^n}{2} \times (G \angle \phi) \right] + \left[\frac{A\overline{Z}^n}{2} \times (G \angle -\phi) \right]$$

$$= \left[\frac{AZ^n}{2} \times Ge^{+j\phi} \right] + \left[\frac{A\overline{Z}^n}{2} \times Ge^{-j\phi} \right]$$

Since $Z = e^{+j\Omega}$ and $\overline{Z} = e^{-j\Omega}$ for an input siusoid, we can use this knowledge to return the output expression to the time domain:

$$y(n) = \left[\frac{A\,e^{+j\Omega n}}{2} \times Ge^{+j\phi} \right] + \left[\frac{A\,e^{-j\Omega n}}{2} \times Ge^{-j\phi} \right]$$

$$= \frac{AGe^{+j(\Omega n + \phi)} + AGe^{-j(\Omega n + \phi)}}{2}$$

$$= AG \cos(\Omega n + \phi)$$

Note that the gain G and phase shift ϕ being applied to the input sinusoid correspond to the response that the system applies to the input $Z = e^{+j\Omega}$. This is the case because both the input sinusoid and the system $h(n)$ are pure-real functions. This situation is very common in practice and allows us to simplify the calculations leading to the system response.

In the case of a system with a pure-real $h(n)$, we can determine the system response to an input sinusoid by evaluating the pole and zero response vectors leading to $Ze^{+j\Omega}$.

The magnitude and phase of each of these response vectors contribute to the overall system response. By examining the length and phase of the individual response vectors, we can visually estimate the contribution of each of the system poles and zeros to the overall system gain and phase shift. This provides an opportunity to design the system response by selectively locating poles and zeros on the Z-plane. In the following sections, we use this technique to design the response of systems.

7.4 SETTING UP THE Z-PLANE

We established in the previous section that each of the system poles and zeros produces a response vector that contributes to the gain and phase shift that is applied to the input signal $X(Z)$. We know that the variable Z can be located anywhere on the Z-plane; however, one set of locations on the Z-plane is of particular interest. This special set belongs to input signals that are *real sinusoids*.

In practice, we are particularly interested in controlling the amplitude and phase of the individual sinusoids that make up *real* input signals. For this reason, this section concentrates on the analysis of system behavior when the input variable Z is confined to adopting values that correspond to the special set of real sinusoidal signals.

7.4.1 POSITIONING INPUT SINUSOIDS ON THE Z-PLANE

When we design the system response, we must consider how the system responds to input sinusoids oscillating at any frequency. Discrete-time sinusoids have the following characteristics:

- They are pure real.
- They are sampled at intervals of $T_S n$.
- They oscillate at an angular frequency $\omega = 2\pi f$, which corresponds to a normalized frequency of $\Omega = \omega T_S$.

When we analyze the system response to input sinusoids, it does not matter if we are dealing with sine or cosine functions. Both use the same conjugate values of Z:

$$A \cos(\Omega n) = \frac{A e^{+j\Omega n} + A e^{-j\Omega n}}{2} \xrightarrow{\quad Z \quad} \frac{A}{2} Z^n + \frac{A}{2} \overline{Z}^n$$

$$A \sin(\Omega n) = \frac{A e^{+j\Omega n} - A e^{-j\Omega n}}{2j} \xrightarrow{\quad Z \quad} \frac{A}{2j} Z^n - \frac{A}{2j} \overline{Z}^n$$

where

$$Z = e^{+j\Omega} \quad \text{and} \quad \overline{Z} = e^{-}$$

According to Section 3.6, discrete-time signals are limited to adopting normalized frequencies in the range $\Omega = 0$ to π. For sinusoid inputs, the conjugate values of Z are therefore confined to the following range of values:

$$Z = 1e^{+j\Omega} \quad \text{and} \quad \overline{Z} = 1e^{-j\Omega}$$

where

$$0 \leq \Omega \leq \pi$$

The values of Z and $\overline{Z}$ therefore correspond to a complex exponentials of magnitude $= 1$ that vary in argument over the range of 0 to π and 0 to $-\pi$, respectively. Figure 7–13 plots the location of these values on the Z-plane.

The top and bottom semicircles illustrated in Figure 7–13 can accommodate the position of an infinite number of normalized input sinusoid frequencies. As an example, the pair of illustrated Z and $\overline{Z}$ values in Figure 7–13 corresponds to a real input sinusoid that has a normalized frequency of $\pi/4$. Fortunately, as the end of Section 7.3.3 noted, there is no need to consider the $\overline{Z}$ part of the input sinusoid when the system has a pure-real impulse response. In this case, we need to calculate only the gain G and phase shift ϕ that the system applies to the input $Z = e^{+j\Omega}$. This means that we have to consider only the response vectors that lead to location that are positioned along the top part of a circle of unit radius.

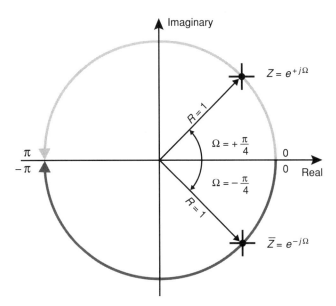

FIGURE 7–13
Set of *Z* values that correspond to input sinusoids.

7.4.2 POSITIONING POLES AND ZEROS ON THE *Z*-PLANE

As Section 7.3 indicated, we saw that the position of the system poles and zeros determines the complete system response. Once we know the location of the poles and zeros of a system, we can build the system Z-transform. In this section, we learn to position sets of poles and zeros anywhere on the Z-plane.

Balancing the Poles and Zeros of a System

We now consider the simplest possible systems that we can design. Such systems are limited to contain one pole and/or one zero.

System Defined with a Single Zero   We begin by examining a system that we define as containing a single *zero*. Such a system has the following Z-transform:

$$H(Z) = (Z - \text{Zero1})$$

where Zero1 can adopt any Z-plane coordinates.

To translate this Z-domain system into a corresponding discrete-time difference equation, we use Equation (7–4) to express the Z-transform as a ratio of output signal over input signal:

$$H(Z) = \frac{Y(Z)}{X(Z)} = (Z - \text{Zero1})$$

The Z-domain output expression becomes

$$Y(Z) = X(Z)(Z - \text{Zero}1)$$
$$= X(Z)Z^{+1} - \text{Zero}1 X(Z)$$

Applying the information presented in Section 7.1.2, we can use the value of the exponent of Z to translate the Z-domain terms to discrete-time domain terms:

$$Y(Z) = X(Z) Z^{+1} - \text{Zero}1 X(Z)$$
$$y(n) = x(n + 1) - \text{Zero}1 x(n)$$

Therefore, the output of this system is not causal since it *lags* one discrete-time unit behind the input signal. A system defined with a single zero requires an input sample located one discrete-time unit in the future. To solve the causality problem, we modify this system by introducing a delay of one discrete-time unit to the output expression. Adding this delay creates a new system that has the following Z-domain output expression:

$$Y(Z) = \left[X(Z) (Z - \text{Zero}1) \right] Z^{-1}$$
$$= X(Z) - \text{Zero}1 X(Z) Z^{-1}$$
$$y(n) = x(n) - \text{Zero}1 x(n - 1)$$

This new causal system has the following Z-transform:

$$H(Z) = \frac{Y(Z)}{X(Z)} = \left(1 - \text{Zero}1 \ Z^{-1} \right) \text{>}$$
$$= \frac{(Z - \text{Zero}1)}{Z}$$

The Z in the denominator represents a pole that is located at the origin:

$$H(Z) = \frac{(Z - \text{Zero}1)}{Z} = \frac{(Z - \text{Zero}1)}{Z - \text{Pole}1}$$

where

$$\text{Pole}1 = 0$$

Note that we had to add a pole at the origin to balance this system. Fortunately, adding a pole (or a zero) at the origin has no effect on the system gain. Figure 7–14 illustrates the response vector of a pole or a zero that is located at the origin. In that figure, we can see that the response vector extends from the origin to the input frequency, which is always positioned on the unit circle. Consequently, the response vector resulting from a pole or a zero located at the origin always has a length of 1. This translates to a gain of 1, which has no impact on the overall system gain.

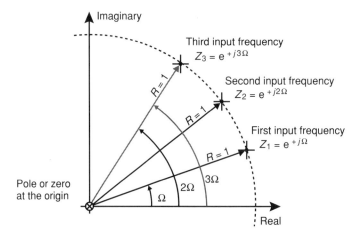

FIGURE 7–14
Response vector of a pole/zero located at the origin.

The phase shift that corresponds to a pole or a zero located at the origin is always *linear*. This means that if the input frequency doubles or triples, the phase shift also doubles or triples. As Figure 7–14 indicates, the response vector extending from the origin always has an angle equal to the normalized input frequency $\Omega = \omega T_S$.

A system that has linear-phase shift characteristics introduces a fixed constant time delay *for all frequencies*. This constant time delay simply increases the amount of time that the signal takes to travel from input to output. Section 8.1.2 discusses the linear-phase shift characteristic in more detail.

System Defined with a Single Pole   We continue by analyzing a new system that we define as containing a single *pole*. Such a system has the following Z-transform:

$$H(Z) = \frac{1}{(Z - \text{Pole 1})}$$

where Pole 1 can adopt any Z-plane coordinates.

We know that the Z-transform is the ratio of the output over the input in the Z-domain

$$H(Z) = \frac{Y(Z)}{X(Z)} = \frac{1}{(Z - \text{Pole 1})}$$

Isolating the Z-domain output $Y(Z)$, we get

$$Y(Z) = \frac{X(Z)}{(Z - \text{Pole 1})}$$

Manipulating to eliminate the denominator, we get

$$Y(Z)(Z - \text{Pole}\,1) = X(Z)$$
$$Y(Z)Z - \text{Pole}\,1\,Y(Z) = X(Z)$$
$$Y(Z)Z^{+1} = X(Z) + \text{Pole}\,1\,Y(Z)$$

We multiply both sides by Z^{-1} to cancel the Z^{+1} shift:

$$[Y(Z)Z^{+1}]Z^{-1} = [X(Z) + \text{Pole}\,1\,Y(Z)]Z^{-1}$$
$$Y(Z) = X(Z)Z^{-1} + \text{Pole}\,1\,Y(Z)Z^{-1}$$

We now translate to the discrete-time domain:

$$Y(Z) = X(Z)\,Z^{-1} + \text{Pole}\,1\,Y(Z)\,Z^{-1}$$
$$y(n) = x(n - 1) + \text{Pole}\,1\,y(n - 1)$$

As we can see, the output of this system *leads* the input expression by one discrete-time unit. A system defined with a single pole uses $x(n - 1)$ instead of the present input sample $x(n)$. To fix this annoying delay problem, we can modify this system by shifting the input samples one discrete-time unit into the future. Adding this shift creates a new system that has the following Z-domain output expression:

$$Y(Z) = \frac{X(Z)\,Z^{+1}}{(Z - \text{Pole}\,1)}$$

We manipulate to eliminate the denominator

$$Y(Z)[Z - \text{Pole}\,1] = X(Z)Z^{+1}$$
$$Y(Z)Z - \text{Pole}\,1\,Y(Z)] = X(Z)Z^{+1}$$
$$Y(Z)Z = X(Z)Z^{+1} + \text{Pole}\,1\,Y(Z)$$

By multiplying both sides by Z^{-1}, we eliminate the Z^{+1} shift, which allows translation to a discrete-time domain difference equation:

$$\left[Y(Z)\,Z\right]Z^{-1} = \left[X(Z)\,Z^{+1} + \text{Pole}\,1\,Y(Z)\right]Z^{-1}$$
$$Y(Z) = X(Z) + \text{Pole}\,1\,Y(Z)\,Z^{-1}$$
$$y(n) = x(n) + \text{Pole}\,1\,y(n - 1)$$

Note that this well-balanced system has one pole and one zero. We can generalize the results of this section as follows:

A system defined with an *equal number* of poles and zeros results in an output expression that neither leads nor lags the input signal.

Designing a Pure-Real System

The applications in this book are limited to systems that have a pure-real impulse response. In such systems, the difference equation coefficients are all real, and this presents the advantage of simplifying the convolution procedures used to process pure-real signals.

Until now, all examples have used poles and zeros positioned along the real axis of the Z-plane. In practice, we can locate poles and zeros at complex Z-plane coordinates, but we must determine how this affects the system.

Figure 7–15 illustrates the characteristics of pure-real versus complex Z-plane coordinates. As we can see, complex Z-plane coordinates are characterized by a nonzero imaginary part in the Cartesian system and by an angle other than zero or π when expressed as complex exponentials.

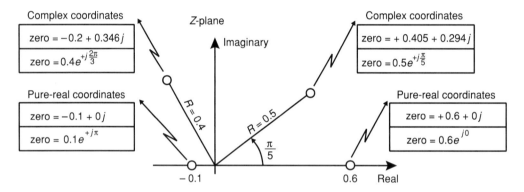

FIGURE 7–15
Pure-real versus complex coordinates.

We begin by analyzing what happens when we create a first-order system:

$$H(Z) = \frac{Y(Z)}{X(Z)} = \frac{(Z - \text{Zero}1)}{(Z - \text{Pole}1)}$$

We eliminate the positive exponents of Z:

$$\frac{Y(Z)}{X(Z)} = \frac{(Z - \text{Zero}1)}{(Z - \text{Pole}1)} \times \frac{Z^{-1}}{Z^{-1}} = \frac{(1 - \text{Zero}1\,Z^{-1})}{(1 - \text{Pole}1\,Z^{-1})}$$

We find the output equation in the Z-domain and translate to the discrete-time domain:

$$Y(Z) - \text{Pole}1Y(Z)\,Z^{-1} = X(Z) - \text{Zero}1X(Z)Z^{-1}$$

$$Y(Z) = X(Z) - \text{Zero}1X(Z)Z^{-1} + \text{Pole}1Y(Z)\,Z^{-1} \qquad \textbf{(7–6)}$$

$$y(n) = x(n) - \text{Zero}1x(n-1) + \text{Pole}1y(n-1)$$

Equation (7–6) provides a standard formula to translate any causal first-order system from its pole/zero form to or from its difference equation.

If we position the zero and/or the pole at a complex location in a first-order system, the output expression calls for multiplying the input samples by a complex value. For example, if $Zero1 = (0.4 + 0.3j)$ and $Pole1 = 0.5$ then

$$y(n) = x(n) + 0.4x(n - 1) + 0.3jx(n - 1) + 0.5y(n - 1)$$

In this case, the output expression is not pure real, and we cannot send it to the DAC as such.

To get pure-real coefficients the pole and zero of a first-order system must be located at pure-real locations on the Z-plane.

In fact, if the numerator or the denominator of the Z-transform is not pure real, the output expression cannot be pure real. To get a pure-real output expression, we must design the numerator and the denominator of the system Z-transform to contain pure-real expressions.

To do this, we recall that the addition and/or the product of complex conjugates yield pure-real results. Both the numerator and the denominator of the Z-transform consist in the product of factored expressions. The following second-order system, for example, contains two factors in the numerator and denominator:

$$H(Z) = \frac{(Z - Zero1)(Z - Zero2)}{(Z - Pole1)(Z - Pole2)}$$

Expanding the numerator and denominator expressions, we get

$$H(Z) = \frac{Z^2 - (Zero1 + Zero2)Z + (Zero1 \times Zero2)}{Z^2 - (Pole1 + Pole2)Z + (Pole1 \times Pole2)}$$

If we locate the zeros and the poles so that they are complex conjugates of each other, then

$$Zero2 = \overline{Zero1} \text{ and } Pole2 = \overline{Pole1}$$

In this case, all of the expressions become pure real:

$$(Zero1 + Zero2) = (Zero1 + \overline{Zero1}) = 2|Zero1|$$
$$(Pole1 + Pole2) = (Pole1 + \overline{Pole1}) = 2|Pole1|$$
$$(Zero1 \times Zero2) = (Zero1 \times \overline{Zero1}) = |Zero1|^2$$
$$(Pole1 \times Pole2) = (Pole1 \times \overline{Pole1}) = |Pole1|^2$$

For example, if

$$Zero1 = 0.4 + 0.3j \text{ and } Pole1 = -0.2 + 0.5j$$

we must position $Zero2$ and $Pole2$ at the following locations:

$$Zero2 = \overline{Zero1} = 0.4 - 0.3j \text{ and } Pole2 = \overline{Pole1} = -0.2 - 0.5j$$

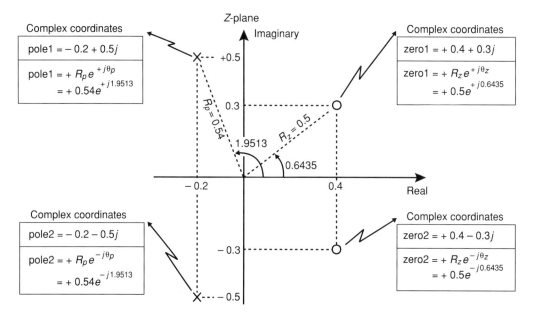

FIGURE 7–16
Conjugate zeros and poles.

Figure 7–16 illustrates the location of these poles and zeros on the Z-plane. For the poles and zeros illustrated

$$\text{Zero}1 + \text{Zero}2 = 0.8 \qquad \text{Zero}1 \times \text{Zero}2 = (0.5)^2 = 0.25$$
$$\text{Pole}1 + \text{Pole}2 = -0.4 \qquad \text{Pole}1 \times \text{Pole}2 = (0.54)^2 = 0.29$$

The Z-transform yields a pure-real result:

$$H(Z) = \frac{Y(Z)}{X(Z)} = \frac{Z^2 - (\text{Zero}1 + \text{Zero}2)\, Z + (\text{Zero}1 \times \text{Zero}2)}{Z^2 - (\text{Pole}1 + \text{Pole}2)\, Z + (\text{Pole}1 \times \text{Pole}2)}$$

$$= \frac{Z^2 - (0.8)\, Z + (0.25)}{Z^2 - (-0.4)\, Z + (0.29)}$$

Use of the complex exponential notation for the poles and zeros yields an equivalent result that adopts a different form:

$$H(Z) = \frac{Y(Z)}{X(Z)} = \frac{Z^2 - \left(R_Z e^{+j\theta_Z} + R_Z e^{-j\theta_Z}\right) Z + \left(R_Z e^{+j\theta_Z} \times R_Z e^{-j\theta_Z}\right)}{Z^2 - \left(R_P e^{+j\theta_P} + R_P e^{-j\theta_P}\right) Z + \left(R_P e^{+j\theta_P} \times R_P e^{-j\theta_P}\right)}$$

Recognizing that $Re^{+j\alpha} + Re^{-j\alpha} = 2R\cos(\alpha)$, we can simplify to the following expression for a system that has conjugate zeros and conjugate poles:

$$H(Z) = \frac{Y(Z)}{X(Z)} = \frac{Z^2 - \left(2R_Z\cos(\theta_Z)\right) Z + R_Z{}^2}{Z^2 - \left(2R_P\cos(\theta_P)\right) Z + R_P{}^2} \qquad (7\text{–}7)$$

Equation (7–7) expresses the location of the poles and zeros using angles and magnitudes. The angle notation lets us establish the relationship between the pole/zero location and the normalized input frequency $\Omega = \omega T_S$ on the Z-plane. This will prove quite useful when we learn to design the system frequency response by positioning poles and zeros.

Equation (7–7) is a second-order Z-transform that yields the following practical advantages:

- It defines a system that is pure real.
- We can combine second-order sections to design a system with an elaborate frequency response.
- We can simplify the programming by modularizing the Z-transform of sophisticated systems into multiple second-order sections.

We substitute the values for the pole and the zero magnitudes and angles of the example:

$$H(Z) = \frac{Y(Z)}{X(Z)} = \frac{Z^2 - \left(0.5 \times 2 \cos(0.6435)\right) Z + \left(0.5 \times 0.5\right)}{Z^2 - \left(0.54 \times 2 \cos(1.9513)\right) Z + \left(0.54 \times 0.54\right)}$$

$$= \frac{Z^2 - \left(0.8\right) Z + \left(0.25\right)}{Z^2 - \left(-0.4\right) Z + \left(0.29\right)}$$

To program a digital signal processor with the double zero/pole system described by Equation (7–7), we need to determine its difference equation. We begin by eliminating the positive exponents of Z:

$$\frac{Y(Z)}{X(Z)} = \frac{Z^2 - \left(2R_Z \cos(\theta_Z)\right) Z + \left(R_Z^2\right)}{Z^2 - \left(2R_P \cos(\theta_P)\right) Z + \left(R_P^2\right)} \times \frac{Z^{-2}}{Z^{-2}}$$

$$= \frac{1 - \left(2R_Z \cos(\theta_Z)\right) Z^{-1} + \left(R_Z^2\right) Z^{-2}}{1 - \left(2R_P \cos(\theta_P)\right) Z^{-1} + \left(R_P^2\right) Z^{-2}}$$

We then manipulate to eliminate the denominator:

$$Y(Z)[1 - 2R_P\cos(\theta_P)Z^{-1} + R_P^2 Z^{-2}] = X(Z)[1 - 2R_Z\cos(\theta_Z)Z^{-1} + R_Z^2 Z^{-2}]$$

Finally, we isolate the output expression in the Z-domain and translate it to the discrete-time domain:

$$Y(Z) = X(Z) - 2R_Z\cos(\theta_z)X(Z)Z^{-1} + R_Z^2 X(Z)Z^{-2}$$
$$+ 2R_P\cos(\theta_P)Y(Z)Z^{-1} - R_P^2 Y(Z)Z^{-2}$$
$$y(n) = x(n) - 2R_Z\cos(\theta_z)x(n-1) + R_Z^2 X(n-2)$$
$$+ 2R_P\cos(\theta_P)y(n-1) - R_P^2 y(n-2)$$

$$(7\text{–}8)$$

Equation (7–8) is the standard difference equation that describes a causal second-order system using *conjugate* zeros and poles. The impulse response of such systems consists of *pure-real* terms. Given a pure-real input signal, such a system produces pure-real output sample values that we can transfer to a DAC. Chapter 9 makes extensive use of second-order systems when it presents the implementation of digital filters. Pure-real second-order systems have the following difference equation coefficient values:

Coefficients that correspond to a pair of conjugate zeros:

$$b_0 = 1 \qquad b_1 = -2\,R_Z\cos(\theta_Z) \qquad b_2 = R_Z{}^2$$

Coefficients that correspond to a pair of conjugate poles:

$$a_1 = +2\,R_P\cos(\theta_P) \qquad a_2 = -R_P{}^2$$

For the pole/zero locations of the example, we calculate the following coefficient values for the difference equation. The zeros have an $R_Z = 0.5$ value and an angle value of $\theta_Z = 0.6435$ radians. This yields

$$b_0 = 1 \qquad \begin{aligned}b_1 &= -2(0.5)\cos(0.6435) \\ &= -0.8\end{aligned} \qquad \begin{aligned}b_2 &= (0.5)^2 \\ &= 0.25\end{aligned}$$

The poles have an $R_P = 0.54$ value and an angle value of $\theta_p = 1.9513$ radians. This yields:

$$\begin{aligned}a_1 &= +2(0.54)\cos(1.9513) \\ &= -0.4\end{aligned} \qquad \begin{aligned}a_2 &= -(0.54)^2 \\ &= -0.29\end{aligned}$$

The example system therefore has the following difference equation:

$$y(n) = x(n) - 0.8x(n-1) + 0.25x(n-2) - 0.4y(n-1) - 0.29y(n-2)$$

7.5 DESIGNING THE FREQUENCY RESPONSE USING POLES AND ZEROS

As Section 7.3 discussed, the system poles and zeros define response vectors that combine to produce the overall system gain and phase shift. Since the location of poles and zeros completely defines the system, we can adjust their position to design the system response.

7.5.1 CONTRIBUTION OF ZEROS TO THE SYSTEM RESPONSE

This section investigates the contribution of zeros to the overall system gain and phase shift. As presented in Section 7.1, an LTI system is completely defined by its Z-transform. Section 7.2 built the numerator of the Z-transform by multiplying the *zero factors:*

$$H(Z) = \frac{(Z - \text{Zero 1})(Z - \text{Zero 2})(Z - \text{Zero 3})(Z - \text{Zero 4}) \ldots}{\text{Denominator pole factors}}$$

Each of the factored expressions in the $H(Z)$ function represents a response vector, which starts at some zero location and ends at Z, the position of the system input. An input sinusoid corresponds to a pair of conjugate Z inputs that lie on opposite sides of a circle of unit radius (refer to Section 7.4.1). Because of the symmetry of the pure-real sinusoid input, we need to consider only the response vectors that lead to the location of Z on the top part of the semicircle for pure-real systems. In this case, each of the numerator response vectors that lead to the input location Z contributes a gain and a phase shift. Figure 7–17 illustrates the response vector of *one* of the system zeros. The response vector of the zero of Figure 7–17 corresponds to a complex exponential of the form

$$\text{Zero1 response vector: } \overrightarrow{V_{Z1}} = (Z - \text{Zero 1})$$
$$= e^{j\Omega} - R_{Z1}e^{j\theta_{z1}}$$
$$= G_{Z1}e^{j\phi_{Z1}}$$

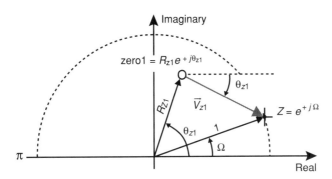

FIGURE 7–17
Response vector associated with a zero.

where G_{Z1} is the gain and ϕ_{Z1} is the phase shift that Zero1 contributes to the system. The graphical representation of zeros in the Z-plane allows us to visualize the contribution made by each of the system zeros:

Each zero contributes a gain of G_Z.
Each zero contributes a phase shift of ϕ_Z.

Remember that the practical range of discrete-time input sinusoid frequencies corresponds to positions of Z along a semicircle of unit radius. As the input frequency changes, the zero response vectors all change to point to the new location of Z.

When response vectors change their length and angle to follow the input frequency, their respective gain and phase shift contributions also change. The contribution made by each zero to the overall response of the system therefore varies with the input frequency.

Figure 7–18 illustrates the changes in the response vector as we change the value of the input frequency along the semicircle. The three response vectors of Figure 7–18 illustrate the response of the same system zero to different input frequencies. If we examine the variation of the response vector closely, it is obvious that the length of the vector becomes shorter as the input frequency approaches to the zero location. Since the length of the response vector represents the gain associated with this zero, we can conclude that positioning a zero close to a particular frequency location will attenuate the amplitude of that frequency.

Positioning system zeros close to a range of input frequencies creates a system that attenuates these frequencies.

FIGURE 7–18
Zero response vector changes with the
input frequency.

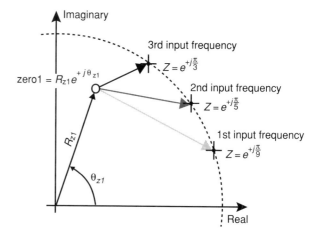

We now consider a system that contains a number of zeros. We can write the numerator of that system Z-transform as a product of all the zero vectors:

$$H(Z) = \frac{\overrightarrow{V_{Z1}} \times \overrightarrow{V_{Z2}} \times \overrightarrow{V_{Z3}} \times \overrightarrow{V_{Z4}} \times \cdots}{\text{Denominator vectors}}$$

$$H(Z) = \frac{\left|\overrightarrow{V_{Z1}}\right| \angle\phi_{Z1} \times \left|\overrightarrow{V_{Z2}}\right| \angle\phi_{Z2} \times \left|\overrightarrow{V_{Z3}}\right| \angle\phi_{Z3} \times \cdots}{\text{Denominator vectors}}$$

The magnitude of each vector corresponds to a gain, and the vector phase corresponds to a phase shift. Separating the gain and phase shift, we get

$$\text{System gain: } |H(Z)| = \frac{\left|\overrightarrow{V_{Z1}}\right| \times \left|\overrightarrow{V_{Z2}}\right| \times \left|\overrightarrow{V_{Z3}}\right| \times \cdots}{\text{Denominator vectors}} =$$

$$\frac{G_{Z1} \times G_{Z2} \times G_{Z3} \times \cdots}{\text{Denominator vectors}}$$

$$\text{System phase shift: } \angle H(Z) = \frac{\angle\phi_{Z1} + \angle\phi_{Z2} + \angle\phi_{Z3} + \cdots}{\text{Denominator vectors}}$$

Since every single zero creates a response vector and since we control the position of the zeros, we can use their position to engineer the system response. For example, consider what happens if we position one of the system zeros exactly on the unit circle. In this case, the response vector associated with that zero will have a length of zero when the input frequency is located under that zero. When any of the numerator response vectors reach a magnitude of 0 at a particular input frequency, the entire numerator evaluates to zero and the overall system gain becomes zero at that frequency. In other words, that frequency is eliminated from the output signal.

Figure 7–19 illustrates what happens when the input frequency coincides with the location of a zero. As the figure indicates, the length of the response vector V_{Z1} is reduced to zero when the input frequency is located under the zero. In this case, the overall system gain becomes zero:

$$|H(Z)| = \frac{|0| \times |\overrightarrow{V_{Z2}}| \times |\overrightarrow{V_{Z3}}| \times \cdots}{\text{Denominator vectors}} = 0$$

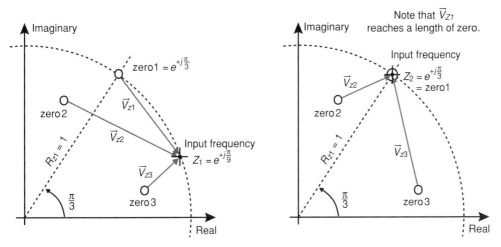

FIGURE 7–19
Frequency eliminated by overlaying it with a zero.

Positioning zeros is very important when we design systems that have special characteristics such as *linear-phase* filters and *all-pass* filters. The section in Chapter 8 on Positioning Zeros for Linear-Phase Response discusses which patterns of zeros allow us to design such filters.

7.5.2   CONTRIBUTION OF POLES TO THE SYSTEM RESPONSE

The position of the system poles determines the denominator factors of the Z-transform. According to the section on the Z-Transform of Infinite Impulse Response Systems, the presence of poles in the system creates an IIR system.

The denominator of a system Z-transform consists in the product of all pole factors:

$$H(Z) = \frac{\text{Numerator zero factors}}{(Z - \text{Pole 1})(Z - \text{Pole 2})(Z - \text{Pole 3})(Z - \text{Pole 4})(\cdots}$$

Each of the pole factors results in a response vector that has a magnitude and a phase:

$$\overrightarrow{V_P} = \left|\overrightarrow{V_P}\right| \angle \phi_P$$

$$H(Z) = \frac{\text{Numerator vectors}}{\overrightarrow{V_{P1}} \times \overrightarrow{V_{P2}} \times \overrightarrow{V_{P3}} \times \overrightarrow{V_{P4}} \times \cdots}$$

Figure 7–20 illustrates the magnitude and phase of a pole response vector.

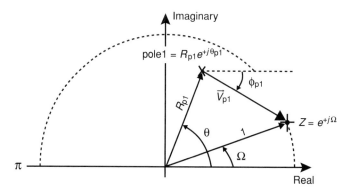

FIGURE 7–20
Gain and phase shift associated with a pole response vector.

Since each of the system poles results in a denominator vector, we can engineer the denominator part of the system by positioning poles on the Z-plane. We now investigate the contribution of poles to the overall system response. Since all pole vectors are located in the denominator of the system Z-transform, the net effect of each pole on the system response is the inverse of each pole vector:

$$H(Z) = \frac{\text{Numerator vectors}}{\left|\overrightarrow{V_{P1}}\right| \angle \phi_{P1} \times \left|\overrightarrow{V_{P2}}\right| \angle \phi_{P2} \times \left|\overrightarrow{V_{P3}}\right| \angle \phi_{P3} \times \cdots}$$

$$= \text{Numerator vectors} \times \frac{1}{\left|\overrightarrow{V_{P1}}\right|} \angle -\phi_{P1} \times \frac{1}{\left|\overrightarrow{V_{P2}}\right|} \angle -\phi_{P2} \times$$

$$\frac{1}{\left|\overrightarrow{V_{P3}}\right|} \angle -\phi_{P3} \times \cdots$$

As we can see, inverting a pole vector results in a response vector whose magnitude is the inverse of the original pole vector magnitude and whose phase is the negative of the original pole vector phase.

The system poles therefore contribute the following gain and phase shift to the overall system response:

Each pole contributes a gain of

$$G_P = \frac{1}{|V_P|}$$

Each pole produces a phase shift of

$$-\phi_P = -\angle V_P = \frac{1}{\angle V_P}$$

Notice that the gain is inversely proportional to the pole vector magnitude. This means that the shorter the pole vector becomes, the greater the gain becomes.

The closer the system poles are located to a range of input frequencies, the more the system amplifies these frequencies.

As an example of using poles in a design, recall that the DAC has a frequency response (see Section 6.4.4) that distorts the amplitude of the output signal frequencies. Most systems contain an equalizer to compensate for the frequency response of the DAC. Ideally, the equalizer cancels the output distortion by providing a response that is the exact inverse of the DAC response.

Building a *perfect* equalizer is impossible, but, fortunately, it is possible to design a variety of systems that closely *approximate* a perfect equalizer response curve. The simplest solution consists of a system that contains a single pole and a zero at the origin. As we will see, this system does a fair job of flattening the system gain curve, but, unfortunately, it slightly distorts the phase. Figure 7–21 illustrates the ideal gain response for a DAC equalizer.

Notice that the high frequencies in Figure 7–21 need more gain than the low ones. Placing a pole close to the high frequencies in the Z-plane boosts the high frequencies. The issue is at what exact location the system pole should be positioned. Since we want the system to yield a pure-real response, the single pole must be located somewhere along the real axis. To balance the system, we must also add a zero at the origin but, fortunately, poles or zeros at the origin do not influence the gain response. Figure 7–22 illustrates the change in the system response when a single pole is relocated at different positions along the real axis.

Examining Figure 7–22 helps to understand the way that the position of the pole influences the system gain. Notice that the gain grows quickly as the pole approaches the unit circle. The system with a pole at –0.2 has a response very close in shape to the inverse of the DAC sinc(x) curve. Actually, it is possible to calculate that a pole located around –0.14 will provide a close approximation to the shape of a 1/sinc(x) curve.

The Z-transform for such a system is

$$H(Z) = \frac{Z}{Z - (-0.14)} = \frac{Z}{Z + 0.14}$$

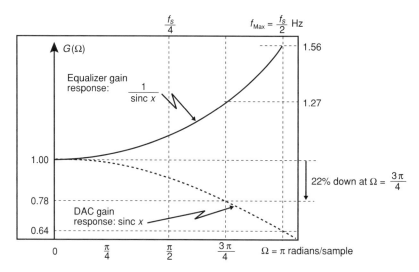

FIGURE 7–21
Ideal gain response for a DAC equalizer.

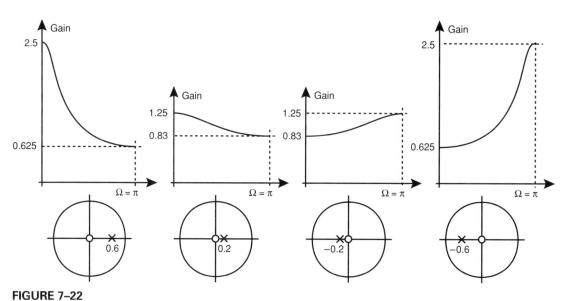

FIGURE 7–22
Relocation of a pole along the real axis.

Calculating the gain at the lowest normalized input frequency $\Omega = 0$ and at the highest normailzed input frequency $\Omega = \pi$, we obtain

Gain at $\Omega = 0$:

$$\left. |H(Z)| \right|_{Z=e^{j0}} = \left| \frac{1}{1 + 0.14} \right| = 0.877$$

Gain at $\Omega = \pi$:

$$\left. |H(Z)| \right|_{Z=e^{j\pi}} = \left| \frac{-1}{-1 + 0.14} \right| = 1.16$$

Review Figure 7–21. The equalizer should have a gain close to 1 at the lowest input frequency $\Omega = 0$ and of 1.27 at a normalized input frequency of $\Omega = 3\pi/4$. To accommodate this, we scale the system by multiplying the Z-transform by a *constant* value. We calculate this value to yield the least deviation in the power levels for the frequencies processed by the system. For example, scaling the system by a value of 1.13 yields good results for systems that process normalized frequencies ranging from zero to $3\pi/4$. In this case, the system Z-transform becomes

$$H(Z) = 1.13 \times \frac{Z}{Z + 0.14}$$

To find the difference equation of this system, we eliminate the positive exponents of Z:

$$H(Z) = \frac{Y(Z)}{X(Z)} = 1.13 \times \frac{Z}{Z + 0.14} \times \frac{Z^{-1}}{Z^{-1}} = 1.13 \times \frac{1}{1 + 0.14\,Z^{-1}}$$

We then solve for an output expression and convert to the discrete-time domain:

$$Y(Z)[1 + 0.14Z^{-1}] = 1.13X(Z)$$
$$Y(Z) = 1.13X(Z) - 0.14Y(Z)Z^{-1}$$
$$y(n) = 1.13x(n) - 0.14y(n - 1)$$

Implementing this equalizer in practice on a signal processor requires only two MAC instructions. Even if the performance of this simple equalizer is not perfect, it keeps the gain deviations within 2% of what they should ideally be for normalized frequencies ranging from zero to $3\pi/4$. Compare this in Figure 7–21 to the 22% error without the equalizer.

Figure 7–23 illustrates the gain response for a system using this equalizer. Unfortunately, the one-pole equalizer slightly distorts the system phase response. All DSP systems that include poles result in nonlinear-phase response. In the case of the one-pole equalizer, the damage is small since the worst-case phase shift amounts to a lead of 7.3 degrees. In many applications, such small deviation can be neglected. Figure 7–24 illustrates the phase response of the one-pole equalizer.

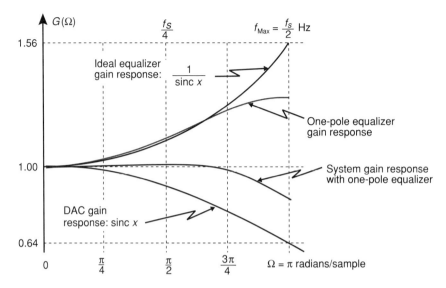

FIGURE 7–23
Gain response with a one-pole equalizer.

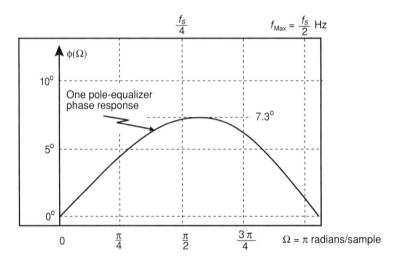

FIGURE 7–24
Phase response with a one-pole equalizer.

The pole in the case of the one-pole equalizer was located quite far from the unit circle. Some applications, however, call for poles to be located very close to the unit circle. In such cases, the gain can become very high, and the phase can shift very suddenly. Such systems are on the edge of being unstable. The next section addresses this issue.

7.6 ENSURING STABLE SYSTEM OPERATION

Section 4.5.2 analyzed the stability in discrete-time systems. It found that FIR systems *always* provide stable system operation while IIR systems may or may not be stable. In the case of IIR systems, unstable operation results occur when the output feedback becomes too great. Unstable IIR systems accumulate and amplify input disturbances, which render the output useless.

The Z-transform of any IIR system contains one or many poles on the Z-plane. Since IIR systems may or may not be stable, there must be areas of the Z-plane where we should not position poles. This section analyzes the Z-plane to determine in which areas it is safe to position poles.

7.6.1 STABILITY IN FIRST-ORDER INFINITE IMPULSE RESPONSE SYSTEMS

The stability of a system can be determined by examining its impulse response. If the impulse response decays toward zero, the system will ultimately achieve stable operation. When the input signal changes, the system goes through a transitory period. The time that the impulse response takes to decay to zero or near-zero values determines the duration of the transitory period. Once the transients have passed, the system reaches stable steady-state operation.

We now analyze a first-order system to determine the way that the position of poles influences the duration of the system impulse response. As this chapter has discussed, the following relationships define the output of a first-order system that contains one pole:

$$Y(Z) = X(Z) + (\text{Pole}1)\, Y(Z)\, Z^{-1}$$

$$y(n) = x(n) + \text{Pole}1\, y(n-1)$$

Figure 7–25 illustrates the impulse response of various single pole systems as the location of a single pole is moved along the real axis. Note that the impulse response in Figure 7–25 takes longer and longer to decay to zero as the pole gets closer to the unit circle:

- When the pole is inside the unit circle, the impulse response decays toward zero.
- When the pole touches the circle, the impulse response does not decay anymore, and the system never returns to a steady-state operation. The system is riding a fine line between stable and unstable operation. Note that the system becomes an oscillator at a frequency of $f_S/2$ when the pole reaches the location -1.
- When the pole moves *beyond* the confines of the unit circle, the impulse response grows exponentially and the system becomes unstable and useless.

We must conclude that the pole of a first-order IIR system must reside *inside* the unit circle to ensure stable operation.

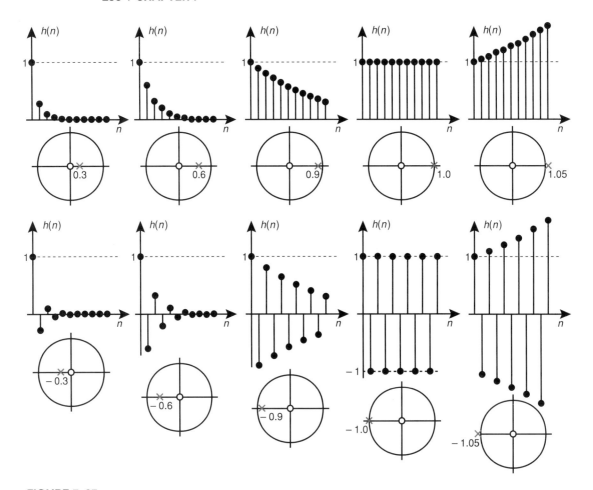

FIGURE 7–25
Impulse response for a single-pole system.

First-order section $= \dfrac{(Z - \text{zero1})}{(Z - \text{pole1})}$

Second-order section $= \dfrac{(Z - \text{zero2})(Z - \overline{\text{zero2}})}{(Z - \text{pole2})(Z - \overline{\text{pole2}})}$

$X(Z) \longrightarrow$ First-order section $\longrightarrow$ Second-order section $\longrightarrow$ Second-order section $\longrightarrow Y(Z) =$ Fifth order

FIGURE 7–26
Cascading first- and second-order sections.

7.6.2 STABILITY OF HIGHER-ORDER INFINITE IMPULSE RESPONSE SYSTEMS

Section 7.6.1 discussed that a system can be stable only if its impulse response decays toward zero. We already know that we can assemble any *linear* system by cascading a number of first- and second-order sections. For example, Figure 7–26 illustrates how a first-order section can be cascaded with two second-order sections to produce a fifth-order system.

If any of the first- or second-order sections is not stable, that section corrupts the entire system and makes it unstable. Since FIR systems are always stable, the steady-state stability is not related to the position of the system zeros. The stability of any system therefore completely depends on the location of its poles.

We next analyze the behavior of second-order IIR sections to determine how the position of conjugate poles influences the system stability. The simplest pure-real second-order IIR section consists of conjugate poles balanced by two zeros at the origin. The following relations define the output of such a section:

$$Y(Z) = X(Z) + 2R_P\cos(\phi_P)Y(Z)Z^{-1} - R_P{}^2Y(Z)Z^{-2}$$
$$y(n) = x(n) + 2R_P\cos(\phi_P)y(n-1) - R_P{}^2y(n-2)$$

When R_P is made to adopt larger values, the conjugate poles move toward the unit circle. As we can see from the output equation, larger values of R_P increase the amount of output feedback. Figure 7–25 illustrates the impulse response of different conjugate pole systems as the value of R_P is increased to move the conjugate poles along an angle of $\pi/4$ on the Z-plane.

An examination of Figure 7–27 reveals that the system takes an increasing amount of time to reach steady-state operation as the poles are moved closer to the unit circle. When the poles touch the unit circle, the system oscillates as it hangs on the edge of instability. If the poles move outside the unit circle, the system becomes unstable and its impulse response grows out of control.

The behavior illustrated in Figure 7–27 is true no matter the angle along which the conjugate poles are located. We must conclude that the conjugate poles of a pure-real second-order IIR system must reside *inside* the unit circle to ensure stable operation. In fact, since all the sections of an IIR system must be stable to ensure an overall stable system, we can conclude the following:

To ensure a stable system, all the poles of an IIR system must reside *inside* the unit circle.

7.6.3 Quantization Error

Many applications call for positioning poles very close to the unit circle. In these delicate cases, we must ensure that the system poles stay inside the unit circle. Unfortunately, the rounding operations and/or the tolerance of the processing often nudge the poles and the zeros to positions that are slightly offset from the intended ones.

In addition, some applications call for processing very narrow frequency bands. For example, imagine that the processing is designed to remove a 60 Hz hum from the output

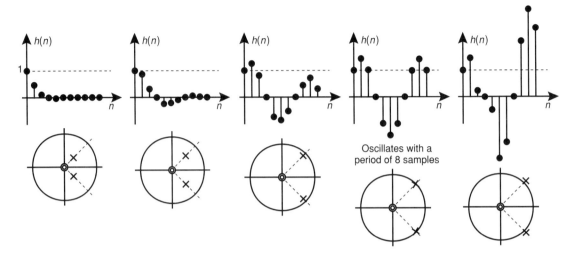

FIGURE 7–27
Impulse response of a conjugate pole section.

of an audio system. In this case, we must ensure that the rounding operations and/or the tolerance of the processing does not change the range of the targeted frequency band.

Imagine that we position poles and zeros along the same input frequency angle. From Equation (7–8) the difference equation of a second-order system that has poles located at $R_P e^{\pm j\Omega}$ and zeros located at $R_Z e^{\pm j\Omega}$ is as follows:

$$y(n) = x(n) - 2R_Z\cos(\Omega)x(n-1) + R_Z^2 x(n-2) + 2R_P\cos(\Omega)y(n-1) - R_P^2 y(n-2)$$

To program the difference equation on a digital signal processor, we quantize the weighting factors (refer to Appendix B). The precision of this translation depends on the number of fractional bits that we use in the conversion. The smaller the Q factor, the worse the quantization error becomes.

For example, using a small Q factor such as Q4 means that the fractional part of the weighting factors can adopt only one of $2^4 = 16$ possible values. There are, therefore, a finite number of locations on the Z-plane where we can locate poles or zeros.

Figure 7–28 illustrates the locations that are possible for a second-order system, using Q4 weighting factors positioned inside the unit circle, and in the first quadrant of the Z-plane.

Quantizing the weighting factors actually alters the practical location of the poles and zeros. For example, consider what happens when we try to locate poles and zeros at the following coordinates:

$$\Omega = \pm\frac{\pi}{8} \qquad R_Z = 1 \qquad R_P = 0.9$$

The difference equation of this second-order system is

$$y(n) = x(n) - 1.84776x(n-1) + x(n-2) + 1.66298y(n-1) - 0.81y(n-2)$$

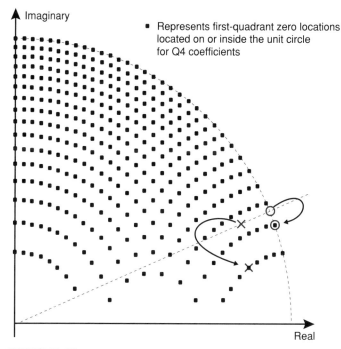

FIGURE 7–28
Possible locations of Q4 factors in the first quadrant.

If we now quantize the values using Q4 factors (rounding down), the equation becomes

$$y(n) = x(n) - 1.8125x(n-1) + x(n-2) + 1.625y(n-1) - 0.75y(n-2)$$

Notice that the fractional values have changed to become multiples of $1/2^4 = 0.0625$:

$$-1.8125 = -29 \times 0.0625 \qquad 1.625 = 26 \times 0.0625 \qquad -0.75 = -12 \times 0.0625$$

Since this system uses Q4 coefficients, it cannot implement the exact equation required by the design. The quantization process therefore alters the intended locations of the system poles and zeros. The effect is obvious in this example since the quantization error is rather large for Q4 factors. Figure 7–28 illustrates that the intended locations have been shifted to quantized locations.

Note a number of important effects in Figure 7–28:

- The possible pole/zero locations are not evenly distributed on the Z-plane. This means that not all Z-plane areas are created equal. Some areas contain a higher density of point than others do. Placing a pole/zero in a "rich" area means that there will be less quantization effect. For example, the locations around $f_S/2$ produce the least amount of error.
- The quantization effect does not necessarily bring the pole or the zero to the closest possible location. The implemented location may be significantly different from the intended one. In the example, examine how the pole has shifted quite far from the intended location.

- Rounding the coefficients *down* moves the pole/zeros to lower locations. This means that they move and remain *inside* the unit circle. Rounding *up* could move poles on or outside the unit circle; therefore, there is a risk that quantization will make the system unstable.
- The poles/zeros are shifted off the intended angular radiant. This means that the system will be processing frequency bands that are slightly different from the intended ones.

Fortunately, the number of possible locations increases quickly when the Q factor becomes larger. In spite of this, if you are trying to position some poles very close to the unit circle, you should check that the quantization error is not making the system unstable. All software systems used to design digital filters allow you to visualize the quantized locations so that you can check this.

7.7 MAKING AN OSCILLATOR

Oscillators are used in DSP systems to provide reference signals or to generate tones. An example is the dual tone multiple frequency (DTMF) tones (refer to Practice Question 7-10 at the end of this chapter) that are used to dial on the telephone network. A number of ways to build oscillators using a digital signal processor exist. The following subsections look at two ways to implement an oscillator.

7.7.1 BUILDING AN OSCILLATOR BY PLACING POLES ON THE UNIT CIRCLE

You may have noticed in Figure 7–25 and Figure 7–27 that the system oscillates when the poles touch the unit circle. When a system has conjugate poles touching the unit circle, the poles are at a distance of $R_P = 1$ from the origin. Note that the two zeros that balance this system are located at the origin with an $R_Z = 0$. In this case, Equation (7–8), which defines the difference equation of this second-order system, simplifies to

$$y(n) = x(n) + 2\cos(\phi)y(n{-}1) - y(n{-}2)$$

Since this system oscillates, the presence of an input signal $x(n)$ is irrelevant, and we can remove it completely from the output relationship:

$$y(n) = 2\cos(\phi)y(n - 1) - y(n - 2) \tag{7–9}$$

The oscillation at the system output $y(n)$ certainly corresponds to a periodic signal. What, however, is the frequency of this oscillation? We assume that the output signal is a sinusoid of the form $y(n) = A\cos(\theta n)$. We can test the validity of this solution by substituting it in the difference equation. If we find that this sinusoid satisfies the $y(n)$ relationship, we should be able to solve for the angular frequency of oscillation θ:

$$y(n) = A\cos(\theta n) = 2\cos(\phi)\,A\cos(\theta(n - 1)) - A\cos(\theta(n - 2))$$

Choosing $n = 1$ helps us to simplify the relation:

$$y(n) = A \cos(\theta \times 1) = 2 \cos(\phi) \, A \cos(\theta(1-1)) - A \cos(\theta(1-2))$$
$$A \cos(\theta) = 2 \cos(\phi) \, A \cos(0) - A \cos(-\theta)$$

We know that a cosine is an even function; therefore, $\cos(\theta) = \cos(-\theta)$. We also know that $\cos(0) = 1$. Substituting these equivalencies, we obtain

$$A \cos(\theta) = 2A \cos(\phi) - A \cos(\theta)$$

Finally, regrouping the $\cos(\theta)$ terms, we can write

$$2A \cos(\theta) = 2A \cos(\phi)$$

The relationship holds when the angular frequency of oscillation θ is equal to the angle ϕ, which sustains the poles of our system.

The pure IIR second-order system described by

$y(n) = 2\cos(\phi) \, y(n-1) - y(n-2)$

generates the sinusoid $y(n) = A\cos(\phi n)$

For example, the conjugate poles in Figure 7–27 have an angle of $\pm\pi/4$. In this case, the system should oscillate at

$$y(n) = A \cos\left(\frac{\pi}{4} n\right)$$

Since π corresponds to a normalized frequency of $f_S/2$, this system oscillates at a frequency of $f_S/8$. Examining Figure 7–27 closely, we notice that the period of oscillation contains eight samples, which confirms a frequency of $f_S/8$. However, what about the amplitude and phase of the oscillation?

Note in Equation (7–9) that the difference equation that describes this oscillator calculates the present output from the previous two output values. To start this system, we must provide it with these two *seed* values. These two seed values must be two samples located on the sinusoid that the system generates. Choosing these values adequately allows us to define the amplitude and the phase of the system oscillations:

$$y(n-1) = A \cos(\phi(n-1) + \Phi) \quad \text{and} \quad y(n-2) = A \cos(\phi(n-2) + \Phi)$$

As an example, consider that we want to use a system that samples at a rate of 10 k-samples/second to design an oscillator. We arbitrarily choose to generate the following sinusoid:

$$y(n) = 3 \cos((2\pi \times 1k)n + 80°)$$

This sinusoid oscillates over an amplitude range of ±3 at a frequency of 1 kHz, and has a phase shift of 80 degrees. In this case, we need to position conjugate poles on the unit circle at a normalized angle of

$$\phi = \omega\, T_S = 2\pi \times 1\mathrm{k} \times \frac{1}{10\mathrm{k}} = \frac{\pi}{5}$$

Converting the phase shift into radians yields the following output relationship:

$$80° \xrightarrow{\text{Radians}} \frac{80}{180}\pi = 1.396$$

$$y(n) = 3\cos\left(\frac{\pi}{5}n + 1.396\right)$$

The difference equation that defines this system is

$$y(n) = 2\cos\left(\frac{\pi}{5}\right)y(n-1) - y(n-2)$$

We must provide the two initial values (the two previous output values) that allow the system to calculate the first output value (at time $n = 0$). These two values define the amplitude and the phase of the output sinusoid:

$$y(0-1) = 3\cos\left(\frac{\pi}{5}(0-1) + 1.396\right) = +2.158$$

$$y(0-2) = 3\cos\left(\frac{\pi}{5}(0-2) + 1.396\right) = +2.971$$

Figure 7–29 illustrates the details of this system.

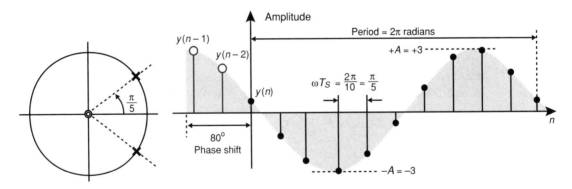

FIGURE 7–29
Oscillator producing a 1 kHz output.

7.7.2 USING TRIGONOMETRIC IDENTITIES TO BUILD AN OSCILLATOR

This section discusses the way to build an oscillator that simultaneously produces sine and cosine waveforms. Such oscillators have applications outside the scope of this introductory book. The technique is shown here to illustrate a different approach to create an oscillator.

In this case, we want to create the following two sinusoid signals oscillating at a normalized frequency of Ω_0:

$$y_{cos}(n) = \cos(n\Omega_0) \text{ and } y_{sin}(n) = \sin(n\Omega_0)$$

We use the following well-established trigonometric identities, the proof of which can be found in most math books dealing with trigonometry:

$$\cos(\alpha + \beta) = \cos(\alpha)\cos(\beta) - \sin(\alpha)\sin(\beta)$$
$$\sin(\alpha + \beta) = \sin(\alpha)\cos(\beta) + \cos(\alpha)\sin(\beta)$$

Now assume that $\alpha = n\Omega_0$ and that $\beta = \Omega_0$. Substituting these in the cosine identity yields

$$\cos(n\Omega_0 + \Omega_0) = \cos(n\Omega_0)\cos(\Omega_0) - \sin(n\Omega_0)\sin(\Omega_0)$$
$$\cos([n+1]\Omega_0) = \cos(n\Omega_0)\cos(\Omega_0) - \sin(n\Omega_0)\sin(\Omega_0) \qquad \textbf{(7–10)}$$

We started by defining that $y_{cos}(n) = \cos(n\Omega_0)$ and $y_{sin}(n) = \sin(n\Omega_0)$. We can increment n to write $y_{cos}(n+1) = \cos([n+1]\Omega_0)$. Substituting these in Equation (7–10) yields

$$y_{cos}(n+1) = y_{cos}(n)\cos(\Omega_0) - y_{sin}(n)\sin(\Omega_0)$$

Decrementing n yields:

$$y_{cos}(n) = y_{cos}(n-1)\cos(\Omega_0) - y_{sin}(n-1)\sin(\Omega_0) \qquad \textbf{(7–11)}$$

If we know the value of Ω_0, the expressions $\cos(\Omega_0)$ and $\sin(\Omega_0)$ are simple numbers whose value may be determined using a calculator. For example, $\Omega_0 = \pi/3$ corresponds to a frequency of $f_S/6$. In this case

$$\cos\left(\frac{\pi}{3}\right) = 0.5 \text{ and } \sin\left(\frac{\pi}{3}\right) = 0.866$$

Equation (7–11) therefore states that the sample of y_{cos} at time n is found by multiplying the previous samples at time $n-1$ by the value of $\cos(\Omega_0)$ and by subtracting $\sin(\Omega_0)$ times the previous sample of y_{sin}.

Similarly, we can use the sine identity to write

$$\sin(n\Omega_0 + \Omega_0) = \sin(n\Omega_0)\cos(\Omega_0) + \cos(n\Omega_0)\sin(\Omega_0)$$
$$y_{sin}(n+1) = y_{sin}(n)\cos(\Omega_0) + y_{cos}(n)\sin(\Omega_0)$$
$$y_{sin}(n) = y_{sin}(n-1)\cos(\Omega_0) + y_{cos}(n-1)\sin(\Omega_0) \qquad \textbf{(7–12)}$$

Figure 7–30 illustrates how we can couple Equation (7–11) and Equation (7–12) to create an oscillator. The boxes in Figure 7–30 represent memory locations, the circles are addition operations, and the triangles are multiplication operations. The oscillator works by initializing the two memory locations that hold the previous samples. For example, using $n = 1$ yields

$$\cos([n-1]\Omega_0) = 1 \quad \text{and} \quad \sin([n-1]\Omega_0) = 0$$

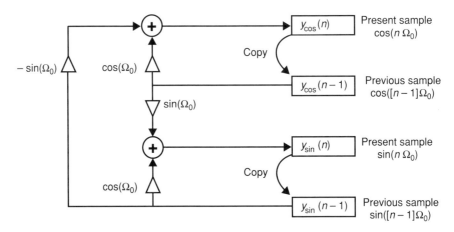

FIGURE 7–30
Coupled form oscillator.

The processor then executes a program that repeatedly executes the following steps:

- The processor computes the sample value of $y_{\cos}(n)$ by performing the operations described by Equation (7–11). This value is stored in memory location $y_{\cos}(n)$.
- The processor computes the sample value of $y_{\sin}(n)$ by performing the operations described by Equation (7–12). This value is stored in memory location $y_{\sin}(n)$.
- The new sample values of $y_{\cos}(n)$ and $y_{\sin}(n)$ may now be outputted to a DAC or may be used by the system.
- The sample values of $y_{\cos}(n)$ and $y_{\sin}(n)$ are copied to make them one time unit older. The process is repeated to produce the next set of samples.

SUMMARY

- The Z-transform describes LTI systems in terms of the *type* of input signal:

$$H(Z) = \sum_{k=-\infty}^{\infty} h(k)Z^{-k}$$

- In the Z-domain, the output function is found by multiplying the input by the Z-transform:

$$Y(Z) = X(Z) \times H(Z)$$

- Any Z-domain signal multiplied by Z^{shift} suffers a discrete-time shift, which is equal to the value of the exponent of Z.
- FIR systems are defined by the numerator part of their Z-transform.
- IIR systems are defined by the denominator part of their Z-transform.
- Finding the roots of the Z-transform allows us to use poles and zeros to completely define the system.

A *zero* is defined as the position of Z in the Z-plane that makes $H(Z) = 0$.

A *pole* is defined as the position of Z in the Z-plane that makes $H(Z) = \infty$.

- The system zeros, system poles, and system input signal can all be plotted on the Z-plane. From this plot, we can draw vectors linking the poles and zeros to the input signal. Each of these vectors contributes a part of the overall system response.
- $\overrightarrow{V_{Zero}}$ provides the following contributions to the overall system response:

$$\left|\overrightarrow{V_{Zero}}\right| = \text{Gain} \quad \text{and} \quad \angle\,\overrightarrow{V_{Zero}} = \text{Phase lead}$$

The magnitude (length) of a zero response vector directly represents the *gain*.

The angle represents a phase *lead*.

- $\overrightarrow{V_{Pole}}$ provides the following contributions to the overall system response:

$$\left|\overrightarrow{V_{Pole}}\right| = \frac{1}{\text{Gain}} \quad \text{and} \quad \angle\,\overrightarrow{V_{Pole}} = \text{Phase lag}$$

The magnitude (length) of a pole response vector represents the inverse of the *gain*.

The angle represents a phase *lag* (delay).

- Solving the Z-transform for a particular input signal amounts to evaluating the gain and phase contributions from every system pole and zero. Multiplying each of these gain contributions yields the overall system gain G. Adding each of the phase contributions yields the phase shift ϕ that the system applies to the input signal.
- When both the system and the input signal are pure-real functions (complex conjugate poles, zeros, and input sinusoids), a symmetry exists that simplifies the solution to the Z-transform.
- A system that contains an identical number of poles and zeros is balanced. Its present output is generated only from present and past inputs. This means that the system output samples neither lead nor lag the input samples.
- The pole and zero of a pure-real first-order system must be located at pure-real locations on the Z-plane.

- A balanced first-order section has the following difference equation:

$$y(n) = x(n) - \text{Zero}\,1x(n-1) - \text{Pole}\,1y(n-1)$$

- Pure-real second-order sections must contain pairs of conjugate zeros and/or conjugate poles.
- A balanced second-order section has the following difference equation:

$$y(n) = x(n) - 2R_Z \cos(\phi_Z)x(n-1) + R_Z^2 x(n-2)$$
$$+ 2R_P \cos(\phi_P)y(n-1) - R_P^2 y(n-2)$$

- We can assemble a sophisticated system by concatenating any number of first- and second-order sections.
- Positioning system zeros close to a range of input frequencies creates a system that attenuates these frequencies.
- The closer the system poles are located to a range of input frequencies, the more the system will amplify these frequencies.
- A stable system has all of its poles residing *inside* the unit circle.
- Quantizing the weighting factors slightly alters the practical location of the poles and zeros. Because of this, we must ensure that the quantization effect does not move poles on or outside the unit circle.
- When conjugate poles are placed on the unit circle, the system becomes an oscillator. The normalized frequency of oscillation corresponds to the angle that sustains the poles.

PRACTICE QUESTIONS

7-1. A system is defined by the following impulse response:

$$h(0) = 1 \qquad h(1) = 1.456 \qquad h(2) = 0.81$$

(a) Calculate the Z-transform that defines this system.
(b) How many poles and zeros does this system have?
(c) Find the roots of this system and plot the poles and zeros on the Z-plane.

7-2. A system is defined by the following Z-transform:

$$H(Z) = \frac{1 - 0.4Z}{Z}$$

(a) How many poles and zeros does this system have?
(b) Find the difference equation that defines this system.
(c) If we multiply this system by Z^{-1}, what happens to the difference equation?

7-3. A system is defined by the following Z-transform:

$$H(Z) = \frac{(Z - 0.4)(Z + 0.6)}{Z(Z - 0.1)}$$

(a) Is this system FIR or IIR? Explain.

(b) Plot the poles and zeros of this system on the Z-plane.

(c) Give the difference equation that describes this system.

7-4. The operation of a DSP system is defined by the following difference equation:

$$y(n) = x(n) + 0.3x(n-1)$$

(a) Draw the poles and zeros of this system on the Z-plane.

(b) If $f_S = 10$ k-samples/second, find the gain of this system at 0 Hz and 2.5 kHz.

(c) If $f_S = 10$ k-samples/second, find the phase shift of this system at input frequencies of 0 Hz and 2.5 kHz.

7-5. A system has conjugate zeros at $1e^{\pm\frac{\pi}{5}j}$ and conjugate poles at $0.6e^{\pm\frac{3\pi}{5}j}$.

(a) Give the difference equation that describes this system.

(b) What gain does this system apply to a DC input?

(c) What gain does this system apply to a normalized input frequency of $\pi/4$?

(d) What phase shift does this system apply to an input frequency of $f_S/4$?

7-6. A system has a sampling rate of 20 k-samples/second.

(a) At what location would you position system zeros on the Z-plane to remove an input frequency of 4 kHz from the system output?

(b) Give the Z-transform of the system in part (a).

(c) Give the difference equation of the system in part (a).

(d) Add a one-pole equalizer to the system of part (a) and repeat parts (b) and (c).

7-7. A DSP system implements the following difference equation:

$$y(n) = x(n) - 1.622x(n-1) + 0.81x(n-2) - 1.892y(n-1) - 1.1025y(n-2)$$

Is this system stable? Explain why.

7-8. A telephone system uses a sampling rate of 8 k-samples/second. You want to position conjugate zeros *on* the unit circle and conjugate poles at 0.9 from the origin along a normalized angle that corresponds to a target frequency of 60 Hz.

(a) Calculate the value of the normalized target frequency.

(b) Give the value of the b_0, b_1, b_2, a_1, and a_2 coefficients that define this system. Provide at least five significant digits of precision for the coefficients.

(c) Quantize the coefficients into 16-bit Q12 numbers (round down).

(d) Give the Z-transform of the quantized system.

(e) Determine the new quantized location of the poles and zeros by finding the quadratic roots of the numerator and denominator.

(f) Using the location of the roots, determine the normalized frequency along which the poles and zeros are positioned. Are the poles and zeros along the same angle? Are the poles and zeros still aimed at a frequency of 60 Hz?

(g) Repeat for a 16-bit system using Q14 quantization factors.

7-9. A system uses a sampling rate of 44.1 k-samples/second.

(a) Position two poles on the Z-plane to create a 100 Hz oscillator. Give the difference equation of this system.

(b) How precise is the frequency of oscillation if 16-bit Q10 factors are used to implement this system?

7-10. The telephone system uses DTMF to identify the keys that are pressed on a telephone (see Figure 7–31). You want to design a DTMF system that can generate any DTMF tone. You will need a 16-bit system, using Q14 coefficients and sampling at 8 k-samples/second; two distinct oscillator subsystems whose outputs are added to generate one of the 16 DTMF tones; and a table that contains the eight sets of coefficients necessary to generate the DTMF frequencies.

(a) Design a block diagram showing the subsystems and the adder that generates the output.

(b) Give the content of the table that contains the eight sets of coefficients.

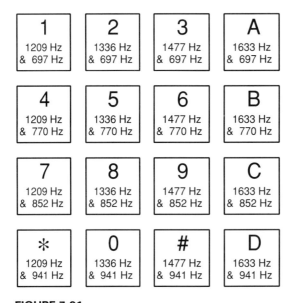

FIGURE 7-31
DTMF tones.

8

DESIGNING DIGITAL FILTERS

This chapter defines the different parameters and characteristics used to design filters. It examines the most popular types of filters used in filtering applications. Each filter type possesses some advantages and exhibits some disadvantages. The positive and negative attributes of each type of filter are outlined to present enough information to choose the filter that is appropriate to satisfy a particular functional specification.

The chapter then discusses some of the most popular techniques used to implement digital filters. In particular, analog filters can be converted into digital filters. The chapter also discusses ways to design filters in the Z-domain by laying special patterns of poles and zeros. Using these techniques, we develop the skills necessary to control the amplitude and/or the phase of signals. The chapter stresses the typical applications and environments in which particular techniques are used.

8.1 DEFINITION OF FILTERS

Filters are systems that allow the frequency response to be shaped. They provide control over both the amplitude and the phase characteristics of the different frequency bands in a signal.

We can use filters in a multitude of applications to emphasize, attenuate, or correct the amplitude and phase characteristics of a signal. Consider the following examples:

- Filters detect the presence of power at particular frequencies such as the DTMF tones used in telephone signaling.
- Filters remove the hum from audio systems.
- Filters extract your favorite radio station frequency band from all the other radio frequencies.
- Filters equalize the gain and phase response of a stereo system.
- Filters process television images to make them sharper.

8.1.1 FILTERS THAT CONTROL THE SIGNAL AMPLITUDE

Controlling the gain over frequency ranges called *bands* allows us to manipulate the amplitude of the different signal frequencies as they travel through a filtering system. We do this by defining the shape of the *gain curve* of the filtering system. This curve describes the way that the filtering system alters the amplitude of frequencies as they travel from the input to the output. The following characteristics define the main parts of a filter system that controls the gain:

- The *passband* defines a range of frequencies that are allowed to travel through the filtering system.
- The *stopband* defines a range of frequencies that are significantly attenuated or blocked by the filtering system.

Since filtering systems are not perfect, the passbands and stopbands are linked through

- The *transition band,* which rolls off the gain from the passband to the stopband.

Figure 8–1 illustrates the different frequency bands making up a gain filter system.

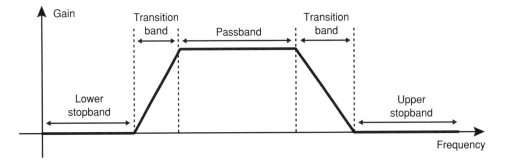

FIGURE 8–1
Filter bands.

Many filtering applications call for a uniformly flat gain in the passbands and in the stopbands. However, achieving a perfectly flat gain response curve is not possible in practice. This means that the gain response curve describing the filter bands exhibits either some *ripple* or some *droop*. Filters using sophisticated designs can reduce the ripple or the droop to small amounts, but some imperfections always remain. Figure 8–2 illustrates the gain fluctuations that are present in the frequency bands of a filter.

It is important to understand that the total signal power is distributed over different frequency bands. The *power* carried by each frequency band defines the relative importance of each of the bands. We can illustrate the power distribution of a signal by drawing the signal power spectrum.

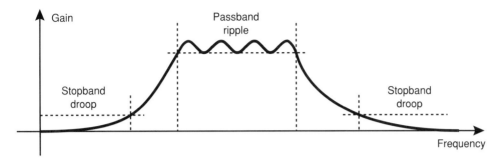

FIGURE 8–2
Filter gain response exhibits fluctuations.

Remember that we define the characteristics of a filter by designing the gain curve, which gives us control over the *amplitude*. This means that we control the power spectrum in an *indirect* way since a signal power is proportional to the square of its amplitude:

$$\text{Power} \propto \text{Amplitude}^2$$

or

$$\text{Amplitude} \propto \sqrt{\text{Power}}$$

For example, *doubling* the amplitude over a particular frequency band *quadruples* the amount of power carried by that band of frequencies.

Since we are mainly interested in controlling the power carried by frequency bands, using power boundaries to delimit the filter bands will be useful. These boundaries are called the *cutoff* frequencies and are usually annotated ω_c.

When we trace power boundaries, the passband defines the frequency range that contains sinusoids that travel through the system while retaining more than 50% of their original input power. The transition band therefore defines the frequency range that contains sinusoids that exit the system with less than 50% of their original power level. Finally, the stopband defines the range of frequencies that carry power levels that we choose to consider negligible in the design context.

The 50% power level is therefore an important reference point that we clearly need to identify on the gain curve. We can express the filter gain response curve by using either linear of logarithmic scales. We calculate the amplitude that corresponds to the 50% power boundary as

$$\sqrt{0.5 \text{ Power}} \cong 0.707 \text{ Amplitude}$$

The 50% boundary therefore corresponds to a level of 0.707 on the gain curve. Note that when the gain is smaller than 1, we refer to it as an *attenuation*. When we use a linear gain scale, a signal that we attenuate to 70.7% of its original amplitude contains one-half of its original power. In the case of systems whose power curve describes large amplitude changes, we find it more convenient to express the power curve using the logarithmic *decibel* (dB) scale. In this case, following the definition of a decibel, a power level change of 50% corresponds to

$$\text{Power in decibel} = 20 \log \left(\frac{\text{Output amplitude}}{\text{Input amplitude}} \right)$$

$$= 20 \log \left(\frac{0.707}{1} \right)$$

$$= -3 \text{ dB}$$

Figure 8–3 illustrates the attenuation that is necessary to reduce a signal power by a factor of 50%.

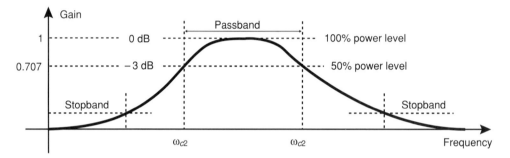

FIGURE 8–3
The 50% power level.

We can classify gain filters using

- The frequency band that they control.
- The types of control (equalizing, passing, or rejecting) they exercise over frequency bands.

Equalizing filters allow a variable adjustment of the gain over frequency bands. Note that the gain is not constant but follows some correction curve. Figure 8–4 illustrates the gain curve of equalizing filters.

Low-pass filters allow the low frequencies that range from 0 Hz (DC) to a –3 dB cutoff frequency (ω_c) to pass through the system. This filter removes the frequency components that are higher than ω_c (high reject).

High-pass filters allow frequencies higher than ω_c to pass through the system. This filter removes the frequency components that are lower than ω_c (low reject). Figure 8–5 illustrates the gain curve of low-pass and high-pass filters.

Band-pass filters allow a range of frequencies from ω_{c1} to ω_{c2} to pass through the system. This filter removes the frequency components that lie outside the selected range.

Band-reject filters prevent a range of frequencies from ω_{c1} to ω_{c2} from passing through the system. This filter allows only the frequency components that lie outside the selected bands to pass through the system. Figure 8–6 illustrates the gain curve of band filters.

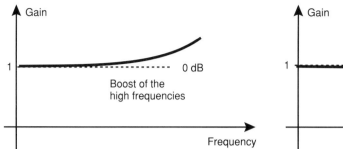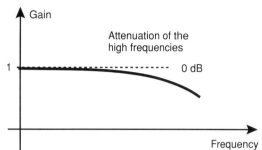

FIGURE 8–4
Gain curve of equalizing filters.

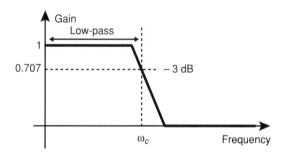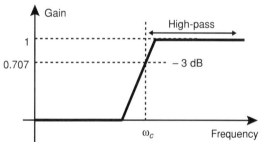

FIGURE 8–5
Gain curve of low-pass and high-pass filters.

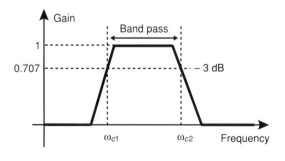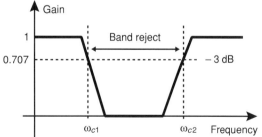

FIGURE 8–6
Gain curve of band filters.

8.1.2 FILTERS THAT CONTROL THE PHASE

When sinusoids travel through a filtering system, each suffers some inevitable time delay. The delay that each sinusoid suffers depends on that sinusoid's frequency. Since sinusoids are periodic, this time delay has the effect of shifting the phase of each of the sinusoids. Figure 8–7 illustrates the relationship between a system time delay and the phase shift of a 1 kHz sinusoid.

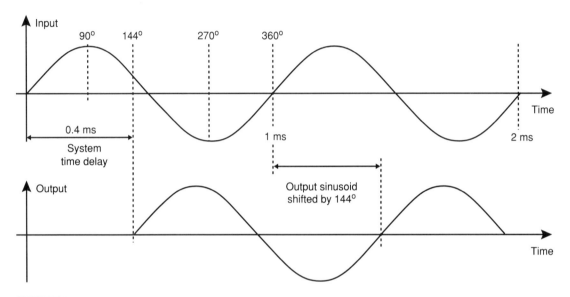

FIGURE 8–7
Time delay corresponds to a phase shift.

Something special happens when *all* the sinusoids in a frequency band suffer an identical time delay. Figure 8–8 illustrates three different sinusoids that suffer the exact same time delay going through a system. Notice that the phase shift in each of the three cases on Figure 8–8 is directly proportional to the sinusoid frequency. If the frequency doubles, the phase shift doubles. If the frequency triples, the phase shift triples.

A frequency band has *linear-phase* shift characteristics when the time delay is constant for all the sinusoids in that band.

Nonlinear-phase shifts produce different system time delays for sinusoids of different frequencies. Figure 8–9 illustrates the difference between linear- and nonlinear-phase response curves. Note that the linear-phase shift graph in Figure 8–9 corresponds to a straight line. This means that we can calculate the amount of phase shift that is applied to any sinusoid by multiplying the sinusoid frequency by a constant. That constant is equal to the slope of the phase response line:

$$Phase\ shift = Constant \times Frequency$$

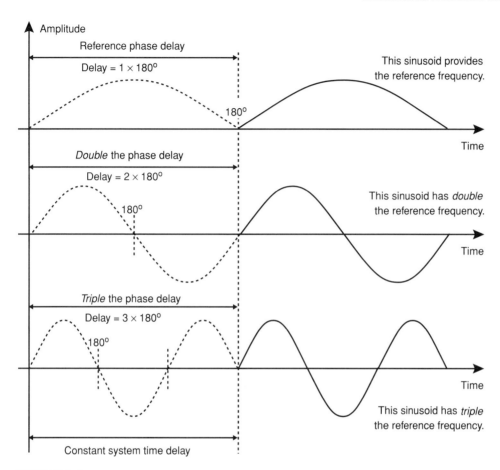

FIGURE 8–8
Constant time delay.

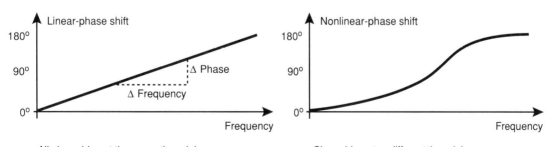

All sinusoids get the *same* time delay.

Sinusoids get a *different* time delay.

FIGURE 8–9
Linear- versus nonlinear-phase delay.

where

$$\text{Constant} = \frac{\Delta \ phase}{\Delta \ frequency}$$

A filter that applies linear-phase shifts to sinusoids produces phase shifts that are directly proportional to the frequency of the sinusoids. Filters that exhibit a linear-phase response over a frequency band maintain the phase relationship of all sinusoids in a band. This phase relationship is desirable and even essential for many applications. Maintaining the phase relationship means that the system will not change the *shape* of the overall signal. For example, preserving the quality of the music is important in audio applications. To illustrate this, Figure 8–10 shows what happens when the system alters the phase of the sinusoids in a signal.

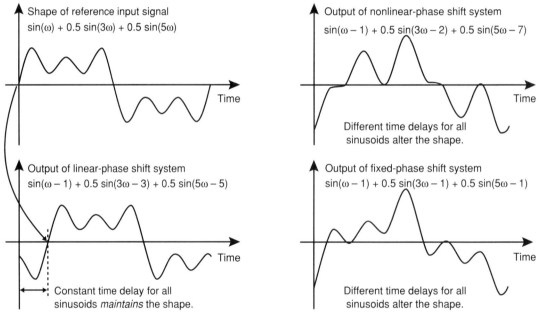

FIGURE 8–10
Effect of changing the phase.

Notice that only the bottom left case in Figure 8–10 results in the signal frequencies being delayed by the same amount of time. Also note that this is the only case for which the phase delay for all three components is directly proportional to the frequency of the individual sinusoid components. In other cases, when the phase shift is not linear, the system alters the shape of the signal as is illustrated in the two cases depicted on the right side of Figure 8–10.

When the phase shift is linear over a particular frequency band, all sinusoids in that band sustain an identical *time* delay as they travel through the system. The frequencies in the band are said to have a constant *group* (or *envelope*) delay.

8.1.3 THE PROCESS OF COMBINING FILTERS

To simplify the design of a complete filtering system, we can break the overall response into a number of filtering *sections*. We then can design each filter section separately. This simplification is possible because we are dealing with *linear circuits,* which allow us to concatenate a number of separate sections to produce the overall desired response. Figure 8–11 illustrates the combination of two sections, a low pass and a band reject, to design a special filter.

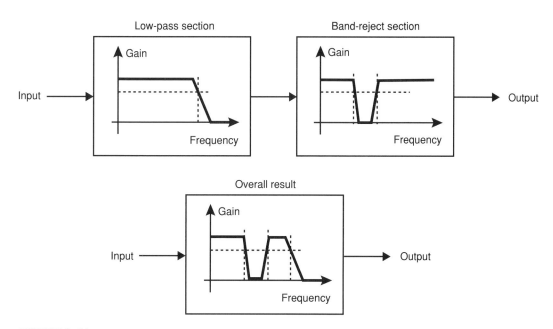

FIGURE 8–11
The process of combining filter sections.

In fact, when we learn to implement DSP systems in Chapter 9, we break them down even further by segmenting filter sections into subsections, each implementing a part of the overall filtering required by the section.

8.2 FILTERS USED IN DIGITAL SIGNAL PROCESSING SYSTEMS

A signal consisting of discrete-time samples corresponds, in the frequency domain, to a spectrum that infinitely repeats itself around multiples of f_S (refer to the section in Chapter 3 on Investigating Periodicity in the Frequency Domain). A DSP system is fully described by a discrete-time impulse response $h(n)$ (Section 4.3), and we developed the discrete-time Fourier transform to examine the frequency domain equivalent of $h(n)$ (Section 6.2.1). At that time, we observed that the discrete time $h(n)$ corresponds, in the frequency domain, to a system response that infinitely repeats itself around multiples of f_S. In fact, it is a characteristic of all discrete-time representations to correspond, in the frequency domain, to a pattern that repeats itself at infinity around multiples of f_S. Since digital filters are discrete-time systems, this means that they possess frequency characteristics that repeat themselves, at infinity, around multiples of f_S.

Figure 8–12 illustrates the periodicity of a filter frequency response. The useful range of frequencies in Figure 8–12 is limited from zero to $f_S/2$. If the system you are working on has an output reconstruction filter, you need not worry about any frequencies above $f_S/2$. On the other hand, if your system cannot accommodate a reconstruction filter, you must ensure that the periodicity of the frequency response does not interfere with the performance of your system. Remember that increasing the sampling rate f_S pushes the replicates to higher frequencies where they may not be such a nuisance.

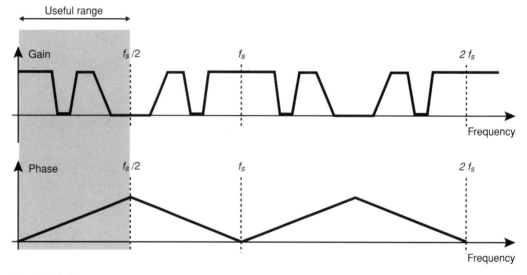

FIGURE 8–12
Digital filters have a periodic frequency response.

We implement digital filters by programming a processor to compute a difference equation. The fact that the difference equation uses time-invariant numerical coefficients (weighting factors) means that the filter maintains *time-invariant performance*. The unvarying characteristics of digital circuits allow us to design filters whose performance is unequaled in the analog world. Digital filters provide superior accuracy, band selectivity, and stability of operation. The digital filter design techniques provide almost unlimited control over the gain and phase response curves. The software tools available allow us to quickly design and simulate filters and, when we practically implement these filters, their performance closely matches the theoretical simulations!

We can use a variety of techniques to design digital filters. The following sections describe the most prevalent filter design techniques.

8.2.1   CONVERTING ANALOG FILTERS INTO DIGITAL FILTERS

Electronics designers have been using filters for a long time in the continuous-time domain. Over the years, researchers have developed approaches to the design of analog filters, and these have been standardized into classes that include the familiar Bessel, Butterworth, Chebyshev, and elliptic filters.

It is possible to transform analog filters into the discrete-time domain through a mathematical process called the *bilinear transformation*. This transformation preserves the gain characteristics and even improves the cutoff characteristics of the transition band. Unfortunately, the transformation has the negative effect of completely distorting the phase response of the analog filters.

**Converting an analog filter into a digital filter
ruins the phase response of the filter.**

Because of this, we should use the bilinear transformation only when the phase characteristics are not important or when the resulting phase distortion stays within reasonable limits.

Developing the bilinear transformation is beyond the scope of this text, but, fortunately, many commercial software packages are available to automatically perform the mathematics of the transformation. When using one of these software packages, you are prompted to specify the filter class (Butterworth, Chebyshev, etc.), the type of filter (low pass, high pass, etc.), and the amount of acceptable ripple. The software then outputs the difference equation coefficients (quantized or not), which are necessary to implement the filter on a digital signal processor. Most filter design packages also allow you to examine the pole/zero plot, the impulse response, and a number of other graphs that will assist you in determining whether the design meets your requirements.

The following section provides the basic information and the terminology necessary to design IIR filters using software that implements the bilinear transformation.

8.2.2   USING STANDARD GAIN-CONTROL FILTERS

When designing gain-control filters, we must define the requirements for the following three parameters:

1. *Sharpness.* This defines the frequency range of the transition band. Sharp filters implement very narrow transition bands. Generally, for a given class of filters, sharpness increases with the order of the filter. As we will see, each filter class has advantages over the others, and it follows that some classes of filters are better than others at achieving sharp cutoff rates.

2. *Flatness.* This is a measure of the gain fluctuations in the passband and stopband. Ideally, the gain should be perfectly constant in these bands. However, achieving a perfectly horizontal gain response curve is not possible in practice. The flatness of the response curve depends on the class of filter and on the order that the designer selects.

3. *Overshoot.* When the input signal changes, the filter system needs time to react and adapt to this change. During this adaptation period, transients appear at the output of the filter. The filter output tends to overshoot the desired values during this interval. The standard way to evaluate the amount of overshoot is to apply the sharpest possible change at the input of the filter and to record the resulting transient at the output. To do this, we apply a *step function,* and the filter responds by outputting a *step response.* Figure 8–13 illustrates a step function and a typical step response.

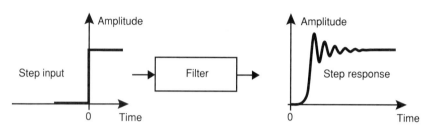

FIGURE 8–13
Step response.

Butterworth Filters

The Butterworth class of filters provides a gain response curve that is maximally flat both in the passband and in the stopband. The cutoff is moderately sharp when compared to other classes of filters. The Butterworth response is ideal in applications requiring very precise gain control since there is no ripple on the gain-response curve. Increasing the order of the filter enhances the sharpness of the transition band. Figure 8–14 illustrates the gain response of Butterworth filters of different orders. The step response yields moderate overshoot, which *increases* with the filter order. Consequently, sharper cutoff is attained at the cost of increased overshoot in the step response.

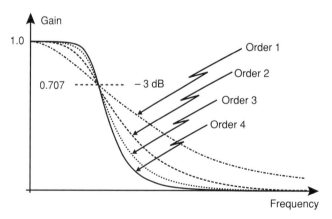

FIGURE 8.14
Butterworth gain characteristics.

Chebyshev Filters

The Chebyshev class of filters provides a sharper cutoff than Butterworth filters of the same order. This sharper cutoff is achieved at the expense of ripples of uniform amplitude (*equiripple*) that appear either in the passband *or* in the stopband. Because of this, there are two types of Chebyshev filters. *Type 1* has equiripple in the passband, and *type 2* has equiripple in the stopband. Applications that call for sharp transitions while tolerating some gain fluctuations either in the passband or in the stopband can be satisfied using Chebyshev filters.

A comparison of filters of the same order indicates that the step response yields a little more overshoot than the Butterworth class. Figure 8–15 illustrates the appearance of typical gain response curves for both types of low-pass Chebyshev filters.

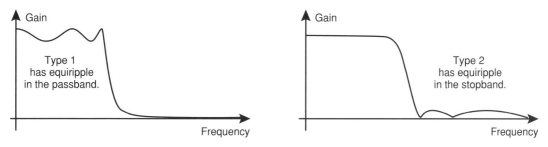

FIGURE 8–15
Low-pass Chebyshev gain characteristics.

Chebyshev filters are defined in terms of the amount of ripple and the sharpness of their transition band. Sharper cutoff and/or lower ripple is achieved by increasing the order of the filter. Again here, increasing the order of the filter increases the amount of overshoot in the step response.

Cauer-Elliptic Filters

The Cauer-elliptic class of filters provides the sharpest cutoff of all classes of filters. This sharper cutoff is achieved at the expense of equiripple in both the passband *and* the stopband. The sharp transition is great to satisfy applications that call for separating frequency bands that are very close to each other. Unfortunately, the application must also be able to tolerate a certain amount of gain ripple. The Cauer-elliptic filters introduce extreme phase distortions and should be avoided in any applications that call for a controlled phase response. Figure 8–16 illustrates the gain response of a low-pass Cauer-elliptic filter.

Bessel Filters

In the continuous time domain, Bessel filters are characterized by a frequency response that exhibits a *linear-phase* response in the passband. These filters therefore produce constant group (envelope) delay. When the input signal has sharp edges (like the steps outputted by a DAC), minimal overshoot results at the output. This means that that there is no

FIGURE 8–16
Low-pass Cauer-elliptic gain characteristics.

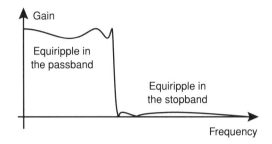

waveform ringing at the output of a Bessel filter. The roll-off in the transition band is unfortunately not as sharp as that of the Butterworth.

Unfortunately, if we use the bilinear transformation to convert Bessel filters to the discrete-time domain, the transformation ruins the linear-phase response. Because of this, they lose their main advantage; consequently, Bessel filters are never used as digital filters. They make a great antialiasing and/or analog reconstruction filter, however, when the phase relationship of the signal frequencies must be preserved. Their minimal overshoot completely smooths the sharp transition steps produced by the DAC.

Figure 8–17 superposes an example from each of the filter classes onto a single gain graph so that the gain characteristics of the three classes of filters worth converting to the discrete-time domain can be compared.

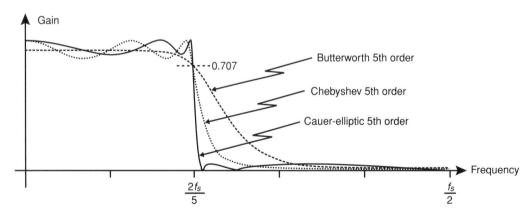

FIGURE 8–17
Comparison of filter classes.

8.2.3 DESIGNING DIGITAL FILTERS WITH LINEAR-PHASE RESPONSE

Filters that possess linear-phase characteristics maintain the phase relationship of the frequencies that are contained in a signal (refer to Section 8.1.2). Linear-phase characteristics mean that the filter maintains the overall shape of the signal as it travels from the input to the output.

Although the proof is outside the scope of this text, it can be shown that a linear-phase response is possible only when all the system's poles are located at the origin. We can only design such systems using zeros; therefore, linear-phase filters are always FIR systems.

Linear-phase filters are always pure FIR systems.

This does not mean that all FIR systems deliver linear-phase characteristics. The linear-phase response may be achieved only when the system zeros are positioned according to special patterns. The good news is that systems that use only zeros always yield stable operation. The bad news is that pure FIR systems are not as efficient as IIR systems. This means that achieving sharp cutoffs typically requires the processor to perform a large number of MAC instructions. In digital systems, the number of zeros (or poles) represents the *order* of the filter. Note that contrary to analog systems, the concept of *order* in discrete-time systems does not necessarily bear relationship to its selectivity. In digital filter systems, some low-order filters may end up being very selective. For an example, refer to the nonlinear-phase notch filters described later in Section 8.2.5.

Another attribute of linear-phase filters is that the coefficients of the difference equation always display symmetrical or antisymmetrical values. This means that plotting the value of the coefficients always yields two halves that are mirror images of each other. Figure 8–18 illustrates this symmetry.

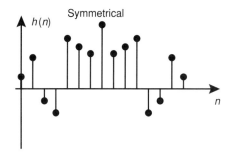
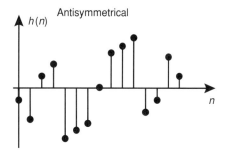

FIGURE 8–18
Coefficient symmetry of linear-phase filters.

Many programming algorithms take advantage of the coefficient symmetry to

- Reduce the memory used to store the coefficients.
- Reduce the high number of MAC instructions required to implement the difference equation of the linear-phase FIR filter.

The next few sections discuss some of the most popular techniques used in practice to design linear-phase filters.

Positioning Zeros for Linear-Phase Response

Researchers have discovered that a real and linear-phased response results only when zeros are positioned according to established patterns. We know that a pure-real response results when the zeros occur either on the real axis or as complex conjugate pairs. To obtain a linear-phase response, all zeros positioned *inside* the unit circle at a distance of R from the origin must be matched by an image zero located *outside* the unit circle, along the same angle at a distance of $1/R$.

The following conditions are therefore necessary to obtain a pure-real linear-phase response. In the case of zeros positioned *along the real axis*

$$Z_1 = R \qquad Z_2 = \frac{1}{R}$$

In the case of zeros positioned *at complex locations*

$$Z_1 = Re^{j\Omega} \quad Z_2 = \frac{1}{R}e^{j\Omega} \quad Z_3 = \overline{Z_1} \quad Z_4 = \overline{Z_2}$$

Note that when the zero is located exactly on the unit circle, $R = 1$, there is no need for the image zero at $1/R$. Figure 8–19 illustrates the possible patterns that yield a system response that is pure real and linear phased. As this figure indicates, the zeros can occur

- As *quads* (A) when they are complex and *not* on the circle.
- As *pairs* when they are real (B) or on the circle (C).
- As *singles* at two special locations when they are both real *and* on the circle (D or E).

It is possible to design linear-phase filters using trial and error by placing zeros according to these basic patterns. There are more efficient ways to design linear-phase filters, however, and the next few sections describe the most popular approaches.

Laying Out the Comb Filter

A special case of practical interest exists when we position a number of zeros *exactly on* the unit circle at positions that are equidistant from each other. In this case, the system exhibits linear-phase response, and the difference equation becomes extremely efficient to implement.

Figure 8–20 illustrates two examples of such a pattern of zeros. Positioning zeros exactly on the unit circle eliminates the normalized frequencies located under the zeros. The resulting gain response resembles the teeth of a comb, and consequently this system is called a *comb filter*. Figure 8–21 illustrates the gain response that results from the pattern of zeros illustrated on Figure 8–20.

The following Z-transform defines a comb filter:

$$H(Z) = \frac{Y(Z)}{X(Z)} = \frac{1}{M} \times \frac{(Z - \text{Zero}1)(Z - \text{Zero}2)(Z - \text{Zero}3)\ldots(Z - \text{Zero}M)}{Z^M}$$

where M is the order of the filter.

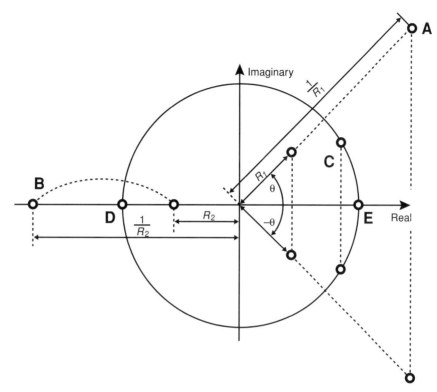

FIGURE 8–19
Zeros positioned for linear-phase response.

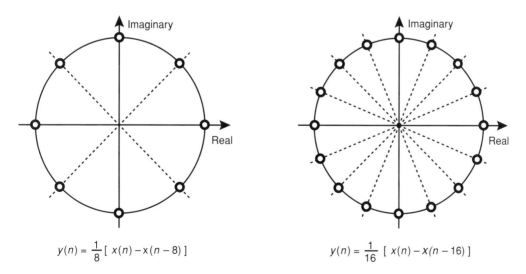

$$y(n) = \frac{1}{8} [\, x(n) - x(n-8) \,]$$

$$y(n) = \frac{1}{16} [\, x(n) - x(n-16) \,]$$

FIGURE 8–20
Equidistant zeros on the unit circle.

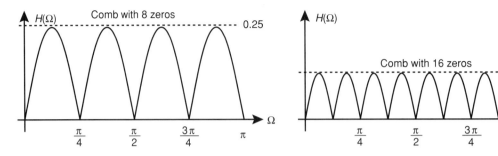

FIGURE 8–21
Gain response of comb filters.

Because of the symmetry of the zero positions, it can be shown that the Z-transform simplifies to

$$H(Z) = \frac{Y(Z)}{X(Z)} = \frac{1}{M} \times \left(1 - Z^{-M}\right)$$

Translating this into the discrete-time domain yields the following difference equation:

$$y(n) = \frac{1}{M}\left[x(n) - x(n - M)\right]$$

The difference equation consists of only two terms. The weighting factors have a value of 1 for both terms, and therefore there is no need to multiply. If M is chosen as a power of 2, the processor can implement the $1/M$ scaling using just a few shift-right instructions. For example, when $M = 8$, the $1/M$ scaling is implemented with three shift-right instructions. Because of its simplicity, any microprocessor can efficiently and easily implement this comb filter.

It is possible to modify comb filters so that they become low-pass, band-pass, or high-pass filters using an unusual approach. This is a "play-with-fire" approach because it involves canceling some of the comb filter zeros by overlaying them with poles. This type of system is a freak of the digital world because it creates an FIR system using poles that are located on the unit circle. Normally, this would make the system into an oscillator, but in this case the system mathematically remains an FIR system since each pole is exactly canceled by the zero it overlays. The process is delicate and dangerous since the pole and the zero must exactly overlay each other. If the computational process is not exact, the system turns into an oscillator. If we manage to position the pole/zero pair with perfect computational accuracy, the system behaves as if the pole and the zero completely vanish. This is possible in practice only when the coefficients that correspond to the pole(s) are kept to the exact value of 1.

This becomes possible in only three cases:

Pole at e^{j0}

$$H(Z) = \frac{1}{M} \times \frac{Z^M - 1}{(Z^{M-1})(Z - 1)}$$

$$H(Z) = \frac{1}{M} \times \frac{1 - Z^{-M}}{1 - Z^{-1}}$$

Pole at $e^{j\pi}$

$$H(Z) = \frac{1}{M} \times \frac{Z^M - 1}{(Z^{M-1})(Z + 1)}$$

$$H(Z) = \frac{1}{M} \times \frac{1 - Z^{-M}}{1 + Z^{-1}}$$

$$\text{Poles at } e^{\pm\frac{\pi}{2}j}$$

$$H(Z) = \frac{1}{M} \times \frac{Z^M - 1}{Z^{M-2}(Z - j)(Z + j)}$$

$$H(Z) = \frac{1}{M} \times \frac{1 - Z^{-M}}{1 + Z^{-2}}$$

In the discrete-time domain, this corresponds to the following difference equations:

Comb with pole at e^{j0}: $y(n) = \frac{1}{M} \times [x(n) - x(n - M)] + y(n - 1)$

Comb with pole at $e^{j\pi}$: $y(n) = \frac{1}{M} \times [x(n) - x(n - M)] - y(n - 1)$

Comb with pole at $e^{\pm\frac{\pi}{2}j}$: $y(n) = \frac{1}{M} \times [x(n) - x(n - M)] - y(n - 2)$

Figure 8–22 illustrates the gain curve of an 8-zero comb filter for each of the three cases of zero cancellation. The attenuation in Figure 8–22 is not considerable, with about –9 dB separating the passband from the stopband. Increasing the number of zeros in the basic comb improves this performance somewhat. For example, a 16-zero low-pass comb filter increases the separation to about –10 dB, but the normalized cutoff frequency is lowered to 0.055π.

This technique is elegant because of its simplicity—not its performance. Its advantage lies in the ease with which it can be programmed using a small number of instructions on almost any type of processor. The only way to adjust the absolute cutoff frequency is to change the sampling rate of the ADC and DAC.

For example, we next create a low-pass filter that has a cutoff frequency of 1000 Hz using an 8-zero comb filter. The low-pass configuration has a normalized cutoff frequency of 0.11π, which corresponds to a frequency of $0.11 \times f_S/2$ (refer to Figure 8–22). In this case, 0.11π must correspond to a frequency of 1000 Hz and therefore

$$f_S = 2 \times \frac{1000}{0.11} = 18{,}182 \text{ Hz}$$

Linear-phase filters that use a comb pattern are used in systems that do not contain a multiplication unit. For example, many microcontrollers do not contain a multiplier but may still require the implementation of crude filtering algorithms. Because low-pass, band-pass, and high-pass comb filters contain one or more poles that are located on the unit circle, they are inherently unstable systems. When we implement such filters, the FIR and IIR sections consist of separate program sections. Because of this, a small round-off error may be introduced as the program moves from one section to another. Since the pole is located on the circle, the system does not damp this error, and it may accumulate to the point that the system becomes unstable after operating over some interval of time. Consequently, this filtering algorithm should be used only for short intervals of time in systems that do not perform operations whose occasional failure could lead to disasters.

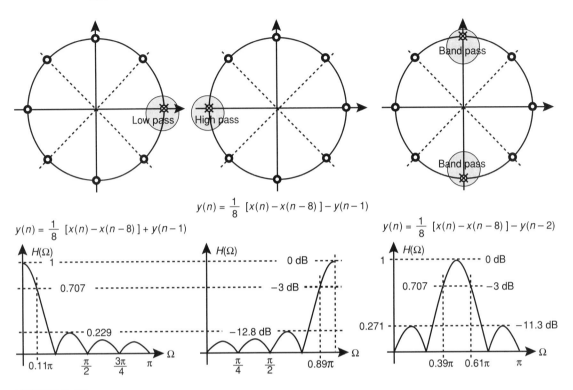

FIGURE 8–22
Cancellation of zeros with poles.

For applications that require complete stability and superior performance, other techniques exist to design linear-phase filters. The next few sections describe the most popular approaches.

Shaping the Reponse with the Windowing Technique

Another approach to design linear-phased filters is to start the design in the frequency domain, where we define the ideal filter characteristics. Figure 8–23 illustrates the ideal frequency gain characteristics of a linear-phase low-pass filter with a normalized cutoff frequency of $\Omega_c = \pi/5$.

Applying the discrete-time Fourier transform to the impulse response of a system, over a range of frequencies, yields the frequency response curve of that system (refer to Section 6.3.2). We know that the frequency response curve is periodic. An inverse to the discrete-time Fourier transform allows us to calculate the impulse response of a system from one period of its frequency response curve. The details of obtaining this transform is beyond the scope of this book; trust that the following mathematical relationship links the frequency response $H(\Omega)$ of a system to its impulse response $h(n)$:

$$h(n) = \frac{1}{2\pi} \int_{-\pi}^{\pi} H(\Omega)\, e^{j\Omega t} d\Omega$$

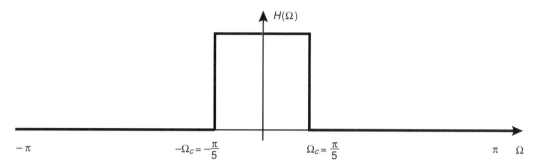

FIGURE 8–23
Ideal frequency response of a low-pass filter.

A number of convenient software packages allow us to calculate the inverse Fourier transform. Figure 8–24 illustrates the inverse Fourier transform of the low-pass filter illustrated in Figure 8–23.

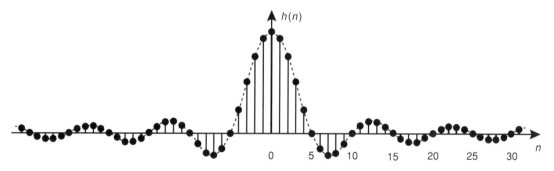

FIGURE 8–24
Inverse Fourier transform is infinite in duration.

Unfortunately, $h(n)$ is infinite and noncausal. Because of this, we cannot implement it as such. Notice, however, that the weighting factors have a tendency to become smaller in value as n increases. It is possible to *approximate* the response of this system by truncating the infinite impulse response at some point. This truncation yields a finite number of weighting factors and, consequently, an FIR system. Figure 8–25 illustrates the impulse response, which has been arbitrarily truncated to 45 terms. The impulse response terms residing outside the rectangular window in Figure 8–25 have been removed. We consequently call this technique *windowing*.

Once the impulse response is truncated to a finite number of terms, it becomes possible to implement it as an FIR system. Making the system causal is simply a matter of adding the appropriate number of poles at the origin. For example, the 45-term impulse response of Figure 8–25 corresponds to a system that has 44 zeros; therefore, adding 44 poles at the origin makes it causal. Figure 8–26 illustrates the impulse response of the resulting causal system.

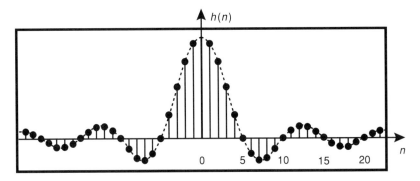

FIGURE 8–25
Truncated inverse Fourier transform.

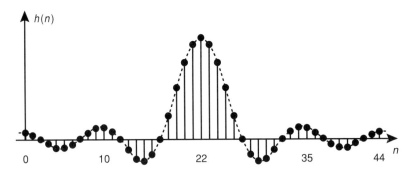

FIGURE 8–26
Causal truncated inverse Fourier transform.

The truncation of the impulse response has the following effects on the actual frequency response:

- Ringing in the passband and in the stopband.
- A transition-band that is not as sharp as the original ideal response.

Figure 8–27 illustrates the frequency response of systems implemented with impulse responses truncated at 45 and 89 coefficients, respectively. As the figure indicates, an increased number of coefficients improves the sharpness of the transition band; unfortunately, the amplitude of the ringing remains the same. This means that the stopband for this type of filter has a fixed attenuation of –21 dB.

The ringing results from the rectangular window that we used to *abruptly* truncate the infinite impulse response. Sharp edges in the time domain correspond to a high-frequency content in the frequency domain. The only way to lower the amount of ringing is to truncate the impulse response using a window that does not cut off as sharply as a rectangular window. These windows gradually reduce the value of the impulse response terms following the shape of the window function.

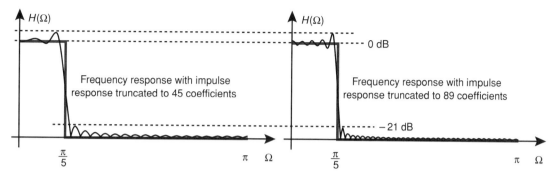

FIGURE 8–27
Frequency response for truncated impulse response.

Over the years, researchers have developed a number truncation window functions that reduce the amount of ringing. These functions allow a deeper stopband attenuation, but the drawback is that it is achieved at the cost of a longer transition band. Figure 8–28 illustrates the shape of some of the most popular truncation windows.

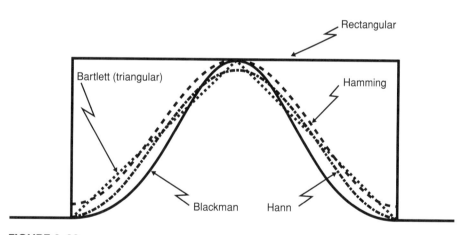

FIGURE 8–28
Truncation windows.

Table 8–1 lists the performance of some of the windows illustrated in Figure 8–28. Note that the Kaiser window in Table 8–1 is a custom type of window that changes its shape when parameters such as stopband attenuation and transition bandwidth are defined. For a given stopband attenuation, the Kaiser window comes close to producing an optimal truncation window.

Numerous software applications are designed to perform all windowing calculations. If you can obtain some of these applications, experiment and see which window satisfies the requirements of the application you have in mind.

TABLE 8–1
Performance of Different Windows

Type of Window	Approximate Attenuation	Normalized Transition Bandwidth
Rectangular	–21 dB	1 (the sharpest possible)
Hann	–44 dB	3.4
Hamming	–53 dB	3.7
Blackman	–74 dB	6.1
Kaiser (custom)	Can be adjusted	Depends on filter parameters

Windowing may be used to create low-pass, band-pass, and/or high-pass filters. Note that because of a mathematical oddity, you should truncate to an *odd number* of coefficients when designing high-pass filters using windows.

Introducing the Parks–McClellan Program

In the early 1970s, Parks and McClellan wrote a computer program that could generate the coefficients for a linear-phase *equiripple* digital filter. Their program uses the *Remez-exchange* algorithm, and, for that reason, it is sometimes referred to by that name. Note that the windowing technique in the preceding section is still preferred when very smooth passbands and stopbands are required. Running the Parks–McClellan program generates a variety of filter types. Its excellent results often make it the best choice to design linear-phase filters. Many versions of the program for different computers are available. Using this program, you can control the transition band cutoff rate, the amount of gain, and the amount of ripple in each of the *individual* bands.

Most versions of the program prompt you to enter the following parameters:

- *NFILT (integer).* This parameter specifies the order of the filter. This defines the number of coefficients in the difference equation. For mathematical reasons that exceed the scope of this book, high-pass filters require NFILT to be an *odd* number; for low-pass filters it does not matter. For example, it is impossible to create a high-pass filter if NFILT = 40.

- *JTYPE.* This entry defines the type of filter. You are likely to select low-pass, high-pass, or band-pass filters. Note that the Parks–McClellan program can generate other filter types such as differentiators and Hilbert transform filters, which are outside the scope of this textbook.

- *NBANDS (integer).* This specifies how many bands are implemented by the filter. For example, a low-pass filter would have two bands: the passband and the stopband.

- *LGRID (integer).* This value is used to define the "grid" density used to interpolate the error function. A value of 16 usually yields good results.

- *EDGES.* This defines the upper and lower frequencies for every transition band. The frequency is usually defined in terms of the sampling frequency f_S where the value of 0.5 defines the maximum possible frequency $f_S/2$.

- *FX.* These entries specify the filter gain in each band in terms of the input amplitude. For example, a value of 0.5 for a band means that the frequencies in that band will lose half of their amplitude when the travel through the filter.

- *WTX.* These values define the amount of ripple allowed in each band relative to the other bands.

8.2.4 IMPLEMENTING PHASE EQUALIZERS (ALL-PASS FILTERS)

When a signal travels through transmission channels, the electrical characteristics of the channels often change the phase relationship of the signal frequencies. To restore the phase relationship, it is necessary to readjust the phase of the individual signal frequencies. The circuit that achieves this phase correction must not change the amplitude of the frequencies; we therefore refer to it as an *all-pass* filter.

A special pattern of poles and zeros achieves the required flat-gain response. This pattern consists of a combination of conjugate poles located inside the unit circle at a distance R from the origin and conjugate zeros that are located along the same angle as the poles but located outside the unit circle at a distance of $1/R$:

$$P_1 = R\, e^{j\Omega} \qquad P_2 = \overline{P_1} \qquad Z_1 = \frac{1}{R}\, e^{j\Omega} \qquad Z_2 = \overline{Z_1}$$

Figure 8–29 is an example of an all-pass filter.

By changing the distance R and the angle Ω that define the all-pass pattern, we can adjust the phase curve while leaving the gain curve flat. By designing a system that uses the required number of such patterns, we can equalize the phase of almost any system.

8.2.5 BUILDING NOTCH FILTERS

Notch-type filters are used to eliminate or select a short frequency band. The technique works by locating conjugate poles and zeros in close proximity. We know that the system

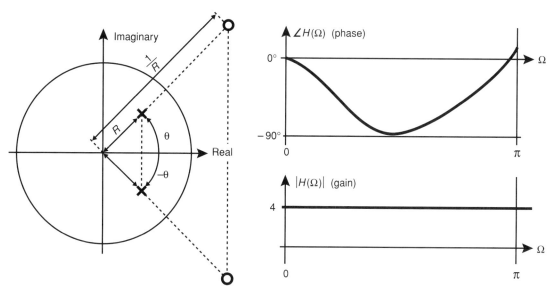

FIGURE 8–29
Pole/zero pattern for an all-pass filter.

zeros have the effect of attenuating any input frequency that occurs close to them. Poles have the exact opposite effect since they amplify the input frequencies that occur close to them.

Each pole and each zero result in a response vector in the system Z-transform. We now consider the Z-transform response when the system consists of a pole and a zero located close to each other:

$$H(Z) = \frac{\overrightarrow{V_Z}}{\overrightarrow{V_P}}$$

Since the pole and the zero are very close to each other, the response vectors V_Z and V_P are of similar length for most of the frequencies that travel through the system. Figure 8–30 illustrates the length of the pole and zero response vectors for two different input frequencies.

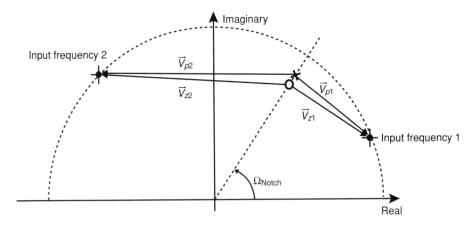

FIGURE 8–30
Response vectors for a close pole and zero pair.

As the Z-transform indicates, the ratio of the length (magnitude) of the two response vectors determines the system gain. As long as we locate the pole and the zero close to each other, the two response vectors are of similar length. Consequently, for most input frequencies, the system gain is close to the value of 1:

$$\left|\overrightarrow{V_Z}\right| \cong \left|\overrightarrow{V_P}\right|$$

therefore,

$$\left|H(Z)\right| = \frac{\left|\overrightarrow{V_Z}\right|}{\left|\overrightarrow{V_P}\right|} \cong 1$$

However, in an important short range of input frequencies, the gain fluctuates considerably. This range spans the input frequencies located in the immediate vicinity of the pole-zero pair. The next few sections analyze the system behavior for that short range of frequencies.

Notch-Stop Filter

In some applications, it is necessary to remove a very short range of frequencies that may be interfering with the signal we are processing. Ideally, we would like to design a filter that removes a very short range of frequencies without disturbing the others.

We now consider a system that contains a close pole-zero pair where the zero is located *exactly* on the unit circle. In this case, to preserve stability, the associated pole is located below the zero just *inside* the circle. Figure 8–31 illustrates this close pole-zero pair.

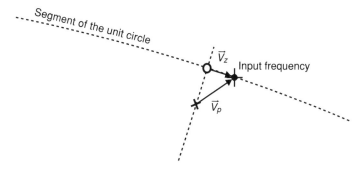

FIGURE 8–31
Notch-stop pole-zero pair.

As the input frequency comes closer to the zero, V_Z gets closer to a length of zero. However, the pole response vector V_P will never reach a length of zero. When the input frequency exactly coincides with the position of the zero, the numerator response vector decreases to zero length. At that point, the system gain reaches zero, and the filter eliminates the input frequency.

At input frequencies *close* to the pole-zero pair

$$\left|\vec{V_P}\right| \gg \left|\vec{V_Z}\right|$$

therefore,

$$|H(Z)| = \frac{\left|\vec{V_Z}\right|}{\left|\vec{V_P}\right|} \ll 1$$

At the input frequency *exactly* overlaping the position of the zero

$$|H(Z)| = \frac{\left|\vec{V_Z}\right|}{\left|\vec{V_P}\right|} = \frac{0}{\left|\vec{V_P}\right|} = 0$$

Figure 8–32 illustrates the gain response of a typical notch-stop filter.

In practice, for the system to exhibit a *real* impulse response, every pole or zero located at a complex location must be matched by its counterpart located at the conjugate location. This means that locating a pole-zero pair at complex locations requires a conjugate pair. A practical notch-stop filter therefore consists of two poles and two zeros. Figure 8–33 illustrates the pole-zero pairs for a practical notch-stop filter.

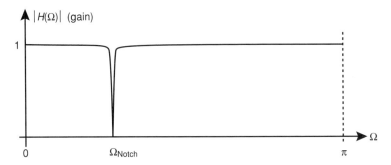

FIGURE 8–32
Notch-stop system gain response.

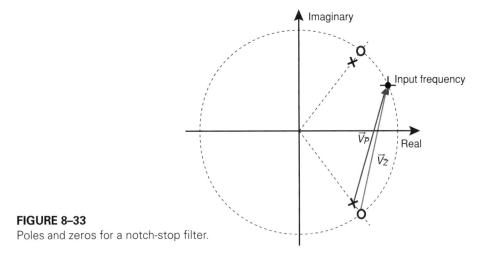

FIGURE 8–33
Poles and zeros for a notch-stop filter.

Notice the response vectors corresponding to the conjugate pole-zero pair in Figure 8–33. The length of both response vectors is approximately the same for all input frequencies; consequently, the presence of this pair hardly changes the system gain:

$$|H(Z)| = \frac{|V_Z|}{|V_P|} \times \frac{|V_{\bar{Z}}|}{|V_{\bar{P}}|} = \frac{|V_Z|}{|V_P|} \times (\sim 1)$$

$$\cong \frac{|V_Z|}{|V_P|}$$

To get the notch effect, we must ensure that the pole and the zero making up the pair are close enough to each other. How close is that? Where does the notch effect stop? First, to obtain a significant effect, we must design the pair *on* or very *close* to the unit circle. As the pair moves away from the unit circle vicinity, the system no longer results in a gain of zero, and the notch effect quickly disappears.

Second, we must consider the distance separating the pole from the zero. Increasing the distance opens the V-shape of the notch. This has the consequence of making the notch filter less selective, and less sharp. Figure 8–34 illustrates the effect of moving the pole-zero pair away from the unit circle and the pole-zero components away from each other.

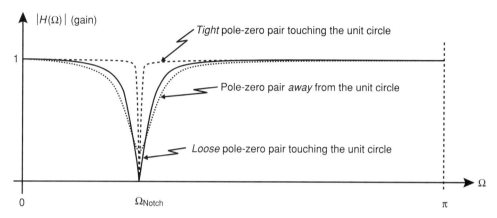

FIGURE 8–34
Change in the relative position of the pole-zero pair.

However, the precision with which the processor handles numbers imposes a limit as to how close we can position the pole and the zero components of the pair. The quantization error actually limits the positions at which we can locate poles and zeros on the Z-plane. As we move the pair components closer together, they may no longer line up properly along the same angle, or they may jump to overlap each other. Refer to Appendix B for details on the quantization process and Section 7.6.3 for a discussion on the effects of quantization error on the location of poles and zeros.

Notch-Pass Filter

Some applications call for a filter that extracts a particular frequency range from a signal that carries many frequencies. Ideally, this filter eliminates all the frequencies except the ones in its narrow target range. Such filters can be approximated using a close pole-zero pair with the pole located just inside the unit circle and the zero located slightly below the pole. Figure 8–35 illustrates this configuration.

In this case, since the pole is the pair member that lies closer to the circle, its effect dominates the gain response curve in the vicinity of the targeted range. When the input signal contains frequencies that lie close to the pair, the pole response vector becomes significantly shorter than the zero vector, producing a gain greater than 1.

For input frequencies located *close* to the pole-zero pair

$$\left| V_P \right| \ll \left| V_Z \right|$$

$$\left| H(Z) \right| = \frac{\left| V_Z \right|}{\left| V_P \right|} \gg 1$$

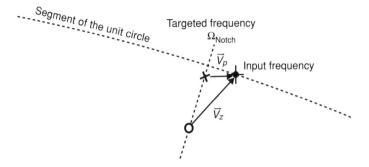

FIGURE 8–35
Notch-pass pole-zero pair.

For input frequencies that are not close to the targeted range, the pole-zero pair response vectors are of approximately the same length, and the system gain is close to a value of 1. Figure 8–36 illustrates the typical response of a notch-pass filter.

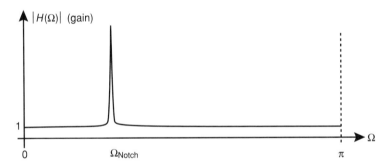

FIGURE 8–36
Notch-pass system gain response.

    The greater gain in the targeted range makes that range of frequencies stand out. Consequently, the other frequencies are relatively low in amplitude, and they sink to insignificant levels. The exact value of the gain in the targeted range depends on the distance separating the pole and the zero from the unit circle.

    For example, consider a system in which the pole is located at 0.99 and the zero at 0.95 from the targeted frequency Ω_{Notch} on the unit circle. When the input frequency is $\Omega = \Omega_{\text{Notch}}$, the response vectors have the following lengths:

$$\text{Pole response vector: } 1 - 0.99 = 0.01$$
$$\text{Zero response vector: } 1 - 0.95 = 0.05$$

In this case, the system gain becomes

$$H(\Omega_T) = \frac{\left|\vec{V_Z}\right|}{\left|\vec{V_P}\right|} = \frac{0.5}{0.1} = 5$$

In this example, the targeted frequency rises to five time its original input amplitude to stand above the other frequencies. This system could achieve greater gain by moving the pole closer to the unit circle or by moving the zero away from the unit circle. The maximum achievable gain is primarily limited in practice by the quantization limits imposed in this system.

8.3   TYPICAL USE OF FILTERS

A signal can carry a very wide bandwidth. When a signal enters a system, the frequencies are processed by the system mechanics, and the system outputs a modified version of the signal.

Signals exist in all shapes and forms: acoustic, electrical, modulated, and control, to name just a few. A large number of applications use filters for a variety of reasons. The next sections discuss some of the environments in which filters are used.

8.3.1   EQUALIZING FILTERS

The sole purpose of many systems is to carry a signal from one location to another. All systems have some frequency response characteristics, and these modify the phase and amplitude of the signal frequencies. Usually these undesired system effects are compensated for with the use of filters.

There are two approaches to compensate for the system frequency response:

1. Filters precondition the signal before it enters the system.
2. Filters correct the signal at the system output.

In both cases, the equalizing characteristics of the filters ideally invert the system frequency response. For example, if the system alters a band of frequencies by *attenuating* the amplitude by a certain factor, the equalizing filters must compensate by *amplifying* by the same factor. If the system makes the phase *lag* by a certain amount, the equalizing filters must compensate by making the phase *lead* by an equivalent amount. An ideal equalizing filter flattens the system gain response and makes the phase response linear.

Equalizing filters typically are used in balancing the acoustics of a room, in compensating the distortions produced by a transmission line, or in restoring the phase relationship of signal frequencies at the output of a transmission system.

8.3.2   BAND FILTERS

Some signals carry information over a number of frequency bands. A *band filter* can extract a specific range of frequencies from the signal. In the case of parasitic frequencies, a tight notch filter eliminates an interfering frequency.

Some typical uses of band filters include recovering carriers in a radio frequency (RF) system, separating the data from the voice in data over voice (DOV) systems, and cleaning up signals that are corrupted by frequencies induced by fields coming from power lines or motors.

8.3.3 CONTROL FILTERS

In control environments, signals are sent to *drive* systems. This could be to control the temperature of a container, the amount of lighting in a room, the speed or position of a motor, or any other system parameter. The intent is to have the system parameter follow a certain behavior curve. Since all systems have a frequency response, they can be mathematically modeled using poles and zeros. Once the system is modeled, it can be controlled.

Often a system inherently possesses some undesired characteristic that we want to reduce or eliminate. We can plot the poles and the zeros of the system model, and this usually reveals that the system contains a naturally occurring unwanted pole or zero. For example, most motors can be modeled as first-order systems that contain a single pole. The presence of this pole usually causes some oscillations in the motor response, and the driving signal must compensate for this unwanted behavior. Making the driving signal go through a filter that implements a zero located on top of the motor pole location performs the compensation. Overlaying the motor pole with a zero effectively eliminates the naturally occurring pole and makes the system behave as if it did not have this natural characteristic.

This technique may be applied to modify the natural characteristics of any system. For example, experimentation allows us to determine the poles and zeros that correspond to the acoustics of a room, the temperature losses in a medium, or the speed at which a motor reacts. By designing control filters that eliminate some system zeros and/or poles, we can usually make systems follow our control signals in spite of their natural characteristics.

SUMMARY

- Filters are used to shape the frequency response.
- Filters may be used to control the gain or the phase of a signal.
- A gain filter consists of passbands, transition bands, and stopbands.
- The passbands and stopbands of a gain filter suffer some response fluctuations.
- By definition, the passband ends where the input frequency is reduced to half of its original power level. This corresponds to 0.707 of the input amplitude (–3 dB).
- Gain filters are generally categorized as low-pass, band-pass, or high-pass filters.
- A frequency band has *linear-phase* shift characteristics when the time delay is constant for all the sinusoids in that band. This ensures that the shape of the signal is maintained.
- Complete filters can be analyzed, designed, and implemented by concatenating a number of subsections.
- Digital filters have a periodic frequency response.
- Converting an analog filter into a digital filter ruins the phase response of the filter.
- Gain filters are defined using sharpness, flatness, and overshoot characteristics.

- The step response of a filter allows us to measure the overshoot and transient characteristics of a filter.
- Butterworth filters provide a gain response curve that is maximally flat.
- Chebyshev filters provide a sharper cutoff than Butterworth filters of the same order. This sharper cutoff is achieved at the expense of equiripple in the passband *or* in the stopband.
- Cauer-elliptic filters provide the sharpest cutoff of all classes of filters. This sharper cutoff is achieved at the expense of equiripple in both the passband and the stopband.
- Bessel filters are linear-phase filters that lose their property when converted into digital filters. They may be used as antialias and/or reconstruction filters.
- Linear-phase filters are always FIR systems.
- Linear-phase filters always display symmetrical or antisymmetrical coefficient values.
- To get a linear-phase response, all zeros at a distance R from the origin must be matched by another zero lying along the same angle at a distance $1/R$.
- A linear-phase comb filter is simple to implement using any type of processor.
- The windowing technique works by converting the frequency domain response to the time domain and then truncating the impulse response using a window to make it finite.
- Windowing results in a ringing response that limits the attenuation of the filter. Different window shapes trade off sharpness for increased attenuation.
- The Parks–McClellan program allows us to design multiband linear-phase filters while controlling the transition and band ripple.
- An all-pass filter allows us to perform phase corrections on a signal.
- Notch filters allow us to eliminate or select a short frequency band.
- Filters may be used to correct or compensate the characteristics of systems.

PRACTICE QUESTIONS

8-1. What is a transition band?

8-2. What is meant by ripple in terms of the gain response of a filter?

8-3. A sinusoid that has an input amplitude of ±5 volts is applied to the input of a filtering system. This sinusoid has an amplitude of ±3 volts at the output of the system.

(a) What attenuation is the system applying to this sinusoid? Express your answer in decibels.

(b) What percentage of the sinusoid power remains at the output of this filter?

(c) If the passband of this filtering system has a gain of 1, is this frequency a component part of the passband?

8-4. The following signal is applied to the input of four different filtering systems:

$$x(n) = \cos(1000n) + \cos(1200n) + \cos(1500n)$$

The four different filters produce the following outputs:

$$y_1(n) = \cos(1000n + 1.35) + \cos(1200n + 1.35) + \cos(1400n + 1.35)$$
$$y_2(n) = \cos(1000n + 1.35) + \cos(1200n + 1.55) + \cos(1400n + 1.75)$$
$$y_3(n) = \cos(1000n + 1.35) + \cos(1200n + 1.62) + \cos(1400n + 1.89)$$
$$y_4(n) = \cos(1000n + 1.35) + \cos(1200n + 2.70) + \cos(1400n + 5.40)$$

Which of the four filtering systems has (have) a linear-phase response?

8-5. A linear-phase digital filter system samples the input at a rate of 8000 samples/second. The frequency response of this filter is illustrated in Figure 8–37:

(a) Is the phase response of this filter linear?

(b) What is the approximate gain of this filter at an input frequency of 2250 Hz?

(c) What is the approximate phase shift of this filter at an input frequency of 2250 Hz?

(d) Assume that this filter system does *not* include an antialiasing filter. What are the gain and phase shift at an input frequency of 10 kHz?

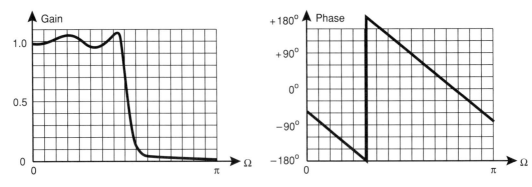

FIGURE 8-37
Filter response.

8-6. We create a Chebyshev type I filter using a software application that implements the bilinear transform.

(a) Does this filter have a linear-phase response?

(b) Is there ripple in the passband?

(c) Is this filter sharper than a Butterworth of the same order?

(d) Is this filter sharper than a Cauer-elliptic of the same order?

8-7. Why is it useless to convert a Bessel filter using the bilinear transform?

8-8. A filter has the following impulse response:

$$h(0) = 1 \qquad h(1) = 1.21 \qquad h(2) = 0.78 \qquad h(3) = 1.21 \qquad h(4) = 1$$

(a) How many poles and zeros are part of this filter?

(b) Does this filter have a linear-phase response?

8-9. A system has a *linear-phase response*. Some of the zeros of this system are at the following locations:

$$0.8 \, e^{j0} \qquad 1 \, e^{-j\pi} \qquad 1 \, e^{j\frac{\pi}{4}} \qquad 1.4 \, e^{-j\frac{3\pi}{4}}$$

   (a) What is the total number of zeros in this system?
   (b) What are the locations of the other system zeros?

8-10. The windowing technique is used to design a filter.
   (a) What is the advantage of using a Blackman window as opposed to a rectangular window?
   (b) What is the advantage of using a rectangular window as opposed to a Blackman?

8-11. A filter has conjugate poles and zeros positioned at the following locations:

$$\text{Poles: } 0.8 \, e^{j\frac{\pi}{4}} \text{ and } 0.8 \, e^{-j\frac{\pi}{4}} \qquad \text{Zeros: } 1.25 \, e^{j\frac{\pi}{4}} \text{ and } 1.25 \, e^{-j\frac{\pi}{4}}$$

   (a) Does this filter have a linear-phase response?
   (b) What special response characteristics does this filter exhibit?

8-12. A system samples at a rate of 8000 samples/second.
   (a) Design a notch-stop filter that completely removes a frequency of 1250 Hz while minimizing amplitude changes at other frequencies. At what location would you position the poles and zeros of this system?
   (b) Design a notch-pass filter that boosts the frequency of 1250 Hz by a factor of 10 with respect to the other frequencies. At what location would you position the poles and zeros of this system?

9

IMPLEMENTING DIGITAL SIGNAL PROCESSING SYSTEMS

Chapters 7 and 8 discussed how to obtain the difference equation coefficients that define filter systems. In this chapter, we examine how to use the value of these coefficients to implement real practical systems.

A large number of techniques that implement filtering systems exist. We examine factors such as computational complexity, precision, memory requirements, speed, noise levels, and cost, which play an important role in selecting a particular implementation scheme. Studying different cases allows us to assess the impact that these factors have on the implementation of the system.

The implementation of digital signal processing (DSP) systems relies on a number of processing structures. Each of the different structures provides some advantages, but, unfortunately, each structure also has weaknesses. We learn to identify the structures that meet the needs of common DSP algorithms. For example, we find that FIR and IIR systems require very different structures.

We then examine the main components of a typical DSP central processing unit to see how to use them effectively when programming the implementation of different system structures. We learn to manipulate and scale the samples as we multiply, add, and shift the numerical samples. The chapter closes with a system checklist that will help you begin to implement your own DSP applications.

9.1   STRUCTURING THE PROCESSING

There are many different types of DSP applications. Convolution (FIR), feedback difference equation (IIR), fast Fourier transforms (frequency analysis), and synthesis are among typical applications implemented using digital signal processors.

A variety of different software and/or hardware structuring schemes is available to implement linear time-invariant signal-processing systems. The different schemes describe how to segment the overall processing function into parts that we can implement as separate subsections. Figure 9–1 illustrates different arrangements of N subsections to configure some of the basic structuring schemes. Combinations of the basic structures illustrated in Figure 9–1 are possible to create more elaborate arrangements of the processing subsections.

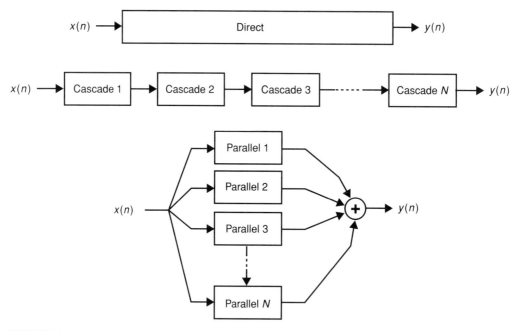

FIGURE 9–1
Some processing structures.

The processing steps performed in each of the separate subsections are described by an *algorithm*. Each algorithm requires a different number of multiplications, additions, and delay (memory location) operations; because of this, each holds some advantage over the others.

The designer must therefore analyze the particular requirements of the application to select the preferred structure and algorithm. The careful evaluation of each of the following factors influences the selection of structure and the choice of an algorithm:

- *Computational complexity.* The number of additions, multiplication operations, memory transfers, and comparisons required by the processing algorithm.
- *Memory requirements.* The amount of memory required to store and delay inputs and intermediate values.
- *Signal-to-noise ratio.* The amount of noise depends on the number of bits that the system uses to store numbers. The finite precision of the calculation requires rounding, truncating, and quantizing operations, which contribute to the amount of noise carried by the output signal.

- *Processing speed*. Processing applications have real-time constraints that limit the amount of processing time. For example, most applications require that all the processing be completed within one sampling interval.
- *Cost*. It would be nice if we could always use the best possible state-of-the-art system, but this unfortunately is not the case. Designing a product that can be marketed successfully requires that the cost and performance of the product be well balanced. The processing structures that yield better performance are often more expensive to implement than less expensive ones.

9.1.1 FINITE IMPULSE RESPONSE STRUCTURES

The difference equation that describes an FIR system calls for summing a finite number of product terms. According to Section 4.2.4, each of the product terms consists of an input sample that is weighted by a constant factor. Although techniques to build FIR systems recursively exist (see the Comb Filters section in Chapter 8 [p. 266]), this section deals only with the nonrecursive (guaranteed stable) system implementations.

Direct Form Structure

A causal FIR system of the Mth order contains $M+1$ weighted input terms. A difference equation limited to $M + 1$ terms (a variation of Equation 4–1) is one way to describe such a system:

$$y(n) = \sum_{k=0}^{M} b_k x(n - k)$$

$$y(n) = b_0 x(n) + b_1 x(n - 1) + b_2 x(n - 2) + \cdots + b_M x(n - M)$$

We can implement the convolution operation directly by storing the last $M + 1$ samples in memory. The processing algorithm then weights each of the stored $M + 1$ input samples and accumulates the results within a single subsection of processing. This direct form implementation is the most straightforward implementation for an FIR system. In a real-time environment, such system must perform the following sequence of events between sampling intervals:

- Acquire and store a new ADC input sample in memory.
- Multiply and accumulate the $M + 1$ product terms of the difference equation.
- Make room for the next input sample by shifting or rotating the stored input samples.

Figure 9–2 illustrates the *direct form* structure, which imposes the following system requirements:

- $M + 1$ memory locations to store the input samples (the boxes).
- $M + 1$ multiply operations to weight the $M + 1$ input samples (the triangles).
- M accumulate operations (the circles).
- M delay operations (these are memory transfer operations).

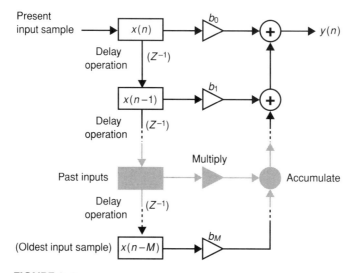

FIGURE 9–2
Direct form implementation for an FIR system.

The algorithm that implements the direct structure starts by computing the oldest M^{th} stage first. To do this, the system multiplies $x(n - M)$ by b_M and stores the result in an accumulator. Once this is done, the content of memory location $x(n - M)$ is no longer required. This allows the previous stage to copy its memory location $x(n - M + 1)$ into $x(n - M)$ to accommodate the delay operation.

The algorithm then calls for computing the previous stages in sequence, working toward the input stage containing $x(n)$. As each stage is processed, the result is accumulated, and the input sample from that stage is copied (delayed) to the next older memory location in preparation for the next pass. Figure 9–3 illustrates the process.

The multiply, the accumulate, and the copy (or rotate) operations described by Figure 9–3 are so commonly used that most digital signal processors provide a special instruction that performs all three operations within a single processor cycle. If we include a small amount of overhead to input and output the samples, a processor equipped with such an instruction requires little more than $M + 1$ processor cycles to compute an Mth order FIR system.

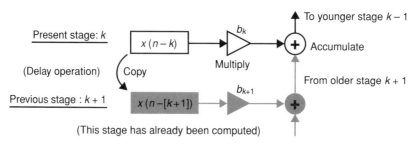

FIGURE 9–3
One stage of the direct form algorithm computed.

For example, imagine an audio system that implements an 80th-order FIR filter. The basic audio system sampling rate is 44.1 k-samples/second. Implementing this system imposes the following system requirements:

- Memory requirements: $80 + 1 = 81$ memory locations to store the input samples.
- Processing speed requirements: The input operation, the $80 + 1$ multiply/accumulate and delay operations, and the output operation must be processed within one sampling interval:

$$\frac{1}{44.1 \text{ k-samples/second}} = 22.6 \text{ µs per output sample}$$

In this case, assume that we use a signal processor that includes a *m*ultiply and *ac*cumulate (MAC) instruction that combines the delay operation. Also assume that the input/output overhead consists of an arbitrary 12 instructions. The processor must then be able to process instructions at the following rate:

$$\frac{(80 + 1) + 12}{22.6 \text{ µs}} \approx 4.1 \text{ million instructions per second}$$

Note that a few million instructions per second is easily achieved even when using the slowest signal processors.

As an example, let's examine the typical output of a software package that designs Parks–McClellan linear-phase filters. In this case, we arbitrarily instruct the software to design a two-band filter that has the following specifications:

- A band-pass filter with a sampling rate of 44.1 k-samples/second.
- Lower transition band: 1 kHz to 1.5 kHz.
- Upper transition band: 2 kHz to 2.5 kHz.
- Ripple: <10%.
- Coefficients quantized to 16 bits.

Table 9–1 illustrates the typical output of the Parks–McClellan linear-phase filter design package. Note the symmetry of the coefficients in the table. As explained in Section 8.2.3, this symmetry exists in all linear-phase filters. Some algorithms may take advantage of this symmetry to minimize the number of multiplication operations.

The direct form structure is the most widespread implementation of FIR filters. However, this structure is not adequate for all situations. In some cases, the direct structure does not implement the desired filter characteristics adequately. The following section discusses the direct structure weaknesses and proposes an alternative to the direct structure.

Cascade Structure

The process that we use to design and implement an FIR system involves the following steps:

1. We determine where to position zeros on the Z-plane to obtain the desired frequency response of the system.
2. Then we write the system Z-transform based on the location of the required zeros. At this point, we can manipulate the Z-transform to shape the structure of the system.

TABLE 9–1
Example of Parks–McClellan linear-phase filter coefficients.

$b_0 = -.02895161 = b_{79}$	$b_{10} = +.01093119 = b_{69}$	$b_{20} = +.00426824 = b_{59}$	$b_{30} = -.02892860 = b_{49}$
$b_1 = +.02729542 = b_{78}$	$b_{11} = +.01572410 = b_{68}$	$b_{21} = -.00325723 = b_{58}$	$b_{31} = -.02141532 = b_{48}$
$b_2 = +.00933937 = b_{77}$	$b_{12} = +.01965610 = b_{67}$	$b_{22} = -.01123120 = b_{57}$	$b_{32} = -.01220048 = b_{47}$
$b_3 = -.00277899 = b_{76}$	$b_{13} = +.02256972 = b_{66}$	$b_{23} = -.01898540 = b_{56}$	$b_{33} = -.00187563 = b_{46}$
$b_4 = -.00924883 = b_{75}$	$b_{14} = +.02424438 = b_{65}$	$b_{24} = -.02606476 = b_{55}$	$b_{34} = +.00881751 = b_{45}$
$b_5 = -.01096522 = b_{74}$	$b_{15} = +.02457206 = b_{64}$	$b_{25} = -.03187779 = b_{54}$	$b_{35} = +.01921620 = b_{44}$
$b_6 = -.00919942 = b_{73}$	$b_{16} = +.02341632 = b_{63}$	$b_{26} = -.03587201 = b_{53}$	$b_{36} = +.02861574 = b_{43}$
$b_7 = -.00519216 = b_{72}$	$b_{17} = +.02076865 = b_{62}$	$b_{27} = -.03784505 = b_{52}$	$b_{37} = +.03632327 = b_{42}$
$b_8 = -.00000526 = b_{71}$	$b_{18} = +.01658634 = b_{61}$	$b_{28} = -.03733018 = b_{51}$	$b_{38} = +.04175790 = b_{41}$
$b_9 = +.00557022 = b_{70}$	$b_{19} = +.01103292 = b_{60}$	$b_{29} = -.03431802 = b_{50}$	$b_{39} = +.04454831 = b_{40}$

3. We follow by converting the system from the Z-domain to the discrete-time domain by writing the difference equation.
4. We quantize the difference equation coefficients and store these in memory locations.
5. We program the processor to compute the output sample values.

Remember that the exact position of the system zeros completely defines the functional characteristics of the system. Unfortunately, when we convert the difference equation constant coefficient values to a binary format, we are limited by the finite bit-size of the memory words (finite word length). This limit results in a quantization error, which means that the quantized coefficient values will not correspond exactly to the values called for by the application.

The coefficient quantization error has the unfortunate effect of nudging the position of the system zeros (see Section 7.6.3). Since the quantized system zeros no longer lie at the required locations, the quantized system will not behave exactly as designed. In some cases, the quantization error results in an insignificant effect on the system behavior and, in extreme cases, the change in zero positioning is enough to yield an unacceptable system response.

The actual amount of zero shifting that is acceptable depends on the position of the system zeros with respect to each other. The following are some of the most common hazardous situations:

- The system function calls for zeros that are *closely clustered* within a small area of the Z-plane. In this case, a small shifting in the position of the zeros results in large variations in the system response.
- The system calls for zeros being positioned in an area of the Z-plane where there is a *low density of available locations,* such as at very high or very low normalized frequencies (see Section 7.6.3). In this case, the quantization error is likely to produce larger shifts in the position of the zeros.

- The system performance depends on zeros being positioned according to very specific *patterns* or at *special locations*. Examples are linear-phase response using quad zero-patterns and zero-gain response by placing zeros exactly on the unit circle. Any deviation from this precise positioning and the desired response is not achieved.

Three main factors affect the amount of zero shifting resulting from the coefficient quantization process:

1. The number of bits used in the quantization operation.
2. The total number of coefficients in the difference equation being quantized.
3. The system structure.

Generally, using more bits to quantize the system coefficients results in less zero shifting. For example, the coefficients of an 8-bit system such as a small microcontroller suffer much more from the quantization process than a 16-bit system does. In general, you should be apprehensive about the amount of zero shifting when using fewer than 12 bits to quantize the system coefficients.

If fewer than 12 bits are used in the coefficient quantization process, you should verify that the zeros have not been shifted to unacceptable locations.

As the system order increases, the difference equation becomes larger, and the system requires a larger number of coefficients. Given a fixed word length in the quantization process, systems defined with a *larger* number of coefficients will suffer a *larger* amount of shifting in the position of their zeros. Although we cannot place an exact limit on a safe number of coefficients, it is generally accepted that quantizing the coefficients of a system having more than 200 zeros is very likely to result in an unacceptable amount of zero shifting.

If the system uses fewer than 12 bits or if it has more than 200 zeros, the direct structure described in the section in this chapter on Direct Form Structure may no longer be adequate. Fortunately, other structures allow us to circumvent the problem. The most popular solution is to implement the system as a cascade of lower-order subsections.

Creating a cascade of subsections works by limiting each subsection to the implementation of two or four of the system zeros. This reduces the number of subsection coefficients to such a small number that the quantization error is dramatically reduced.

Let's examine how we can divide a system into a number of subsections. If the system has a causal, pure-real response, each complex conjugate pair of zeros corresponds to the following Z-transform expression:

$$H_{\text{Subsystem}}(Z) = \frac{Z^2 - 2R\cos(\phi)\,Z + R^2}{Z^2}$$

$$= \frac{b_0 Z^2 + b_1 Z + b_2}{Z^2}$$

Higher-order systems may be created by sequencing a number of such second-order subsections. Figure 9–4 illustrates such cascade of subsections.

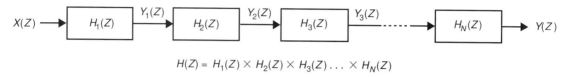

$$H(Z) = H_1(Z) \times H_2(Z) \times H_3(Z) \ldots \times H_N(Z)$$

FIGURE 9–4
Cascading subsections.

For example, let's look at implementing an 80th-order pure-real system on an 8-bit system. In this case, the system may be assembled using 40 individual second-order subsections. The overall Z-transform of this 80th-order system is found by multiplying the 40 second-order Z-transforms that represent the subsections:

$$H(Z) = \frac{Z^2 - 2R_1 \cos(\phi_1)Z + R_1^2}{Z^2} \times \frac{Z^2 - 2R_2 \cos(\phi_2)Z + R_2^2}{Z^2} \times \cdots$$

$$\times \frac{Z^2 - 2R_{40}\cos(\phi_{40})Z + R_{40}^2}{Z^2}$$

$$H(Z) = \frac{b_0 Z^{80} + b_1 Z^{79} + b_2 Z^{78} + \cdots + b_{79}Z + b_{80}}{Z^{80}}$$

These two expressions are Z-transforms of exactly the same system. The second expression corresponds to a *direct* structure; however, because only 8-bit coefficients are used and because there is a relatively large number of coefficients, the quantization error is likely to produce a large amount of zero shifting. Fortunately, the first Z-transform expression represents a cascade of second-order subsections where each subsection requires the quantization of only three coefficients. The cascade expression may therefore be used to provide a system structure that provides more control over the shifting of the system zeros.

Separating the overall system function into a number of second-order subsections yields less quantization error on the system coefficients for these reasons:

- There are only three coefficients in each of the cascaded second-order subsections.
- The value of the three subsection coefficients is likely to span a smaller range of values than the range spanned by the many coefficients of a direct implementation of the system.

This last point results from the fact that the direct structure coefficients are found by expanding the second-order subsections into a single very large order function. The multiplication operations involved in the expansion process can make some coefficients become very small and others become very large. As an example, examine the range of coefficient values for the Parks–McClellan example used in Table 9–1. You will find that the coefficients range from –0.00000526 to +0.04454831, which covers almost four orders of magnitude.

To achieve programming efficiency in fixed-point systems, it is customary to quantize all coefficients using the same Q format. Because of this, the absolute worst-case quantization error is the same for all coefficients.

For example, if we round to the nearest integer during the quantization process, the worst-case error approaches the value of 1/2 of the least significant bit. Consider a simplified example in which we calculate the maximum error that results from an 8-bit Q5 quantization operation:

Absolute value of half of the least significant bit in a Q5 number

$$\frac{1}{2} \times \frac{1}{2^5} = 0.015625$$

This error may be close to insignificant when we quantize coefficients that have rather large values. However, the percentage error could be relatively large for coefficients bearing smaller values. For example, consider the percentage error that results in the following two cases:

Maximum percentage error on a Q5 coefficient with value ≈ 2 :

$$\frac{0.015625}{2} \times 100 = 0.78\%$$

Maximum percentage error on a Q5 coefficient with value ≈ 0.1 :

$$\frac{0.015625}{0.1} \times 100 = 15.62\%$$

As we can see, the quantization error is worse on the smaller-valued coefficients. Cascading second-order subsections reduce the range of coefficient values; consequently, the quantization process is likely to result in a smaller error.

A cascade structure is likely to result in a smaller range of coefficient values. This results in a smaller percentage quantization error.

Sequencing of the Subsections of a Cascade Structure

In DSP applications, the processing involves a large number of MAC operations. The accumulator, used to sum the weighted samples, contains at least twice as many bits as the finite word length of the memory locations. When the accumulated result must be sent to a DAC and/or stored in memory, the contents of the accumulator must be truncated to fit the bit-size of the destination. Truncating bits may change the level of the noise floor; consequently, the processing steps used in subsections may set new noise levels. In a cascade implementation of subsections, the noise level of one subsection imposes noise limits that are transferred to the next subsection.

It is difficult to estimate exactly how much noise propagates in a cascade of subsystems because each subsystem effectively behaves as a filter. The subsection filtering affects not only the input signal but also the input noise level injected from the previous subsection. Although the formal proof exceeds the introductory nature of this text, subsections with large coefficient values will amplify the noise contained in the input samples more than subsections that implement smaller coefficient values.

Subsections with smaller coefficient values propagate less input noise to the system output.

Each subsection has a frequency response defined by the position of the pair of zeros it is implementing. If a subsection frequency response produces a gain for a particular band of frequencies, it will have a tendency to magnify any noise in that band that was inherited from the previous subsection. Fortunately, a proper arrangement of the cascaded subsections can minimize the amount of noise being propagated to the system output.

To minimize the amount of noise, the subsections should be arranged in order of decreasing gain from input to output. For example, imagine a system that has an overall gain of 10, which is split over three second-order subsections having gains of 0.5, 2, and 10. Because the system is linear, the three subsections may be sequenced in any order. Imagine that the input signal has an amplitude of 1 and a noise level of 0.004 and that each second-order subsection has an arbitrary absolute noise floor of 0.01. Figure 9–5 illustrates two of the six possible ways to sequence the three subsections of this system.

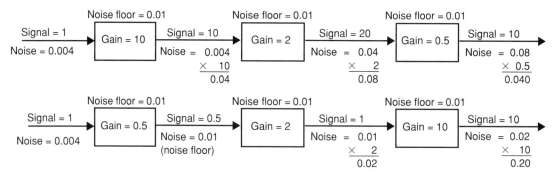

FIGURE 9–5
Ordering the second-order subsections.

Although Figure 9–5 is an oversimplification of what really happens, it clearly illustrates that the input noise is amplified by subsections. When we sequence the subsections in order of decreasing gain, the output signal is much cleaner because the noise floor of the subsections sets a lower limit to the noise level. In this example, the noise level is five times lower if the subsections are arranged in order of decreasing gain.

Sequencing the second-order subsections of a cascaded system in order of decreasing gain produces cleaner output signals.

Implementing a Cascade of Subsections

Let's examine how to implement a cascade of second-order FIR subsections. The first step is to regroup the conjugate zeros making up the system into second-order subsections having the following Z-transform:

$$H(Z) = \frac{\text{Out}(Z)}{\text{In}(Z)} = \frac{b_0 Z^2 + b_1 Z + b_2}{Z^2} \times \frac{Z^{-2}}{Z^{-2}}$$

$$\text{Out}(Z) = b_0 \text{In}(Z) + b_1 \text{In}(Z) Z^{-1} + b_2 \text{In}(Z) Z^{-2}$$

The Z-domain output of a second-order FIR subsection translates to the following difference equation:

$$\text{Out}(n) = b_0\text{In}(n) + b_1\text{In}(n-1) + b_2\text{In}(n-2)$$

Figure 9–6 illustrates how to implement this second-order difference equation.

FIGURE 9–6

Second-order FIR subsection.

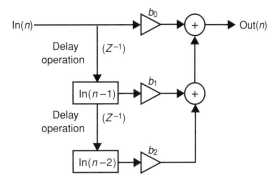

We can create larger systems of the Mth order by concatenating N second-order subsections. The number of individual second-order subsections is simply half of the overall system order:

$$N = \frac{M}{2}$$

When M is an odd number, we simply need to append an extra *first-order subsection* to the system. Figure 9–7 illustrates how to cascade second-order subsections to implement an FIR system of the Mth order (where M is an even number equal to $2N$). As the figure illustrates, the cascade form structure imposes the following system requirements:

- A total of $M + 2$ memory locations.
  $N \times 2 = M$ memory locations to store the delayed input samples.
  One memory location to store the present input.
  One shared memory location to store the intermediate result between subsections.
- $3N = 1.5M$ multiply operations.
- $N \times 2 = M$ accumulate and delay operations in the N subsections.

The cascade form requires approximately 50% more multiplication operations when compared to the direct form structure. This is the price paid to obtain more precise coefficient quantization.

In the frequent case when an FIR system is used to implement a system that has linear-phase characteristics, we must be especially careful. The linear-phase characteristics usually require the use of patterns of four zeros arranged as follows (refer to Figure 8–19):

$$Z_1 = Re^{j\phi} \qquad Z_2 = \frac{1}{R}e^{j\phi} \qquad Z_3 = \overline{Z_1} \qquad Z_4 = \overline{Z_2}$$

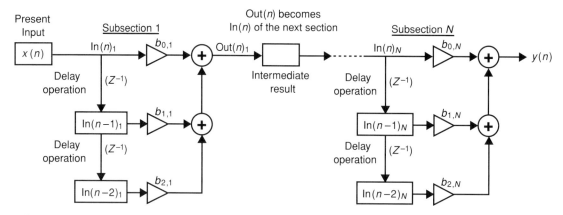

FIGURE 9-7
Cascade implementation of an FIR system.

This quad of zeros translates to two second-order subsystems that combine to yield the following Z-transforms:

$$H(Z) = \frac{Z^2 - 2R\cos(\phi)Z + R^2}{Z^2} \times \frac{Z^2 - \frac{2}{R}\cos(\phi)Z + \left(\frac{1}{R}\right)^2}{Z^2}$$

Expanding these into a fourth-order expression reveals a coefficient symmetry that is essential in all systems with linear-phase characteristics:

$$H(Z) = \frac{b_0 Z^4 + b_1 Z^3 + b_2 Z^2 + b_3 Z^1 + b_4}{Z^4}$$

where

$$b_0 = b_4 \quad \text{and} \quad b_1 = b_3$$

The quantization process must not alter this symmetry to preserve the linear-phase characteristics. Fortunately, quantizing the symmetrical coefficients of a direct structure or the symmetrical coefficients of a fourth-order system does not alter the symmetry. The quantization process produces a symmetrical error. For example, consider quantizing the following quad system, which implements a linear-phase system where $R = 0.5$ and $\phi = \pi/6$:

$$H(Z) = \frac{Z^2 - 0.866025\,Z^1 + 0.25}{Z^2} \times \frac{Z^2 - 3.46410\,Z^1 + 4}{Z^2}$$

$$H(Z) = \frac{1\,Z^4 - 4.33012\,Z^3 + 7.25000\,Z^2 - 4.33012\,Z^1 + 1}{Z^4}$$

Quantizing this system using 8-bit Q4 coefficients yields

$$H_{\text{Quantized}}(Z) = \frac{1\,Z^4 - 4.3125\,Z^3 + 7.2500\,Z^2 - 4.3125\,Z^1 + 1}{Z^4}$$

Notice that $b_0 = b_4$ and $b_1 = b_3$ even after the quantization process. Unfortunately, this may not be true if we quantize the second-order sections separately since the second-order coefficients are not symmetrical:

$$H(Z) = \frac{Z^2 - 2R\cos(\phi)Z + R^2}{Z^2} \times \frac{Z^2 - \dfrac{2}{R}\cos(\phi)Z + \left(\dfrac{1}{R}\right)^2}{Z^2}$$

$$H(Z) = \frac{Z^2 - 0.866025\,Z^1 + 0.25}{Z^2} \times \frac{Z^2 - 3.46410\,Z^1 + 4}{Z^2}$$

Quantizing this system using 8-bit Q4 coefficients yields

$$H_{\text{Quantized}}(Z) = \frac{Z^2 - 0.8750\,Z^1 + 0.25}{Z^2} \times \frac{Z^2 - 3.4375\,Z^1 + 4}{Z^2}$$

$$H_{\text{Quantized}}(Z) = \frac{1\,Z^4 - 4.3125\,Z^3 + 7.2578\,Z^2 - 4.3594\,Z^1 + 1}{Z^4}$$

Notice that b_1 is no longer equal to b_3. This is bad news since it means that we have lost the linear-phase characteristics of this system. To prevent this problem, we can make a scaling adjustment to the coefficients of the second subsections of the Z-transform by multiplying it by R^2:

$$H(Z) = \frac{Z^2 - 2R\cos(\phi)Z + R^2}{Z^2} \times \frac{R^2\,Z^2 - 2R\cos(\phi)\,Z + 1}{Z^2} \times \frac{1}{R^2}$$

Note that we multiply the complete Z-transform by $1/R^2$ to ensure that the overall system is left unchanged. Notice that the scaling forces both second-order subsections to use exactly the same numerical values. Applying this modification to our example yields

$$H(Z) = \frac{Z^2 - 0.866025\,Z^1 + 0.25}{Z^2} \times \frac{0.25\,Z^2 - 0.866025\,Z^1 + 1}{Z^2} \times 4$$

System quantized using 8-bit Q4 coefficients yields

$$H_{\text{Quantized}}(Z) = \frac{Z^2 - 0.875\,Z^1 + 0.25}{Z^2} \times \frac{0.25\,Z^2 - 0.875\,Z^1 + 1}{Z^2} \times 4$$

$$H_{\text{Quantized}}(Z) = \frac{1\,Z^4 - 4.375\,Z^3 + 7.3125\,Z^2 - 4.375\,Z^1 + 1}{Z^4}$$

Notice that the coefficient symmetry now ensures a linear-phase response.

When there are many fourth-order subsections, the resulting scaling factor (which is larger than 1) may be distributed evenly over the cascaded subsections. This is done by applying the appropriate factor to the b_0, b_1, and b_2 coefficients of all second-order parts of the other fourth-order subsections (the section in this chapter on Scaling the System Difference Equation explains this in detail). Figure 9–8 illustrates the cascaded structure with the scaling factor of $\sqrt[N]{1/R^2}$ being applied to each subsection so that it results in $1/R^2$ over the N subsections.

FIGURE 9–8
Distribution of the scaling factor over N subsections.

9.1.2   INFINITE IMPULSE RESPONSE STRUCTURES

The difference equation that describes any practical IIR system calls for summing a finite number of weighted *input* samples and a finite number of weighted *output* samples. Consequently, the difference equation that describes an IIR system contains an FIR part that weights the *input* samples and a recursive part that weights the *output* samples.

In a well-balanced causal system (see the section in Chapter 7 on Balancing the Poles and Zeros of a System), every pole must be balanced by a zero. Consequently, the FIR and feedback parts are of the same order. The following difference equation describes a *causal M^{th}* order IIR system that combines M zeros and M poles:

$$y(n) = \sum_{k=0}^{M} b_k x(n - k) + \sum_{k=1}^{M} a_k y(n - k)$$

This difference equation expands to the following terms:

$$y(n) = b_0 x(n) + b_1 x(n - 1) + b_2 x(n - 2) + \ldots + b_M x(n - M)$$
$$+ a_1 y(n - 1) + a_2 y(n - 2) + \ldots + a_M y(n - M)$$

As we will see, there are two possible ways to implement this system using a direct form. Other structures are possible when we manipulate the system function into second-order subsections. In particular, we will examine the popular cascade structure, the transposed structure, and the parallel structure.

Direct Implementation

The first implementation of an IIR system is unimaginatively called the *direct form I*. It is a straightforward, what-you-see-is-what-you-get direct implementation of all the terms of the difference equation, as Figure 9–9 illustrates. According to the figure, the direct form I structure imposes the following system requirements:

FIR Part Requirements

- $M + 1$ memory locations to store the input samples.
- $M + 1$ multiply operations to weight the $M + 1$ input samples.
- M accumulate operations.
- M delay operations.

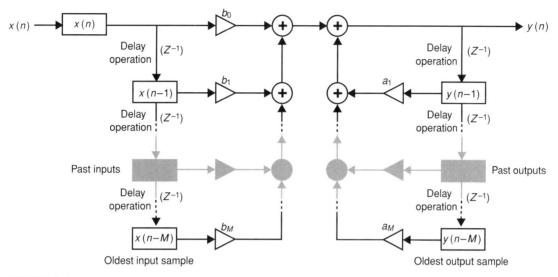

FIGURE 9–9
Direct form I of an IIR system.

Feedback Part Requirements

- M memory locations to store past M output samples.
- M multiply operations to weight the past M output samples.
- M accumulate operations.
- M delay operations.

This structure is avoided in practice because the high gain associated with the IIR part comes *after* the low-gain FIR part. According to the section Sequencing of the Subsections of a Cascade Structure on page 295, the high-gain part should be positioned before the low-gain part to maintain low noise levels.

To fix the noise problem, we can devise a second direct form by taking advantage of the linear properties of the system. Remember that the subsections of a linear system may be sequenced in any order. We contrive the new structure by swapping the FIR and the recursive parts. Figure 9–10 illustrates the resulting structure.

Upon examining Figure 9–10 closely, we note that the structure requires an intermediate value $d(n)$. This intermediary value holds the output samples of the recursive part of the system. Notice that the intermediate value is delayed and stored down two identical columns that duplicate the delay operations and the required memory locations. This duplication is unnecessary, and removing it allows us to simplify the structure. Figure 9–11 illustrates this by merging the identical columns into a single one. Note that the input $x(n)$ and the output $y(n)$ in the figure no longer need to be delayed. This means that we can input directly into the accumulator with no need to store the input sample in memory. Notice, however, that the structure requires us to delay an intermediate result; consequently, we must provide memory locations to store these values. Figure 9–12 illustrates these modifi-

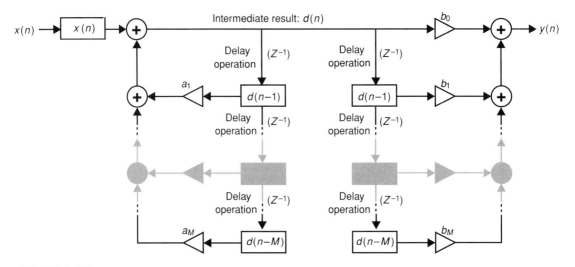

FIGURE 9–10
FIR and recursive parts swapped.

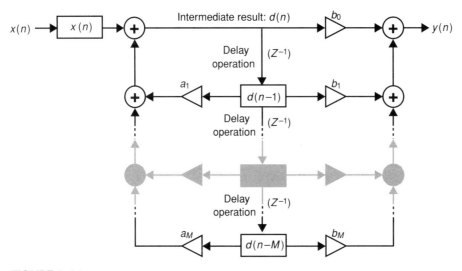

FIGURE 9–11
Central parts merged.

cations, which yield the *direct form II* structure. As the figure indicates, the direct form II structure reduces the system requirements to the following:

- $M + 1$ memory locations to store the delayed intermediate values.
- $2M + 1$ multiply operations to weight the intermediate values.
- $2M$ accumulate operations.
- M delay operations.

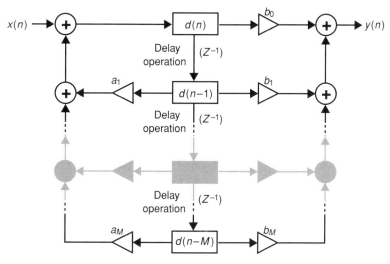

FIGURE 9–12
Direct form II structure of an IIR system.

The direct form II structure achieves lower noise levels by sequencing the IIR and FIR parts so that the high-gain recursive part is applied first (see the section Sequencing of the Subsections of a Cascade Structure [p. 295]). The direct form II structure therefore cuts the memory requirements almost by half. This allows a significant reduction of the resources required in pure hardware implementations using PLDs or ASICs. The saving is also important in software-based systems since digital signal processors provide only a small amount of on-chip memory.

The typical algorithm that computes the direct form II output works by starting at the bottom left part of the structure. From there, it works up the M multiply/accumulate stages, accumulating the delayed intermediate values, which have been weighted using the a_k coefficients. Once it reaches the top, the input is accumulated and the intermediate result $d(n)$ is stored. The algorithm then proceeds to compute from the base of the right part of the structure. Again, it starts at the bottom stage, accumulating delayed intermediate values, but this time they are weighted using the b_k coefficients. Once a stage is processed, the delayed intermediate value of that stage is no longer required, and the delay operation is implemented by overwriting it, with the intermediate result coming from the next "younger" stage. The mechanics here are identical to the ones described in Figure 9–3.

Cascading IIR Systems

As discussed in Chapter 7, we can design IIR systems by positioning poles and zeros on the Z-plane. A well-balanced causal system requires a pair of zeros for every pair of poles (see the section in Chapter 7 on Balancing the Poles and Zero of a System [p. 220]). In most cases, we need to position the poles at complex locations. When this happens, the poles must occur as complex conjugate pairs to obtain a pure-real system response (see the section in Chapter 7 on Designing a Pure-Real System [p. 224]).

The following second-order Z-transform describes a well-balanced second-order IIR system containing a pair of conjugate poles and a pair of conjugate zeros:

$$H(Z) = \frac{Z^2 - 2R_Z \cos(\phi_Z) Z + R_Z^2}{Z^2 - 2R_P \cos(\phi_P) Z + R_P^2}$$

$$= \frac{b_0 Z^2 + b_1 Z + b_2}{Z^2 - a_1 Z - a_2}$$

where

$$b_0 = 1$$
$$b_1 = -2R_Z \cos(\phi_Z) \quad \text{and} \quad a_1 = +2R_P \cos(\phi_P)$$
$$b_2 = R_Z^2 \qquad\qquad\qquad a_2 = -R_P^2$$

Moving this result to the discrete-time domain yields the following second-order difference equation:

$$y(n) = b_0 x(n) + b_1 x(n-1) + b_2 x(n-2)$$
$$+ a_1 y(n-1) + a_2 y(n-2)$$

Figure 9–13 illustrates a scaled second-order IIR subsection implemented using a direct form II structure.

We can create higher-order systems by combining second-order IIR systems. For example, we can assemble an even M^{th} order pure-real causal IIR system using $N = M \div 2$ second-order subsections:

$$H(Z) = \frac{b_{0,1} Z^2 + b_{1,1} Z + b_{2,1}}{Z^2 - a_{1,1} Z - a_{2,1}} \times \frac{b_{0,2} Z^2 + b_{1,2} Z + b_{2,2}}{Z^2 - a_{1,2} Z - a_{2,2}} \times \cdots$$

$$\times \frac{b_{0,N} Z^2 + b_{1,N} Z + b_{2,N}}{Z^2 - a_{1,N} Z - a_{2,N}}$$

$$H(Z) = H_1(Z) \times H_2(Z) \times \cdots \times H_N(Z)$$

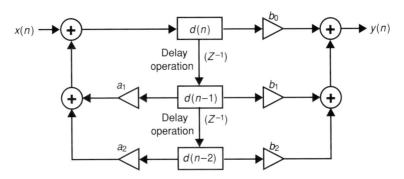

FIGURE 9–13
Direct form II implementation of a second-order IIR subsection.

Figure 9–14 illustrates N-cascaded direct form II subsections making up a pure-real, causal IIR system of the Mth order. As the figure indicates, each second-order subsection behaves as an independent system that receives its input from the previous subsection. The output of one subsection becomes the input of the following subsection. Cascading second-order subsections requires the following system resources:

- $2N + 1 = M + 1$ memory locations to store the delayed intermediate values (remember that the top intermediary value may be a single shared location).
- $5N = 2 \, 1/2 \, M$ multiply operations to weight the intermediate values.
- $4N = 2 \, M$ accumulate operations.
- $2N = M$ delay operations.

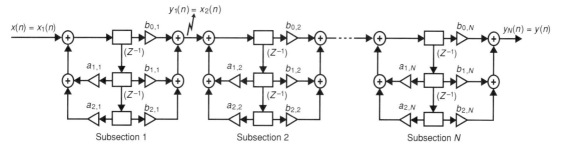

FIGURE 9–14
Cascading direct form II subsections.

Compared with the direct form II illustrated in Figure 9–12, the cascade structure requires 25% more MAC (multiply/accumulate) operations. However, because of the high gain of the IIR sections, the cascade structure is still preferred because it propagates less noise and results in less quantization error on the system coefficients.

As an example, let's examine the typical output of a software package that designs IIR filters. In this case, we instruct the software to design a filter that has the following specifications:

- An elliptic band-pass filter with a sampling rate of 44.1 k-samples/second.
- Lower transition band: 1 kHz to 1.1 kHz.
- Upper transition band: 2 kHz to 2.1 kHz.
- Ripple: <10%.

Table 9–2 illustrates the output of the filter design package. According to the table, the required filter requires four subsections. The software automatically arranges the subsections in order of increasing gain. Consequently, the system input $x(n)$ should feed into subsection 4, which feeds into subsection 3, which feeds into subsection 2, which feeds into subsection 1, which provides the filter output $y(n)$.

In this typical elliptic filter example, the poles are positioned very close to the unit circle. A quick check can determine the distance of the poles from the unit circle. Remember from Equation 7–8 that the IIR coefficient $a_2 = -Rp^2$. For example, the conjugate poles of section 4 are located at the following distance from the origin on the Z-plane:

$$Rp_4 = \sqrt{0.993689} = 0.9968$$

TABLE 9–2
Example of cascaded IIR system coefficients.

Subsection	a_1	a_2	b_0	b_1	b_2
1	+1.883113	−0.944366	+0.136271	−0.249794	+0.136271
2	+1.929371	−0.960208	+0.568084	−1.129568	+0.568084
3	+1.908490	−0.988664	+0.688227	−1.313588	+0.688227
4	+1.969248	−0.993689	+0.959823	−1.898915	+0.959823

This is $1 - 0.9968 = 0.0032$ unit from the unit circle.

Fortunately, because of the small number and small value of the subsection coefficients, the quantization process should produce a minor shift in the positions of the system pole/zero. To help examine the amount of shift in the pole position, most software packages can produce a plot of the quantized pole and zero positions. It is always wise to verify, on this plot, that all poles are still inside the unit circle *after* the software has applied the quantization process.

Incidentally, compare the relatively tight transition band specifications of this IIR elliptic filter with the rather loose specifications of the 80 coefficient FIR filter described in Table 9–1. The IIR elliptic filter achieves the much tighter gain response using only four subsections for a total of 20 coefficients. Unfortunately, the linear-phase characteristics are completely lost with this elliptic IIR system.

Transposed Structure

A cascade of direct form II subsections has one disadvantage. The fact that the high-gain recursive part is processed first means that the delayed intermediary memory locations must hold larger results and therefore they require more integer bits. When using finite word-length arithmetic, having more I-bits means having fewer F-bits. Unless you have very large word lengths, the resulting lower Q formats usually mean higher system noise levels (noise levels are discussed in Section 9.2.4).

An alternative structure exists to spread the gain of the recursive part over the whole subsection. This new structure reduces the value of the samples held in intermediary memory locations, allowing for larger Q formats and consequently a reduction of the noise level.

Let's start with the second-order difference equation that describes the subsection:

$$y(n) = b_0 x(n) + b_1 x(n-1) + b_2 x(n-2) + a_1 y(n-1) + a_2 y(n-2)$$

We can regroup and name the terms that have delays of one and two discrete-time units:

$$\Psi_1(n-1) = b_1 x(n-1) + a_1 y(n-1)$$
$$\Psi_1(n) = b_1 x(n) + a_1 y(n)$$

and

$$\Psi_2(n-2) = b_2 x(n-2) + a_2 y(n-2)$$
$$\Psi_2(n) = b_2 x(n) + a_2 y(n)$$

Note that the new terms $\Psi_1(n)$ and $\Psi_2(n)$ use the input and output samples without any delay. Figure 9–15 illustrates how to build these terms separately. We regroup the terms $\Psi_2(n-1)$ with $\Psi_1(n)$ and delay the result to create the output function $y(n)$. Figure 9–16 illustrates the resulting transposed direct form II structure.

When compared to the direct form II, the two memory locations holding delayed elements will hold smaller sample values. This enables us to use larger Q factors that allow lower noise levels. Unfortunately, when compared to the direct form II structure, the transposed structure requires more memory resources to hold the input and output samples. It also requires an extra addition operation; consequently, the processing will require extra time.

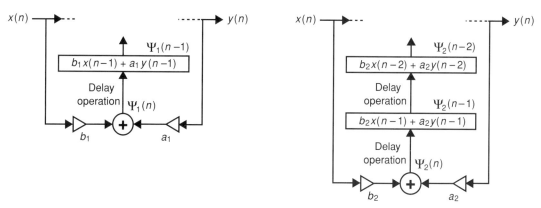

FIGURE 9–15
Delayed terms built separately.

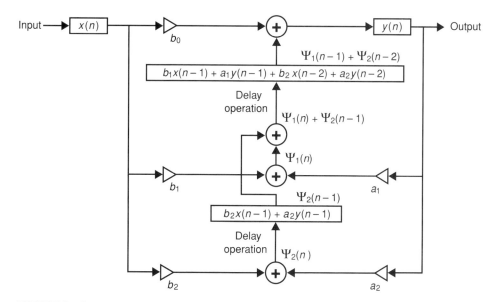

FIGURE 9–16
Transposed form structure.

Parallel Structure

Alternate structures are possible if we manipulate the system Z-transform. The parallel structure allows for the efficient use of systems that have parallel processing units. We start by examining the Z-transform of a well-balanced causal IIR system expressed as a cascade of second-order subsections:

$$H(Z) = \frac{b_{0,1}Z^2 + b_{1,1}Z + b_{2,1}}{Z^2 - a_{1,1}Z - a_{2,1}} \times \frac{b_{0,2}Z^2 + b_{1,2}Z + b_{2,2}}{Z^2 - a_{1,2}Z - a_{2,2}} \times \cdots$$

$$\times \frac{b_{0,N}Z^2 + b_{1,N}Z + b_{2,N}}{Z^2 - a_{1,N}Z - a_{2,N}} \tag{9-1}$$

$$= H_1(Z) \times H_2(Z) \times \cdots \times H_N(Z)$$

Through a mathematical operation known as *partial fraction expansion*, it is possible to rewrite the Z-transform of the system as a *sum* of N second-order subsections. Fortunately, many software applications are available to compute the partial fraction expansion for us. Using such software application yields the following general result:

$$H(Z) = C + \frac{Z\left[b_{0,1P}Z + b_{1,1P}\right]}{Z^2 - a_{1,1P}Z - a_{2,1P}} + \frac{Z\left[b_{0,2P}Z + b_{1,2P}\right]}{Z^2 - a_{1,2P}Z - a_{2,2P}} + \cdots$$

$$+ \frac{Z\left[b_{0,NP}Z + b_{1,NP}\right]}{Z^2 - a_{1,NP}Z - a_{2,NP}}$$

$$= C + H_{1P}(Z) + H_{2P}(Z) + \cdots + H_{NP}(Z)$$

where

$$b_{0,NP} \neq b_{0,N} \qquad b_{1,NP} \neq b_{1,N} \qquad \text{and} \qquad a_{1,NP} = a_{1,N} \qquad a_{2,NP} = a_{2,N}$$

Comparing this subsection with the cascade of second-order subsections (Equation 9–1), you will note that the poles of the subsections remain at exactly the same locations but that the numerator zeros have shifted location. Converting to a parallel system, we keep control over the pole locations but we lose control over the zero locations.

The partial fraction expansion operation always converts the numerator conjugate zeros into one zero at the origin and one zero at some other location. Although the partial fraction expansion alters the individual subsection functions, once the subsection results are combined, the overall system implements exactly the same output function!

The partial fraction expansion results in subsections that have the following form:

$$H_{NP}(Z) = \frac{Z\left[b_{0,NP}Z + b_{1,NP}\right]}{Z^2 - a_{1,N}Z - a_{2,N}}$$

Let's determine the difference equation that describes one of the parallel subsections. We start by manipulating the parallel second-order subsection so that it contains negative exponents of Z:

$$H_{NP}(Z) = \frac{Y_{NP}(Z)}{X(Z)} = \frac{Z\left[b_{0,NP} Z + b_{1,NP}\right]}{Z^2 - a_{1,N} Z - a_{2,N}} \times \frac{Z^{-2}}{Z^{-2}}$$

$$= \frac{b_{0,NP} + b_{1,NP}Z^{-1}}{1 - a_{1,N}Z^{-1} - a_{2,N}Z^{-2}}$$

We follow by establishing the Z-domain output relation $Y(Z)$, which we convert to a discrete-time domain difference equation

$$Y_{NP}(Z) = b_{0,NP}X(Z) + b_{1,NP}X(Z)Z^{-1} + a_{1,N}Y_N(Z)Z^{-1} + a_{2,N}Y_N(Z)Z^{-2}$$
$$y_{NP}(n) = b_{0,NP}x(n) + b_{1,NP}x(n-1) + a_{1,N}Y_N(n-1) + a_{2,N}Y_N(n-2)$$

Figure 9–17 illustrates a form II implementation of the parallel subsection difference equation.

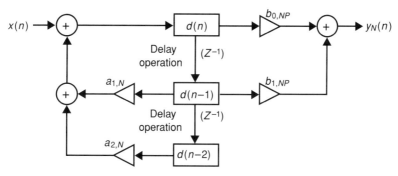

FIGURE 9–17
Direct form II second-order subsection of a parallel system.

Since the system function calls for summing the subsections, this allows us to achieve a parallel structure by computing each of the individual second-order subsections in parallel and by summing the results.

$$H(Z) = C + H_{1P}(Z) + H_{2P}(Z) + \cdots + H_{NP}(Z)$$

There are two possible ways to implement this parallel structure. One way uses a DSP system equipped with a single processing unit and the other uses a DSP system equipped with multiple processing units. Figure 9–18 illustrates both structures.

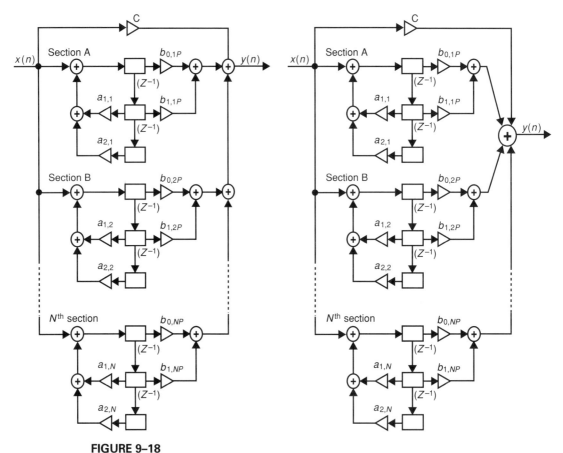

FIGURE 9–18
Parallel structures of an IIR system.

As Figure 9–18 indicates, each second-order subsection behaves as an independent system that receives a common $x(n)$ input. The output of each subsection is then added to produce the overall system output.

The two parallel structures differ only in the method used to sum the outputs of the individual subsections:

- In the left structure of Figure 9–18, a single processing unit computes and accumulates the subsection outputs one at a time. If you compare this structure with the cascade structure of Figure 9–14, you will find that it uses the same system resources. Therefore, nothing is to be gained using such a parallel structure. In fact, because the output regroups the subsections through addition operations, this structure tends to propagate more noise than a cascaded system.

- In the right structure of Figure 9–18, multiple processing units compute the sub-sections in parallel and the outputs are then summed. In this case, paralleling the second-order subsections requires the following system resources:

  $3N$ memory locations to store the delayed intermediate values (we cannot share the top intermediary value since the processing units are distinct).

  $2N = M$ delay operations.

  $4N = 2M$ MAC operations to compute the subsection values.

  $1N = 1/2M$ additional accumulate operations to add the results of the parallel subsections.

The system resources required to implement the parallel structure illustrated on the right side of Figure 9–18 are quite different from the resources required to implement the cascade structure of Figure 9–14:

- First, the parallel structure requires the use of a system equipped with parallel processing units. Such processors are available at a rather substantial increase in cost.
- Second, since the top intermediate memory location cannot be shared among parallel units, the overall parallel system requires $3N$ memory locations as opposed to $2N + 1$ memory locations for the cascaded system.
- Third, the parallel structure requires fewer multiplication operations, but this is not much of an advantage since it still requires the same number of accumulate and delay operations.

Given finite word length, the parallel structure is noisier than its cascade counterpart because it regroups the subsection outputs through an addition operation. The noise generated by each subsection therefore tends to accumulate (this is discussed in more detail in Section 9.2.3). Conversely, the noise in a cascaded subsection tends to be reduced by the filtering effect of the next subsection(s).

If cost is not a concern and speed is desired, a parallel structure will compute the output value much faster since the second-order subsections are processed in parallel.

Great processing speed may be achieved by using many processing units to implement a parallel structure.

The increase in speed is not directly proportional to the number of paralleled processing units. Figure 9–19 depicts a system implemented with different levels of parallelism. We can observe in the figure that the adder becomes larger as the level of parallelism increases. This means that the noise tends to grow worse. If the system uses software to implement the adder, increasing the level of parallelism decreases the time for MACs but increases the time for addition operations. Therefore, extra processing speed may be achieved by using a single *hardware adder* that regroups all the output(s) of the subsections. This hardware adder should have as many inputs as there are parallel subsections.

A: 8 pure cascaded subsections. Processing time of 8 × 5 = 40 MACs

B: 4 cascaded subsections using *double* processing units. Processing time of 4 × 4 = 16 MACs

+ 4 × 2 = 8 Adds

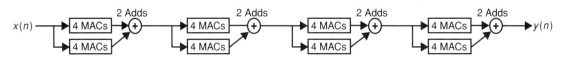

C: 2 cascaded subsections using *quadruple* processing units. Processing time of 2 × 4 = 8 MACs

+ 2 × 4 = 8 Adds

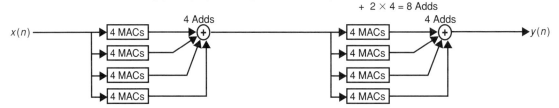

FIGURE 9–19
Different levels of parallelism.

9.2 UNDERSTANDING NOISE ISSUES

Acquiring, storing, and processing digital samples involves manipulating binary numbers stored in finite word-length memory. Consider the following example in which the signal goes through various bit formats when traveling from the input to the output of a system: A 14-bit ADC produces Q13 numbers that are stored in 16-bit memory locations. The processor then multiplies the stored input samples by 16-bit Q14 coefficients, and a 32-bit accumulator sums the Q27 result. Once the processing is finished, the accumulated result is stored in a 16-bit memory location and copied to a 14-bit DAC to generate the system output.

As the input samples journey through the DSP system, the format and word length change to accommodate the different situations. Generally, using more bits achieves a more accurate representation of the signal.

Three major factors influence the levels of noise in a DSP system:

1. The signal acquisition noise, which is determined by the signal level, the number of ADC bits, and the quality of the ADC.
2. The processing noise generated in the system subsections.
3. The propagation of noise as the subsections of a system feed into one another to form a processing structure.

The following section discusses how to control these noise factors in DSP systems.

9.2.1 MEASURING NOISE

Most ADCs and DACs use uniform step sizes when quantizing a signal. In this case, the absolute quantization error is always ±1/2 of the value of the least significant bit.

Let's examine what happens if we use a single bit to encode a signal. In this extreme case, the signal may adopt only one of two values, and the encoded number has an error margin of $\pm^1/_2$ bit. A one-bit number therefore carries quantization noise that spans the amplitude of one quantization level centered on the quantized values.

We can visualize the amount of quantization noise by picturing the ADC quantizing a signal of very small amplitude. The maximum amount of noise occurs when this small signal fluctuates at mid point between two quantization levels just at the threshold of decision. Figure 9–20 illustrates the worst-case quantization noise that results when quantizing a signal of very small amplitude.

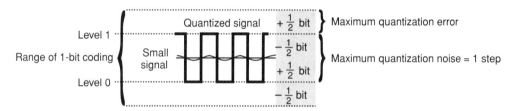

FIGURE 9–20
Quantization noise on a one-bit signal.

The maximum quantization noise is always the size of one quantization step no matter how many steps the ADC uses to encode the signal. We normally use a large quantity of steps to digitize a signal; therefore, it is likely that the signal will be much larger than the noise.

The standard way to evaluate noise issues at the system level is to compare the strength of the signal to that of the noise. This is done by calculating the *signal-to-noise ratio* (S/N), which we calculate by dividing the signal power by that of the noise. Considering that the power is proportional to the square of the amplitude and that the signal is much larger than the noise, this ratio is likely to yield very large results. Hence, we calculate the signal-to-noise ratio using the decibel units as follows:

$$\frac{S}{N} \text{ in dB} = 10 \log\left(\frac{\text{Signal amplitude}^2}{\text{Noise amplitude}^2}\right) = 10 \log\left(\frac{\text{Signal amplitude}}{\text{Noise amplitude}}\right)^2$$

$$= 20 \log\left(\frac{\text{Signal amplitude}}{\text{Noise amplitude}}\right) \text{dB}$$

When dealing with digitized signals, the number of quantization steps defines the maximum amplitude of a digitized signal, and the noise amplitude is that of one step. Substituting this yields the *signal-to-quantizing noise ratio* (SQR):

$$\text{SQR of a digitized signal} = 20 \log\left(\frac{\text{Number of steps used to encode the signal}}{\text{Noise is equal to one step}}\right) \text{dB}$$

$$= 20 \log\left(\frac{2^{\text{Number of bits}}}{1}\right) \text{dB}$$

Applying this to a one-bit signal, we calculate that every bit used to encode a signal contributes the following SQR:

$$20 \log\left(\frac{2^1}{1}\right) \cong 6 \text{ dB}$$

For example, the compact discs in most digital audio systems use 16-bit word lengths; therefore, they provide the following SQR:

$$16 \text{ bits} \times 6 \text{ dB} \cong 96 \text{ dB}$$

Interpreting this result, we can appreciate why it is that we use the decibel unit. A 96 dB SQR means that the signal amplitude is $2^{16} = 65{,}536$ larger than the noise amplitude and the signal power is $(2^{16})^2 = 4{,}294{,}967{,}296$ larger than the noise power. No wonder compact disc digital systems produce such high audio quality.

In theory, all ADCs and DACs achieve 6 dB of SQR and S/N per bit, but in practice, limitations in the fabrication processes introduce additional noise. The circuitry is unlikely to achieve the theoretical $\pm 1/2$ bit of precision that we mentioned earlier. Because of this, it is possible that the ADC or the DAC that you select will not perform exactly to expectations. For example, depending on the quality of the quantization circuitry, a 16-bit ADC may yield the equivalent of only 15 bits of accuracy. In general, using more bits results in a better SQR.

9.2.2 CONTROLLING THE INPUT NOISE

To obtain the best possible SQR performance from the ADC acquisition circuitry, it is necessary to apply an analog signal that swings as close as possible to the input rails. Such signal swings through the largest possible number of quantization steps; consequently, the ADC can use all of its available bits to encode the signal.

Making the analog signal as large as possible (without clipping) achieves the best possible input sample SQR.

Numerous techniques exist to ensure that the acquired digital signal maintains the best possible SQR as the system processes the samples. Following is a description of two of the most popular digital techniques that help maintain a better SQR through the processing steps.

Automatic Gain Control

Automatic gain control (AGC) schemes are designed to adjust the input signal so that it occupies the largest possible range. Modems commonly use such AGC circuitry to adjust the signal level acquired from the telephone line carrier.

A number of ways to implement AGC schemes exists, and we describe one of the most popular approaches. This approach implements an AGC by first acquiring a block of input samples. Once the samples are stored in memory, the processor locates the *largest* input sample value in the block. Based on the largest value, the processor uniformly scales all the samples in the block by a factor calculated to make the largest sample grow to the largest possible value that can be accommodated. Note that this technique does not actually improve the signal SQR because the scaling process also scales the signal noise up by exactly the same factor.

The usefulness of the technique reveals itself when we look at the subsequent processing of these larger-valued input samples. The multiplication operation that is invariably used by the processing always produces results having a number of bits equal to the sum of the bits contained in the multiplier and the multiplicand. For example, multiplying two 16-bit numbers coming from memory locations yields a 32-bit result in the product register. To store this result back into a memory location, we must typically discard half of the resulting bits. The number of remaining bits sets the noise floor level. If the processed signal is low in amplitude, the SQR could be reduced. Fortunately, the AGC scaling process produces larger-valued input samples, therefore, the product will also yield larger results. More significant bits will therefore survive the truncation process. Had the samples not been scaled up by the AGC process, the remaining signal and the resulting SQR would be much smaller after the processing.

A-Law and μ-Law Encoders

These encoders are used throughout the telephone industry. They are based on special ADCs designed to encode signals that contain a large range of natural amplitude variations (such as voice). For example, some people are very soft spoken and, conversely, others have thunderous voices. The encoding technique used by the ADC manages to maintain a reasonably good SQR for low- and high-amplitude signals.

The solution is to design the ADC so that it does not use linear quantization steps. Such devices are called *codec*s because they are used to *co*de and *dec*ode the input signal using a companding (*comp*ressing and ex*panding*) scheme. Two standards exist for codecs:

- European countries use the A-law standard.
- North Americans use the μ-law standard.

The μ-law standard is so common in North America that it warrants an additional discussion. This standard effectively compresses the equivalent of signed 13-bit samples into an 8-bit signed format, which is compatible with the telephone transmission system. The codec makes some compromise when making this conversion. The approach is to use smaller steps at the low levels and larger steps at the higher levels. Figure 9–21 illustrates the 8-bit encoding of μ-law numbers and depicts the different step sizes for positive values.

S	Compressed 8-bit μ-law							S	Biased 13-bit linear input											
0	1	1	1	Q_3	Q_2	Q_1	Q_0	0	1	Q_3	Q_2	Q_1	Q_0	x	x	x	x	x	x	x
0	1	1	0	Q_3	Q_2	Q_1	Q_0	0	0	1	Q_3	Q_2	Q_1	Q_0	x	x	x	x	x	x
0	1	0	1	Q_3	Q_2	Q_1	Q_0	0	0	0	1	Q_3	Q_2	Q_1	Q_0	x	x	x	x	x
0	1	0	0	Q_3	Q_2	Q_1	Q_0	0	0	0	0	1	Q_3	Q_2	Q_1	Q_0	x	x	x	x
0	0	1	1	Q_3	Q_2	Q_1	Q_0	0	0	0	0	0	1	Q_3	Q_2	Q_1	Q_0	x	x	x
0	0	1	0	Q_3	Q_2	Q_1	Q_0	0	0	0	0	0	0	1	Q_3	Q_2	Q_1	Q_0	x	x
0	0	0	1	Q_3	Q_2	Q_1	Q_0	0	0	0	0	0	0	0	1	Q_3	Q_2	Q_1	Q_0	x
0	0	0	0	Q_3	Q_2	Q_1	Q_0	0	0	0	0	0	0	0	0	1	Q_3	Q_2	Q_1	Q_0
1	1	1	1	Q_3	Q_2	Q_1	Q_0	1	1	Q_3	Q_2	Q_1	Q_0	x	x	x	x	x	x	x
1	1	1	0	Q_3	Q_2	Q_1	Q_0	1	0	1	Q_3	Q_2	Q_1	Q_0	x	x	x	x	x	x
1	1	0	1	Q_3	Q_2	Q_1	Q_0	1	0	0	1	Q_3	Q_2	Q_1	Q_0	x	x	x	x	x
1	1	0	0	Q_3	Q_2	Q_1	Q_0	1	0	0	0	1	Q_3	Q_2	Q_1	Q_0	x	x	x	x
1	0	1	1	Q_3	Q_2	Q_1	Q_0	1	0	0	0	0	1	Q_3	Q_2	Q_1	Q_0	x	x	x
1	0	1	0	Q_3	Q_2	Q_1	Q_0	1	0	0	0	0	0	1	Q_3	Q_2	Q_1	Q_0	x	x
1	0	0	1	Q_3	Q_2	Q_1	Q_0	1	0	0	0	0	0	0	1	Q_3	Q_2	Q_1	Q_0	x
1	0	0	0	Q_3	Q_2	Q_1	Q_0	1	0	0	0	0	0	0	0	1	Q_3	Q_2	Q_1	Q_0

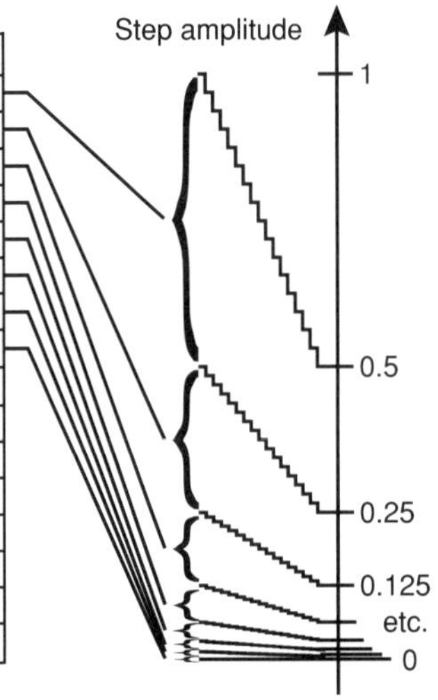

FIGURE 9–21
μ-Law encoding.

According to Figure 9–21, the μ-law encoding standard divides the positive and negative signal amplitudes into eight different ranges. Each range contains 16 equal steps that are half the amplitude of the step size used in the previous range. The lower ranges therefore encode low-amplitude signals using smaller, more precise steps, thereby providing good resolution for low-amplitude signals. Notice, however, that both the entire top half and bottom half of the signal amplitude are encoded using only 16 rather large steps, which provide coarse resolution for high-amplitude signals. This encoding technique has the advantage of providing a reasonably good SQR for all types of signals. In practice, μ-law encoding provides 24 to 30 dB of SQR for most signal amplitudes. This allows the low whispers to get through the telephone transmission with an acceptable SQR.

Since μ-law encoding is not linear, the samples cannot be sent through a DSP algorithm in the coded form. We must expand the μ-law samples to a linear bit-format before applying any processing. Although the expansion uses some processing time, the standard sampling rate in telephone systems is 8 k-samples/second, which provides a huge processing interval of 125 μs to perform the conversion and the processing.

9.2.3 PROCESSING A NOISY SIGNAL

The processing algorithm operates on the input samples by multiplying the signal samples and by accumulating the products. As the signal is processed, the mathematical manipulation of the samples may introduce additional noise. Understanding where the noise comes from will help us to develop noise control techniques.

Accumulator Noise

DSP algorithms call for accumulating weighted samples. If each of the samples contains a certain amount of noise, what happens to the noise level as we add the samples? To answer this question, consider the example of adding two arbitrarily chosen decimal numbers that have an absolute tolerance of ±0.1:

$$6.7 \pm 0.1$$
$$+ \ \ 5.8 \pm 0.1$$
$$12.5 \pm 0.2 \ \textit{(this is the maximum possible tolerance error)}$$

Examining this addition, we note the following:

- The number of integer digits has increased by an extra digit.
- The number of fractional digits is not increasing.
- The maximum possible tolerance error is the sum of the tolerance errors (cumulative).

Noise is similar to a tolerance error; however, the accumulation of noise depends on the *type* of noise that is present. For example, if the noise is completely random (white noise), it tends to statistically cancel out when performing a large number of additions. This is similar to picking 100 numbers at random in the ±5 range and adding all of them. In this case, the result is likely to be close to zero. Nevertheless, there is still a very small probability that the result will be 500.

In digital signal processing, exactly the same rules apply to binary numbers:

$$\texttt{SSII.FFFN} \ \text{(where N is a noisy bit)}$$
$$+ \ \ \texttt{SSII.FFFN}$$
$$\texttt{SIII.FFNN} \ \text{(note that the noisy bits accumulate)}$$

The addition operation therefore results in the following effects:

- The number of integer bits *could* increase.
- The number of fractional bits *does not* increase.
- The noise *could add* up to a large value, depending on the type of noise.

The amount of noise generated by an addition operation depends on the nature of the noise. If the noise is not completely random, repeated addition operations could considerably deteriorate the SQR of the processed signal.

Multiplier Noise

DSP algorithms call for accumulating weighted samples. A weighting operation is performed by multiplying a sample by some constant coefficient value. If the samples contain a certain amount of noise, what happens to the noise as we multiply the samples? To answer this question, consider the example of multiplying the following two decimal numbers:

$$6.7 \ \pm \ 0.1 \ \ \text{(sample)}$$
$$\times \ 5.8 \ \ \ \ \ \ \ \ \ \ \ \text{(constant coefficient)}$$
$$38.86 \pm \ 0.58 \ \textit{(this is the maximum possible tolerance error)}$$

Examining this multiplication, we note the following:

- The number of integers digits is increasing.
- The number of fractional digits is increasing.
- The maximum possible tolerance error is multiplied.

If we consider the tolerance to be the noise on the sample, the SQR *has* not been changed by the multiplication operation since the noise and the sample have increased by the same factor:

$$\text{SQR: } 20 \log\left(\frac{6.7}{0.1}\right) = 36.5 \text{ dB} \qquad \text{SQR: } 20 \log\left(\frac{38.86}{0.58}\right) = 36.5 \text{ dB}$$

Let's now examine multiplication in the binary world:

```
       SII.FFFFN        (sample)
    ×    SIII.FFFF      (constant coefficient with 3 I-bits)
    SSIIIII.FNNN NNNNN  (the noise moves up by three positions)
```

As we can see, the number of sign and integer bits in the result is simply the total number of such bits in the multiplier and multiplicand. However, the resulting noise *level* (the position of the most significant noise bit) depends on the value of the multiplier. In the preceding example, the multiplier has three integer bits so the noise level could grow to become almost eight times larger, which explains why the noise bit is shifted three positions to the left.

It is interesting to note that although the SQR does not change, the noise level depends on the magnitude of the multiplier. Consider the following numerical example, in which the multiplier has a pure fractional (smaller than 1) value of 1/8:

```
      011.1111N             (sample)
                                                   ⎛                         1 ⎞
    ×    0.0010000     ⎜ constant coefficient  =  ─ ⎟
                                                   ⎝                         8 ⎠
    0000.0111111NNNNN   (the noise shifts 3 positions to the right)
```

In this case, the noise level becomes smaller since the noisy bits get shifted three bits to the right of the multiplicand noise level. Although the multiplication operation does not change the SQR, the numerical result and the noise level become smaller when the multiplier bears smaller values. Remember that we need to accumulate products when we compute a difference equation. When summing many products, the repeated accumulation operations are likely to require more integer bits. Given a limited number of bits in the accumulator, we face the prospect of an overflow. Nevertheless, if the individual products bear smaller values, we can accumulate more of them before overflow becomes a concern. Consequently, using constant coefficients bearing smaller values allows us to implement a difference equation that contains more terms.

This makes a case for breaking the system difference equation into subsections that implement coefficients that bear smaller values. In the Cascade Section (p. 291) it was shown that subsections may be arranged so that they contain a small number of coefficients whose value is close to 1. Since the cascaded subsections implement a smaller difference equation, fewer accumulate operations are required to compute each subsection result.

**A system implemented as a cascade of subsections
reduces the risk of accumulator overflow.**

Truncation Noise

As discussed in the analysis of the multiplication and addition operations, the processing accumulates the weighted samples described by the difference equation. As each weighted sample accumulates, the result becomes larger, which means that it requires more integer bits. If an in-depth analysis of the system reveals a probable overflow of the accumulator, some intervention is required to reduce the numerical value of the weighted samples. To prevent the overflow, either the *product* and/or the *samples* must be truncated to a smaller bit-size (the circuitry to perform truncation is described in Section 9.3). Truncating bits from the input samples would reduce the SQR, so we are left with the alternative of truncating the product through shift-right operations. In this case, we face the prospect of shifting valuable noise-free bits out of the product register. If this happens, the SQR will deteriorate.

Truncation also becomes necessary when the final accumulated result is transferred to some memory location and/or to a DAC. The memory location and the DAC are not likely to have enough bits to accommodate all the bits of the accumulated result; consequently, we must discard some of the least significant bits from that result. Again, if noise-free bits are truncated, there will be a reduction of the SQR.

9.2.4 CONTROLLING THE SYSTEM SQR

Every system subsection is associated with a certain level of noise. For example, a 14-bit DAC has an ideal SQR of 14×6 dB = 84 dB. Every subsection of the overall DSP system operates using a limited number of bits; therefore, each subsection is associated with a particular SQR. Once noise insinuates itself, it cannot be removed. Consequently, the SQR of a signal traveling through subsections cannot be improved. Fortunately, in a digital system, the amount of noise does not grow worse so long as the subsections have an SQR that is at least as good as the SQR of the signal coming into the subsection. Because of this, the overall system SQR will be only as good as the SQR of the *worst* subsection. Figure 9–22 illustrates this by showing the SQR of a signal as it travels through a system.

Note in Figure 9–22 that the ADC and the DAC are practical devices that do not achieve the theoretical 6 dB of SQR per bit. Also note that we lose only 1 dB of SQR by using a less expensive 12-bit DAC because this system contains a rather noisy subsection. A conclusion that we can draw is that there is no advantage to using an ADC that has more bit resolution than the DAC.

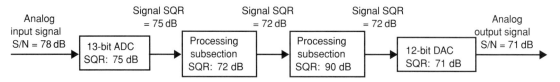

FIGURE 9–22
SQR of a signal traveling through many subsystems.

Filtering Effects on System Noise

In a cascaded system, the signal travels with the noise from subsection to subsection. Each subsection is effectively a filter that acts on both the signal and the noise. The filtering effects of subsections are best illustrated by examining the behavior of a narrowband filter such as a notch-stop filter. As an example, consider a second-order narrowband IIR notch-stop filter that has poles and zeros positioned at the following locations:

$$\text{Conjugate zeros at } 1\, e^{\pm\frac{\pi}{5}j} \qquad\qquad \text{Conjugate poles at } 0.99\, e^{\pm\frac{\pi}{5}j}$$

Figure 9–23 illustrates a plot of the poles and zeros and the gain characteristics of this narrowband filter. The Z-transform of the narrowband notch filter of Figure 9–23 is as follows:

$$H(Z) = \frac{Z^2 - 2 \times 1 \cos\left(\frac{\pi}{5}\right) Z + R_Z{}^2}{Z^2 - 2 \times 0.99 \cos\left(\frac{\pi}{5}\right) Z + R_P{}^2} = \frac{Z^2 - 1.61803\, Z + 1}{Z^2 - 1.60185\, Z + 0.9801}$$

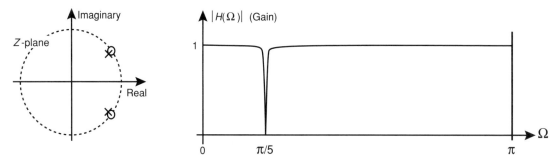

FIGURE 9–23
Characteristics of a narrowband notch filter.

We separate the processing of this second-order system in two parts. One implements the system zeros (FIR part) and the other implements the poles (recursive part). The order in which we process these parts allows us to create two implementation structures:

$$H(Z) = \frac{Z^2 - 1.61803\, Z + 1}{1} \times \frac{1}{Z^2 - 1.60185\, Z + 0.9801} \qquad \text{(Direct form I)}$$

$$H(Z) = \frac{1}{Z^2 - 1.60185\,Z + 0.9801} \times \frac{Z^2 - 1.61803\,Z + 1}{1} \quad \text{(Direct form II)}$$

Note that the direct form I is not used in practice. We are about to witness why this is so. The direct form I structure processes the *FIR* part first as opposed to the direct form II, which processes the *recursive* part first. As we can see, the two processing structures share the exact same coefficient values. Consequently, the coefficient quantization error is the same for both structures. Figure 9–24 compares the processing structure necessary to implement the zeros of the FIR part to that which is necessary to implement the poles of the recursive part.

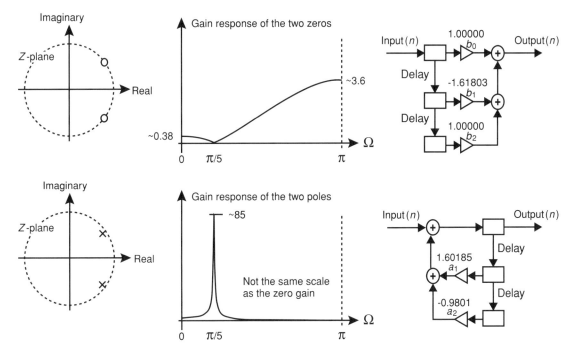

FIGURE 9–24
Comparison of the implementation of the FIR and recursive parts.

As Figure 9–24 indicates, the poles of the recursive section contribute a gain that is much greater than that of the zeros in the FIR section. This is typical of most IIR systems, and we must be careful to arrange the processing so that the high-gain part does not amplify the noise levels. The order in which we cascade the two sections is crucial.

Let's purposely make the mistake of using a direct form I to compute the low-gain FIR part first. The FIR response curve shows that the signal amplitude increases over some range of frequencies but that it is attenuated to zero at the notch frequency. In this example, we dramatize this by assuming that the input signal has a rather poor SQR. We can picture what happens to the signal and to the noise level if we imagine an input signal that has a frequency content of uniform amplitude for all frequencies. Figure 9–25 shows this input signal on the left side, and the central part illustrates the amplitude of the signal and of the noise level after the zeros have been processed.

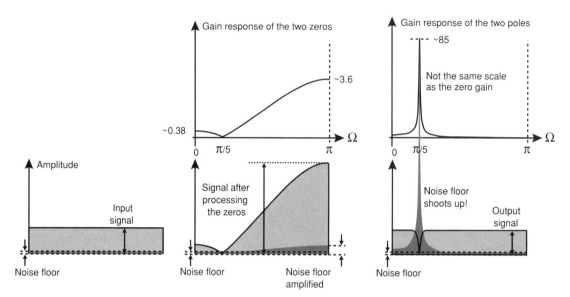

FIGURE 9–25
Effect on the noise levels if we process the zeros first.

According to the central part of Figure 9–25, although the amplitude of the signal drops to zero at the notch frequency, the noise floor is *not* at zero. If we now shift our attention to the high-gain IIR processing that follows, the right part of the figure illustrates that the noise level shoots up when we process the high-gain poles of the system.

The alternative is to compute a direct form II structure in which the poles are computed first. In this case, the recursive part amplifies both the signal and the noise floor proportionally. The central part of Figure 9–26 illustrates the signal and noise levels after processing the poles.

We notice, in the right part of Figure 9–26, that processing the zeros attenuates both the signal and the noise to acceptable levels around the narrowband notch frequency. In this example, since we assumed that the noise floor is constant, the small gain of the zeros slightly increases the noise levels in the high frequency range. This reinforces the conclusion that we reached in the section in this chapter on Cascade Structure, where we showed that it is better to process the high-gain subsections first.

Lower noise levels are achieved when we implement
IIR subsections using a direct form II structure.

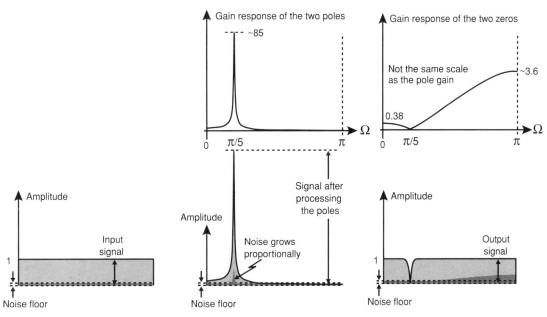

FIGURE 9–26
Effect on the noise levels if we process the poles first.

Scaling the System Difference Equation

Sometimes it is necessary to make scaling adjustments to the system gain response curve. For example, a system that can produce a maximum gain of 9 requires four I-bits to accommodate the magnitude of the output signal $y(n)$. The four I-bits can accommodate a range of $\pm16^-$, but the output signal covers only 9/16 of that range, wasting about 5 dB of SQR. If we could scale the system gain curve *proportionally* so that the maximum occurred at approximately 15, the relative *characteristics* of the system gain curve would be left unchanged. In this case, the output signal would use 15/16 of the available range, gaining us a few dB of SQR.

Scaling is simply a question of making all the output samples $y(n)$ larger or smaller by a certain constant factor.

> **Scaling the system coefficients results in making the gain response curve proportionally larger or smaller. Scaling the system coefficients does not affect the phase response.**

Let's determine how to scale a system. Assume that we want to scale the output samples of a second-order IIR section by a factor of ψ.

$$y(n) = b_0\, x(n) + b_1\, x(n-1) + b_2\, x(n-2) + a_1 y(n-1) + a_2 y(n-2)$$

$$y_{\text{Scaled}}(n) = \psi \times y(n)$$

$$y_{\text{Scaled}}(n) = b_0 \, \psi \, x(n) + b_1 \, \psi \, x(n-1) + b_2 \, \psi \, x(n-2) + a_1 \, \psi \, y(n-1)$$
$$+ a_2 \, \psi \, y(n-2)$$

Since $y_{\text{Scaled}} = \psi \times y(n)$, then

$$\Psi \, y(n-1) = y_{\text{Scaled}}(n-1) \quad \text{and} \quad \Psi y(n-2) = y_{\text{Scaled}}(n-2)$$

Substituting, we can therefore rewrite

$$y_{\text{Scaled}}(n) = b_{0(\text{Scaled})}x(n) + b_{1(\text{Scaled})}x(n-1) + b_{2(\text{Scaled})}x(n-2)$$
$$+ a_1 y_{\text{Scaled}}(n-1) + a_2 y_{\text{Scaled}}(n-2)$$

where

$$b_{0(\text{Scaled})} = \Psi b_0, \; b_{1(\text{Scaled})} = \Psi b_1, \; \text{and} \; b_{2(\text{Scaled})} = \Psi b_2$$

Notice that the value of the a_k coefficients has *not* been altered. Consequently, scaling a system requires a scaling of only the b_k coefficients that correspond to the FIR part of the system. The recursive a_k coefficients of the IIR part are left unchanged by the scaling process.

9.3 PROCESSING ISSUES IN FIXED-POINT SYSTEMS

Whatever processor or structure you elect to use, computing a difference equation invariably requires going through a standard sequence of operations. Figure 9–27 illustrates the operations performed by the typical parts of a digital signal processor's arithmetic logic unit (ALU). Referring to Figure 9–27, let's examine the typical steps involved in the processing of a signal.

Step 1: Fetching the Samples

Computing the difference equation requires a number of multiplication operations that weight samples according to the difference equation coefficients. To prepare for each multiplication, the processing algorithm fetches a sample and stores it in the multiplicand. Note that FIR systems use only *input* samples. Unless the samples have been shifted, the Q format of the multiplicand is likely to be the same as that of the pure-fractional ADC samples.

Conversely, an IIR system needs to weight samples coming from either the input, the output, or intermediary memory locations. In IIR systems, the Q format of the multiplicand typically reflects that of the output sample being fed back.

Step 2: Fetching the Coefficients

The next step is to fetch the multiplier, which is the value of one of the constant coefficients stored in memory. The Q format of the multiplier therefore reflects that of the stored system coefficients. The coefficients typically adopt a Q format that contains enough integer bits to accommodate the largest coefficient value.

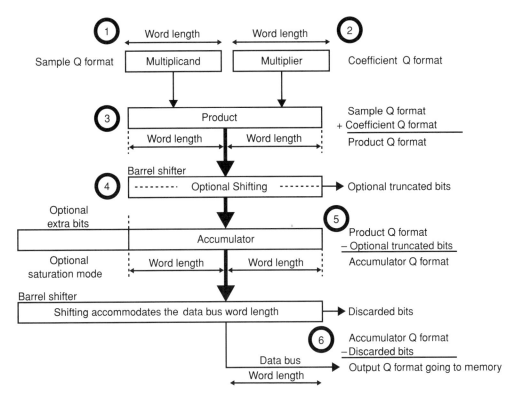

FIGURE 9–27
Signal processing operations.

Step 3: Weighting the Samples

Now that the multiplicand and multiplier are in place, the processor performs the multiplication necessary to compute the value of the weighted sample. The result of this multiplication contains a number of bits equal to the sum of the bits in the multiplicand and multiplier. Accordingly, the Q format of the product is the sum of the multiplicand and multiplier Q formats. For example, a Q14 multiplier and a Q13 multiplicand yield a Q27 product.

Step 4: Optional Shifting of the Product

The product obtained in step 3 is only one of the many weighted samples that the processing algorithm must accumulate. Based on the value of the weighted samples and on the length of the difference equation, we must choose an accumulator Q format that includes enough integer bits to accommodate the sum of all the weighted samples. We know that the accumulated result has the same Q format as that of the numbers being accumulated. For example,

$$Q27 \text{ number} + Q27 \text{ number} + Q27 \text{ number} = Q27 \text{ accumulated result}$$

Because of this, some processors provide the option of a barrel shifter to truncate some fractional bits *before* the product is accumulated. This allows us to adjust the product format to match our choice of accumulator Q format. Shifting effectively truncates the Q format of the product by discarding fractional bits. This runs the risk of removing valuable noise-free bits, but it effectively leaves more room for integer bits in the accumulator.

Step 5: Accumulating the Weighted Samples

Once the processing has finished computing the difference equation, the accumulator contains the sum of all the weighted samples. We must set up the processing so that the Q format of the accumulator minimizes the prospect of an overflow. If the accumulator overflows, the output samples become corrupted for some transitory interval of time. To reduce the possibility of overflow, some processors offer extra accumulator bits. These extra bits also help to maintain a better SQR by allowing the accumulator to sum a larger number of products that use larger Q formats.

Even if you plan every detail properly, it is still possible to experience infrequent accumulator overflows. Any accumulator overflow is not good for the frequency content of the signal being processed. An overflow can seriously modify the frequency content of the output signal. In the eventuality of overflow, the damage can be reduced by an optional saturation mode (see Section 1.3.2). Note that since FIR systems normally use modulo addition (see p. 376), the saturation mode should be used only with IIR systems.

Step 6: Outputting the Result

The processing of the difference equation is finished when all the weighted samples have been summed in the accumulator. At this time, the accumulator contains an output sample $y(n)$, and this value must be transferred either to the DAC or to the next processing subsection and/or to a memory location. In any case, some bits must be discarded from the accumulated result to accommodate the smaller word length handled by these destinations. A barrel shifter effectively produces the same effect as shift-right operations to discard some of the least significant bits from the accumulated result.

9.3.1 SELECTING APPROPRIATE Q FORMATS

The Q format and the number of bits used to encode any variable in fixed-point binary systems determines the following:

- The level of precision attained in coding the variable.
- The range of values that may be accommodated by the fixed-point variable.

Q Format of the Accumulator

Selecting particular Q formats for the different parts of the system usually begins by evaluating a range of values that is acceptable for the *output* samples. Since the output sample is assembled in the accumulator, the Q format selected for the accumulator *must* hold enough integer bits to accommodate some maximum numerical value. For example, a 32-bit accumulator containing results that cover a range of ±25 must bear five I-bits and hence a Q26 format:

SIIIII.FF FFFF FFFF FFFF FFFF FFFF FFFF (range of ±32⁻)

Q Format of the Multiplicand

The multiplicand holds the samples that are weighted by the multiplication operations. The Q format of these samples is determined as follows:

- If the processor is weighting input samples $x(n - k)$, the multiplicand should reflect the pure-fractional Q format of the ADC input samples. For example, a 14-bit ADC produces pure-fractional Q13 samples:

   `S.F FFFF FFFF FFFF` (range of $\pm 1^-$)

   Note that the number of significant bits used to encode the samples has a direct impact on the system SQR.
- If the processor is computing the recursive part of a difference equation, the multiplicand must accommodate the number of integers contained in the output samples being fed back. These samples are either output values $y(n - k)$ or intermediary values $d(n - k)$. For example, a signed 16-bit multiplicand loaded with intermediary samples containing five I-bits must carry a Q10 format:

   `SIII II.FF FFFF FFFF` (range of $\pm 32^-$)

Q Format of the Multiplier

The multiplier holds the value of the weighting factors. These are the system coefficients, and the multiplier Q format therefore accommodates the range of values covered by the a_k or b_k coefficients. Note that the number of significant bits used to encode the coefficients has a direct impact on the amount of poles/zeros shift produced by the quantization error. However, the constant coefficient values in the multiplier have no impact on the system SQR.

Q Format of the Product

The format of the product is the sum of the multiplier and multiplicand Q formats:

$$Product \text{ Q Format} = Multiplier \text{ Q Format} + Multiplicand \text{ Q Format}$$

Note that there are two possibilities here:

1. Weighted input samples, $Product = b_k \times x(n - k)$, which are used by FIR and IIR systems.
2. Weighted recursive samples, $Product = a_k \times y(n - k)$ or $Product = a_k \times d(n - k)$, which are used only by IIR systems.

System parameters such as the word length of various registers and of the ADC, the maximum system gain, and the value of the system coefficients set an upper value for the Q format of the product. These constraints cause the Q format of the multiplicand and multiplier to become linked as follows when using fixed word length:

- *Increasing* the multiplicand (samples) Q format increases the system SQR. Unfortunately, to achieve this, we must *decrease* the multiplier Q format.
- *Increasing* the multiplier (coefficients) Q format results in a more precise positioning of the system poles and zeros. Unfortunately, to achieve this, we must *decrease* the multiplicand Q format.

The next step is to compare the Q format of the weighted samples held in the product register to the Q format required by the accumulated sample. This comparison determines how many bits, if any, to truncate from the product register before performing the accumulate operation. As an example, consider a system that accumulates Q28 products and that requires seven I-bits to accommodate the magnitude of the output samples. Figure 9–28 illustrates how extra accumulator bits may help to avoid product-shifting operations. Each of the two cases illustrated in Figure 9–28 uses Q28 products that must be adjusted to comply with an accumulator format that contains seven integer bits. In Case 1, the presence of four extra accumulator bits provides enough room to directly accumulate the products with no need for product shifting. In Case 2, four fractional bits must be truncated from the product to accommodate the Q24 accumulator format that provides the required seven I-bits. Note that the truncation, which is necessary in Case 2, removes four bits of precision, which may reduce the system SQR.

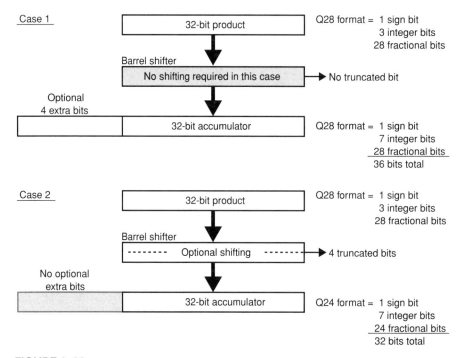

FIGURE 9–28
Product shifting to comply with the accumulator Q format.

9.3.2 OVERFLOW IN FIXED-POINT FINITE IMPULSE RESPONSE SYSTEMS

The output of an FIR system is based on the addition of $N + 1$ difference equation terms:

$$y(n) = b_0x(n) + b_1x(n-1) + \cdots + b_Nx(n-N)$$

Each difference equation term consists of a weighted input sample, which the processor sums in the accumulator when computing the value of $y(n)$. To prevent overflow conditions, we must find ways to predict the magnitude of the results being accumulated. The following sections describe two approaches that may be used to determine the required number of I-bits and hence the appropriate accumulator format.

Selecting the Q Factors of a Finite Impulse Response System Based on the Value of the System's Coefficients

Since the $x(n - k)$ input samples are generated by an ADC, they bear a pure-fractional Q format. Multiplying any number by a fraction can only decrease the value of that number. The individual weighted terms of the summation must therefore be equal to or smaller than the value of the weighting factor:

$$|b_k x(n - k)| \le |b_k|$$

Since the input samples can be positive or negative, its possible that all weighted samples end up being positive no matter what the sign of the weighting factor. This allows us to state that the value of the output sample $y(n)$ cannot exceed the absolute sum of the weighting factors that define the system:

$$y(n) \le \left| \sum_{k=0}^{N} b_k \right| \quad (\text{since } |x(n - k)| < 1)$$

Knowing the absolute upper limit on the accumulated weighted input samples allows us to format the accumulator so that it contains enough integer bits to prevent an overflow. For example, let's plan the direct implementation of a linear-phased FIR system defined by the following difference equation:

$$y(n) = x(n) - 6.010437x(n - 1) + 17.062636x(n - 2) - 6.010437x(n - 3) + x(n - 4)$$

In this case, the largest possible value for $y(n)$ is

$$y(n)_{\text{Max}} \le \left| \sum_{k=0}^{4} b_k \right|$$

$$= 1 + 6.010437 + 17.062636 + 6.010437 + 1$$

$$= 31.08351$$

The accumulator therefore requires five I- bits to prevent an overflow.

Imagine now that we are using a 32-bit accumulator, a 14-bit signed ADC, and 16-bit memory locations to store the system coefficients. Based on this, we can make a first estimate at the required Q formats for this system:

- Q26 accumulator: 1 sign, 5 *integers* ($\pm 32^-$ range), 26 fractional bits.
- Q13 samples: The ADC produces signed pure-fractional samples ($\pm 1^-$ range).
- Q10 coefficients: 1 sign, 5 integers ($\pm 32^-$ range), 10 fractional bits.

Note that a Q10 format accommodates the largest coefficient: $b_2 = +17.062636$. The product of a coefficient and an input sample $b_k x(n-k)$ result in a Q23 product Q10 × Q13. Directly accumulating the Q23 products yields a Q23 result in the accumulator. This Q format in the accumulator allows 23 fractional bits, which leaves room for eight I-bits and one sign bit. The eight I-bits provide more than the required five I-bits that satisfy this particular system. We could therefore alter the estimated accumulator Q format from Q26 to Q23 with no loss of precision or SQR.

The fourth-order system that we just analyzed carried conditions that were rather easy to meet. Systems that have a large number of coefficients and/or large gains and/or many frequency bands will impose tighter programming constraints.

The following example describes a linear-phase FIR system that has stringent operational requirements. This system uses a signed 16-bit ADC, a 32-bit accumulator, and word lengths of 16 bits. In this case, let's say that the difference equation of this system contains a large number of coefficients that result in the following upper limits:

$$y(n) \leq \left| \sum_{k=0}^{N} b_k \right| = 72.38344 \qquad \text{Largest } b_k \text{ has a value of 5.32654}$$

Based on this, we can make a first estimate at the required Q formats for this system:

- Q24 accumulator to accommodate the absolute maximum gain of 72.38344: 1 sign, *7 integers*, 24 fractional bits ($\pm 128^-$ range).
- Q15 samples produced by the pure-fractional signed ADC.
- Q12 coefficients to accommodate the largest coefficient value of 5.32654: 1 sign, *3 integers* ($\pm 8^-$ range), 12 fractional bits.

In this example, the weighted samples $b_k x(n-k)$ result in a Q27 product Q12 × Q15. We cannot accumulate the Q27 products directly since this would yield a Q27 result in the accumulator. This would not leave enough room for the required seven I-bits; consequently, the accumulator would overflow.

A number of alternatives exist to solve this problem condition.

EXTRA ACCUMULATOR BITS The first step is to verify that the system is equipped with optional extra accumulator bits. In our example, the accumulator bit-size needs to be increased by three bits so that the Q27 result has room for seven I-bits and a sign. If the system hardware provides three extra accumulator bits, there is no overflow problem.

PRODUCT SHIFTER If the accumulator is not equipped with extra accumulator bits, the next step is to determine whether the system allows us to shift the product *before* performing the accumulate operation. If this is so, the shifter can be used to truncate three fractional bits from the product. This effectively converts the product to a Q27 − 3 = Q24 format, which matches the required accumulator format. Note that this could reduce the SQR.

REDUCING THE MULTIPLIER AND/OR MULTIPLICAND Q FORMAT If the system is not equipped with extra accumulator bits or a product shifter, the next step is to examine the possibility of reducing the number of bits used to code either the multiplicand or the multiplier. This will alter some part of the system performance. Reducing the bit-size of the multiplicand effectively means reducing the bit-size of the ADC. This has the immediate effect of reducing the system SQR.

Reducing the bit-size of the multiplier is likely to increase the zero quantization error. This means that the system zeros may experience increased shifts in positions when the coefficients are quantized (see Section 7.6.3). Consequently, the practical system response could move farther from the desired response.

In our example, we could use a 14-bit ADC (Q13 samples) and Q11 coefficients. The product would then result in a Q13 × Q11 = Q24 format, matching the accumulator constraint. Unfortunately, this compromise introduces an additional 12 dB of noise (we lost two ADC bits), and the system response runs the risk of moving farther from the target response curve.

Selecting the Q Factors of a Finite Impulse Response System Based on the System Response Curve

In the previous section, we established that the maximum value of any output sample may be calculated by summing the absolute value of the system coefficients:

$$y(n) \leq \left| \sum_{k=0}^{N} b_k \right|$$

This approach of calculating the maximum value of $y(n)$ is very conservative because it covers absolute worst-case conditions. In practice, the value of most output samples is likely to be well below the absolute maximum limit. If your system can tolerate the occasional rare accumulator overflow, its possible to lower the limit value for $y(n)$. A lower limit on $y(n)$ means that we can increase the Q format of the input samples and system coefficients. This allows a better SQR and a more accurate positioning of the system zeros. In any case, FIR systems are inherently stable; consequently, the rare event of an overflow produces only short-lived output transients.

We can use the system gain response curve to establish a reasonable practical limit on the maximum value for $y(n)$. This approach is especially valid for the great majority of FIR systems that implement linear-phase systems. As an example, Figure 9–29 illustrates the gain response curve of a linear-phase multiband filter.

According to the gain response curve illustrated on Figure 9–29, the example system has a maximum gain of 5. Since the input samples $x(n)$ coming from the ADC are pure fractional, the maximum output of this system should not exceed the following range of values:

$$y(n) = 5x(n) \quad \text{where } x(n) \text{ covers a range of } \pm 1^-$$
$$y(n) = 5 \times (\pm 1^-) = \pm 5^-$$

Using this approach sets a practical limit for $y(n)$ that is certainly lower than the absolute sum of all the system coefficients. Consequently, it allows us to lower the number of integer bits required in the accumulator. In this case, an accumulator containing three I-bits

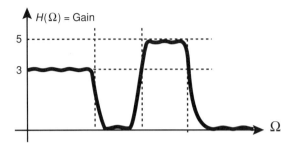

FIGURE 9–29
Gain response curve of an FIR multiband filter.

easily accommodates the maximum gain of 5. If our system has a 32-bit accumulator, we obtain a Q28 format:

$$\texttt{SIII.FFFF FFFF FFFF FFFF FFFF FFFF FFFF}\ (\text{Range of } \pm 8^-)$$

It is important to note that for this approach to work properly, we must be careful when setting up one of the ALU features. Although the output samples are unlikely to exceed the range of $\pm 5^-$, this is true only when the processor has finished computing all the terms of the difference equation. It does not imply that there will not be any intermediary accumulator overflow during the calculation (see *using modulo addition* on p. 376 in Appendix B). If we use modulo addition, it is normal for the accumulator to produce intermediary overflows, which do not affect the final result. For modulo addition to work properly, we must ensure that we disable the accumulator *saturation* mode.

9.3.3 OVERFLOW IN FIXED-POINT INFINITE IMPULSE RESPONSE SYSTEMS

Programming IIR systems is more tricky than dealing with FIR systems. All IIR systems contain poles and this means that

- The recursive subsections could generate very high gains.
- The a_k coefficients, which control the position of the poles, may be very sensitive to the quantization error.
- The phase is nonlinear and therefore the altered shape of the signal could yield rather high-valued output samples.

The Gain in IIR Systems

The gain produced by any one system pole is inversely proportional to its distance from the unit circle on the Z-plane. Consequently, this gain grows exponentially as the pole moves closer to the unit circle. For example, if a pole is located at a distance of 0.99 from the origin, it is 0.01 away from the unit circle. The maximum gain that this particular pole contributes to the system is:

$$\frac{1}{1 - 0.99} = 100$$

This does not mean that the overall system gain will reach 100. However, it does mean that the recursive part of the subsection that contains this pole will reach a gain of 100 when processing an input containing that particular frequency.

The Quantization Error on the a_k Coefficients

Typically, the poles of IIR systems occupy positions located relatively close to the unit circle on the Z-plane. This means that a small shift in the position of poles could produce large deviations from the expected system response and possible unstable operation. To reduce the quantization error, you must ensure that the Q format of the a_k coefficients is large enough to prevent an unacceptable shifting of the pole locations. Note that the Q format of the b_k coefficients is not so sensitive since they are associated with the system zeros, which may safely be located anywhere on the Z-plane. Consequently, you may use different Q formats for the a_k and b_k coefficients.

The Nonlinear Phase Response

Remember that the input signal contains a mix of sinusoid components. If the phase response is nonlinear, the alignment of these components will change as the signal travels from input to output. Figure 9–30 illustrates what could happen when a signal containing two sinusoid components travels through an all-pass phase equalizer. In this case, the system has a flat gain response of 1, and the phase of the component with the highest frequency shifts by 180°.

Since the gain in Figure 9–30 is 1, the amplitude of both sinusoid components remains the same. However, the phase equalizer realigns the two components in such a way that the resulting output signal reaches significantly higher amplitudes.

Let's learn how to handle the IIR constraints by working out a practical example. We will use a direct form II structure to implement a narrowband notch-stop filter that has poles and zeros positioned at the following locations (this uses the filter described in Figure 9–23):

$$\text{Conjugate zeros at } 1\,e^{\pm\frac{\pi}{5}j}$$

$$\text{Conjugate poles at } 0.99\,e^{\pm\frac{\pi}{5}j}$$

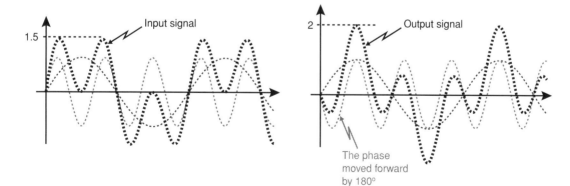

FIGURE 9–30
Change of the phase can change the amplitude.

The direct form II Z-transform of this narrowband notch filter is as follows:

$$H(Z) = \frac{1}{Z^2 - 1.60185\,Z + 0.9801} \times \frac{Z^2 - 1.61803\,Z + 1}{1}$$

Implementing this system requires us to select Q formats for the different parts of the processing. In this example, we arbitrarily try to accommodate a signed 14-bit ADC, a 32-bit accumulator, and 16-bit word lengths.

The direct form II structure requires that the recursive part be processed first. The processing requirements of the recursive part lead to the following arguments for choosing the Q formats used in the processing:

- The pure-fractional signed 14-bit ADC produces Q13 samples.
- According to the gain curve of Figure 9–24, the two poles of the recursive part combine to produce a large gain of ~85. This requires an accumulator format containing at least seven I-bits (±128⁻ range). The *maximum* accumulator Q format is

  1 sign, 7 integers, 24 fractional = Q24 format

- We hope that the Q24 format will provide enough headroom to accommodate the extra amplitude swing that is possible due to the phase-shifting effect. The phase shifting should not cause problems in this case because of the narrowband performance of this filter.

  Important note: The accumulator saturation mode should be enabled to reduce the negative effects of the occasional accumulator overflow.

- Because this is an IIR system, the accumulated output samples of this section must be stored in $d(n)$, the intermediary 16-bit memory locations. These memory locations must therefore bear a Q8 format to accommodate the gain of ~85.

  16-bit Q8: 1 sign, 7 integers, 8 fractional bits (±128⁻ range)

- The a_k and b_k coefficients require a Q14 format to accommodate the largest coefficient value of −1.61803.

  16-bit Q14: 1 sign, 1 integers, 14 fractional bits (±2⁻ range)

- The $y(n)$ output sample has a word length of 16 bits and its Q format depends on the overall gain of the complete second-order subsection plus some subjective fudge factor required to provide room for the inevitable amplitude fluctuations inherent in nonlinear-phase IIR systems. The particular notch-stop filter we are implementing has a maximum gain of 1; if we arbitrarily allow 1 I-bit to the result, this accommodates output values in the ±2⁻ range.

  Q14: 1 sign, 1 integer, and 14 fractional bits.

Figure 9–31 illustrates a direct form II that uses the selected Q factors. Note the following in the figure:

- The system accumulates Q8 × Q14 = Q22 products. The Q22 format in the accumulator leaves plenty of room for the seven I-bits that are required.
- The Q13 ADC input samples need to be shifted left by nine positions so that they are aligned with the Q22 accumulator.

- Fourteen F-bits are truncated Q22 result to store Q8 numbers in the intermediary memory locations. This results in truncating noise-free bits, which reduces the SQR. The ADC samples were first shifted left nine positions and then right 14 positions for a net loss of five fractional bits. Using a 14-bit ADC is a waste in this case since only nine bits are effectively being used.
- The Q22 final result needs to be shifted right eight positions to output a Q14 result. Note that because we lost some SQR in the intermediary locations, there is no point in using a 14-bit DAC.

Note that a better SQR could have been achieved if we had used a transposed structure (see the section *Transposed Structure* on p. 306) to implement this system. In the example in Figure 9–31, the limiting factor is the Q8 sample that has to be stored in the direct form II intermediary positions. If we implement a transposed structure, the intermediary direct form II values will tend to bear much smaller values; therefore, a larger Q format could be used for the intermediary memory locations. Determining the correct Q format to use in a transposed structure is not straightforward. Overflow could occur at any of the four summing nodes. As the following paragraphs explains, you will need to run some simulation to determine the correct Q format to use.

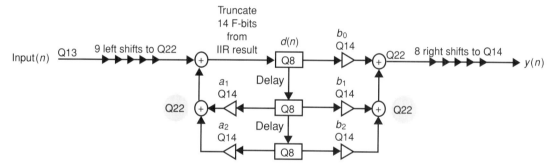

FIGURE 9–31
Q formats chosen for the complete direct form II structure.

Selecting an appropriate Q format for a particular node (memory location or accumulator) is based on the maximum value of the numbers that could be stored at that location. The section on Selecting the Q Factors of a Finite Impulse Response System Based on the Value of the System's Coefficients (p. 329) showed that summing the absolute value of the system coefficients yields the worst-case maximum computed value. Since the impulse response of a system actually yields the system coefficients, taking the absolute value of the impulse response at a node yields

$$\text{Maximum value at a node} \leq \left| \sum_{k=0}^{\infty} h_{Node}(k) \right|$$

In the case of IIR systems, the impulse response is infinite, but, fortunately, the impulse response terms decay toward zero in stable IIR systems. We take advantage of this by summing the absolute value of the node's impulse response terms over a reasonable finite number of M terms. This yields

$$\text{Estimated maximum value at a node} \leq \left| \sum_{k=0}^{M} h_{\text{Node}}(k) \right|$$

In the case of a transposed structure, this estimated maximum value should be evaluated at all four summing nodes using some simulation program that inputs a discrete-time impulse. In most practical applications, the estimated maximum is too conservative since the system nodes reach this maximum only under very exceptional circumstances.

A second technique that stems from analyzing the statistical variance of a list of numbers may be used. This yields a

$$\text{Liberal estimate of the maximum value at a node} \leq \sqrt{\sum_{k=0}^{M} h_{\text{Node}}(k)^2}$$

The reason that we use the square of the impulse response terms is linked to the statistical variance of a list of numbers and exceeds the scope of this book. In practical applications, this technique usually produces more realistic estimates of the maximum value at a particular summing node. However, it should be noted that the liberal estimate resulting from this technique is not an absolute maximum value. The node will occasionally exceed this value and produce overflows. Fortunately, the overflow events will be exceptional.

9.4 ANALYZING DSP SYSTEM REQUIREMENTS

This section provides a quick checklist that will help you determine the right type of system for the application that you seek.

9.4.1 CHOOSING BETWEEN FINITE IMPULSE RESPONSE AND INFINITE IMPULSE RESPONSE SYSTEMS

Consider the following:

- *Stability:* If your application cannot tolerate glitches or if it has to recuperate very quickly from glitches, you should seriously consider an FIR system. Pure FIR systems are completely stable and cannot oscillate.
- *Robustness:* FIR systems are robust systems. Since they are implemented exclusively with zeros, they do not suffer as much from coefficient quantization. FIR systems will recover from any temporary crazy input condition. FIR systems are typically implemented using a direct form I structure and occasionally using a cascaded structure.
- *Linear-Phase:* Only FIR systems can provide linear-phase characteristics (except comb filters). Although IIR systems may be phase equalized, this correction is never perfect and is usually relatively hard to achieve.
- *Efficiency:* IIR systems are much more efficient than FIR systems. Compared with equivalent FIR systems, an IIR implementation will drastically lower the order of

the system. This means that the program will execute much faster, allowing for larger sampling rates.

- *Flexibility:* IIR systems contain both poles and zeros. This means that you can easily combine these to achieve large gains and/or great attenuations. You can also use the bilinear transform to quickly convert any analog filter into a digital filter.

9.4.2 SELECTING A PROCESSOR

Solving your DSP needs by running software on a digital signal processor is the most popular solution. However, there are so many processors to choose from that selecting a particular one may prove to be quite a challenge. Consider the following questions when you choose a processor:

- *Do I really need the flexibility of a software-based system?* Many specialized hardware chips or macros for *programmable logic devices* (PLDs) or *field programmable gate arrays* (FPGAs) are available to implement systems such as FIR or IIR filters, echo cancellation, and signal synthesis. If your system is used only to implement a single simple feature, then a pure hardware solution may provide you with an inexpensive and simple solution.
- *How long is the difference equation that must be implemented?* This will help you to estimate the number of processing steps that must be executed within one sampling interval. It will help to determine the amount of memory that the system requires. The different processors come with various memory resources, and overkill could take you into an overbudget situation.
- *Can the processing be completed within the time the next sample (or group of samples) comes in?* Combined with the answer to the previous question, answering this question will help you to determine the processor's performance in MACs/second.
- *How fast is the required sampling rate?* This determines the ADC-to-processor transfer rates. Not all processors can accommodate the high speed of some ADCs or DACs.
- *Does the processor require special features?* This includes evaluating the need for extra accumulator bits, special addressing modes, and perhaps a parallel architecture.
- *What are the power constraints in my system?* Most processors consume a large amount of power, and if your system works on battery power, power conservation will be a priority.

SUMMARY

- Factors such as computational complexity, memory requirements, signal-to-noise ratio, processing speed, and cost must be considered when choosing a particular implementation scheme.
- A direct structure implements the system difference equation in one single pass that uses all coefficients.

- A direct structure implementation suffers from a large amount of coefficient quantization error. This results in some shifting of system zero positions, which alters the response of the system. This can result in poor system performance if the system zeros are closely clustered or are located in certain areas of the Z-plane or if a specific pattern of zeros is essential to the proper operation of the system.
- A cascade implementation usually provides better coefficient quantization than a direct implementation.
- By properly sequencing cascaded subsections, we can achieve an output signal that has lower noise levels.
- A cascade of second-order subsections unfortunately requires additional multiplication operations.
- When implementing a cascaded system that provides a linear-phase response, the cascaded subsections must be fourth-order subsections.
- IIR systems are implemented with a cascade structure to minimize the quantization error on the coefficients. This is because poles are very sensitive to coefficient quantization.
- The cascaded subsections of an IIR system use a direct form II structure. This structure yields much lower noise levels by processing the high-gain recursive part before the low-gain FIR part.
- Parallel structures provide faster processing than do cascaded structures.
- When compared to cascaded structures, parallel structures unfortunately produce higher levels of noise and require a more expensive parallel arrangement of processors.
- There are three major sources of noise: the ADC acquisition noise, the processing noise, and the system noise.
- Each bit in a word length provides 6 dB of SQR.
- Processing techniques such as AGC and special hardware such as codecs help to lower the acquisition noise levels.
- The addition operation has a tendency to increase the noise level, which deteriorates the SQR.
- The multiplication operation scales the noise and the signal equally. Consequently, it does not affect the SQR.
- To prevent overflow, it is sometimes necessary to shift bits out of the product register. If noiseless bits are shifted out, the SQR deteriorates.
- The overall system gain may be scaled by multiplying the b_k coefficients by a given factor. Scaling does not affect the value of the a_k coefficients.
- The subsections of a cascaded system may be sequenced to act as noise filters.
- Selecting appropriate Q formats starts by establishing a maximum value for the output samples. This sets the accumulator format. The maximum value may be determined either by taking the absolute sum of the coefficients (in FIR systems) or by examining the gain response curve (in FIR or IIR systems).
- Selecting appropriate Q formats for the samples and coefficients depends on the required pole/zero positioning precision (the Q format of the coefficients) and the required noise level (the Q format of the samples).

- Shift operations may be used on some processors to adjust the format of the $b_k x(n-1)$ and $a_k x(n-1)$ products to meet the required accumulator format.
- If no product shifter is available, the Q format of the product (sample Q format + coefficient Q format) must match the Q format of the accumulator.
- The saturation mode should be disabled when computing linear-phase FIR filters.
- When implementing an IIR system, you must consider the gain of the recursive part separately to determine the Q format of the intermediary memory location $d(n)$.
- The nonlinear-phase response of IIR systems is likely to produce the occasional overflow. The accumulator should be formatted so that overflow conditions are rare events. The saturation mode should be enabled to minimize the negative effects of such overflows.
- Go through the checklist in Section 9.4 before you begin to design a DSP system.

PRACTICE QUESTIONS

9-1. A linear-phase FIR system consists of a zero quad and a pair of conjugate zeros on the unit circle.
   (a) Draw a direct form I implementation of this system.
   (b) How many MAC instructions are required when you use a direct form I?
   (c) Draw a cascaded version of this system.
   (d) How many MAC instructions are required when you use a cascade?
   (e) Is there a danger if we use 16-bit word length to quantize the coefficients of the direct form I implementation? What if we use 8-bit word length?

9-2. A linear-phase FIR system is defined by the following difference equation:

$$y(n) = -0.064x(n) + 0.17x(n-1) + 0.31x(n-2) + 0.39x(n-3)$$
$$+0.39x(n-4) + 0.31x(n-5) + 0.17x(n-6) - 0.064x(n-7)$$

   (a) What is the maximum value of any output sample?
   (b) What is the maximum Q format of the 32-bit accumulator used to compute the output sample?
   (c) What is the Q format of the 16-bit coefficients?
   (d) The ADC provides signed Q13 samples. Draw a direct form I implementation of this system.

9-3. An IIR system contains 12 poles and 12 zeros.
   (a) What are the dangers of using a direct form I structure?
   (b) How many second-order subsections are required if we use a cascade implementation?

9-4. A filtering system has the following Z-transform:

$$H(Z) = \frac{Z^2 - 1.41Z + 1}{Z^2 - 1.273Z + 0.81} \times \frac{Z^2 + 0.618Z + 1}{Z^2 + 0.556Z + 0.81}$$

(a) Give a schematic showing a cascade of direct form II second-order subsections.

(b) How many MAC instructions are required to compute one output sample?

(c) Approximate the value of the largest gain in one of the subsections.

(d) Approximate the value of the largest gain of this system.

9-5. An IIR system implements a cascade of two subsections. The processing of the subsections is defined by the following difference equations:

$$y_1(n) = x(n) - 1.41x(n-1) + x(n-2) + 1.273y(n-1) - 0.81y(n-2)$$
$$y_2(n) = x(n) + 0.618x(n-1) + x(n-2) - 0.556y(n-1) - 0.81y(n-2)$$

(a) Scale the frequency response of subsection 1 by a factor of 0.7.

(b) What does the scaling of part (a) do to the maximum gain of subsection 1?

(c) What does the scaling of part (a) do to the phase response of the whole system?

(d) Draw a parallel structure of this system. (Show only the value of the a_k coefficients.)

9-6. (a) What is the theoretical SQR of a 14-bit ADC?

(b) What is the theoretical SQR of a 12-bit DAC?

9-7. An analog signal with an amplitude of ±5 v is applied to the input of a 14-bit ADC. This ADC can accept input voltages in the ±12 v range.

(a) What is the SQR on this data acquisition?

(b) What could be done to improve the noise performance of the system?

9-8. An ADC provides 12-bit samples that have an SQR of 70 dB. These samples are multiplied by 16-bit Q14 coefficients that have values that are close to 1.

(a) What is the Q factor of the product?

(b) Does the multiplication operation change the SQR of the signal?

(c) If we are to avoid an overflow condition, approximately how many of these products could be accumulated in a 32-bit accumulator?

(d) Consider that the products are right shifted twice before being accumulated.

   1. Does the shifting reduce the SQR?

   2. Approximately how many of these shifted products could be accumulated in a 32-bit accumulator?

(e) Ten of these products are added to compute the value of an output sample. If the noise is a random function, will the addition process change the SQR?

9-9. An IIR system implements a notch-pass filter that has an approximate gain of 10 at a normalized frequency of $\Omega = \pi/2$. The conjugate poles are located at 0.01 from the unit circle and the conjugate zeros are at 0.1 from the unit circle.

(a) What is the maximum gain of the recursive section?

(b) If 16-bit word lengths are used to implement a direct form II structure, what is the Q factor of the intermediary samples stored in $d(n)$?

(c) What is the Q factor of the 16-bit output samples?

Appendix A

COMPLEX NUMBERS AND VECTORS

This book simplifies the mathematical approach that is traditionally used to present digital signal processing (DSP). The mathematical presentations used in this book are generally limited to exploiting complex numbers, basic vectorial analysis, and exponential properties. These rather elementary mathematical concepts are typically introduced during the first year of undergraduate college or university science programs.

Because of this simplified approach, the concepts and techniques introduced in this book are within reach of students entering the second year of most science programs. This appendix provides a concise review of the mathematical concepts and operations involving complex numbers, vectors, and exponentials to ensure that the reader is familiar with these topics.

A.1 UNDERSTANDING IMAGINARY NUMBERS

If we square a *real* number x, the result is always positive:

$$x^2 = \text{Positive result}$$

This is true even if the *real* number x happens to be a negative number. Consider the following examples:

$$9^2 = +81$$
$$(-9)^2 = +81$$

By definition, the square root of x is found by determining which number will yield x when it is squared. Since squaring the positive or the negative of a real number yields the same positive result, the square root of a *real* number always yields two distinct results:

$$\sqrt{+9} = +3 \text{ or } -3$$
$$= \pm 3$$

This creates a problem if we want to determine the square root of the *negative* number $-x$. This problem occurs because squaring any *real* number always results in a positive answer. We must conclude that the square root of a negative number cannot yield a *real* number:

$$\sqrt{\text{Negative number}} \text{ cannot yield a } real \text{ number}$$

This leads us to define a new type of number, which we call *imaginary.* This number will be referred to as *j* and is defined such that

$$j = \sqrt{-1} \text{ or that } j^2 = -1$$

This new number allows us to find the square root of negative numbers if we use the following procedure:

$$\sqrt{- \text{ number}} = \sqrt{(-1)\text{ number}}$$
$$= j\sqrt{+ \text{ number}}$$

Consider the following example:

$$\sqrt{-9} = \sqrt{(-1) \times 9}$$
$$= j\sqrt{+9}$$
$$= \pm 3j$$

A.2 USING COMPLEX NUMBERS

It is possible to create numbers that contain both a *real* and an *imaginary* part. We call such numbers *complex numbers*:

Complex number = Real + *j*Imaginary

A.2.1 SQUARING COMPLEX NUMBERS

Consider finding the square of a complex number such as $4 + 5j$. Since complex numbers consist of two separate parts, it is necessary to cross-multiply both the real and the imaginary parts of the number as illustrated. $(4 + 5j)^2$ becomes

$$\begin{array}{r} 4 + 5j \\ \times\ 4 + 5j \\ \hline (5j \times 5j) + (5j \times 4) + (4 \times 5j) + (4 \times 4) \end{array}$$

Expanding the terms and regrouping, we find that a new complex number results from the operation

$$(4 + 5j)^2 = (25j^2) + 20j + 20j + 16$$
$$= (25 \times -1) + 40j + 16$$

$$= -25 + 40j + 16$$
$$= -9 + 40j$$

A.2.2 USING COMPLEX CONJUGATES

In general, the electronic devices that we use in our real world produce *real* results and *real* waveforms. In our real world, we cannot perceive results that have an imaginary part. If a DSP system produces an answer that has a complex part, it cannot be used directly to produce a pure real output. It is therefore necessary to find a technique that will allow us to eliminate the imaginary part of a complex answer.

To eliminate the imaginary part of complex numbers, we define the notion of a *complex conjugate*. We find the complex conjugate of a number simply by changing the sign of the imaginary part. Complex conjugate of

$$\text{Real} + j\text{Imaginary} = \text{Real} - j\text{Imaginary}$$

Since complex conjugate numbers have imaginary parts of opposite sign, adding these numbers eliminates the imaginary part completely.

The complex conjugate of a number is expressed mathematically by adding a bar on top of the number:

$$\text{Complex conjugate of } x = \bar{x}$$

and

$$x + \bar{x} = \text{Pure real}$$

Consider the following example:

$$\begin{aligned} \text{if:} \quad & z = 4 + 5j \\ \text{then:} \quad & \bar{z} = 4 - 5j \\ \text{and:} \quad & z + \bar{z} = 8 + 0j \end{aligned}$$

Adding a number with its complex conjugate eliminates the imaginary part of that number.

Note that the real part of the number is doubled in the process:

$$z = \text{Real} + j\text{Imaginary}$$
$$\bar{z} = \text{Real} - j\text{Imaginary}$$
$$= 2\,\text{Real} + 0\,\text{Imaginary}$$

A.3 PLOTTING COMPLEX NUMBERS

We can represent complex numbers on a diagram that closely resembles the *Cartesian* coordinate system invented by French mathematician René Descartes in 1637. When we use such system to represent complex numbers, the real part corresponds to the *x*-axis coordinate and the imaginary part corresponds to the *y*-axis coordinate. Because of this, we change the notation on the axes and the diagram becomes an *Argand diagram* after Jean-Robert Argand (1768–1822), a French mathematician who published a paper in 1806 on the subject. Figure A–1 illustrates an Argand diagram.

FIGURE A–1
Argand diagram.

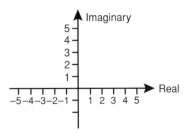

We can plot complex numbers on an Argand diagram using two different approaches. The following subsections describe how this is done.

A.3.1 PLOTTING COMPLEX NUMBERS USING RECTANGULAR COORDINATES

When a complex number is represented using a combination of a real and an imaginary part, we can look at this notation as corresponding to rectangular coordinates. The real part corresponds to the real-axis coordinate and the imaginary part corresponds to the imaginary-axis coordinate. Using this approach, we can plot points on the Argand diagram. Figure A–2 illustrates a complex number plotted as an × on an Argand diagram.

FIGURE A–2
Rectangular coordinates on an Argand diagram.

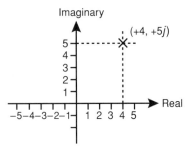

You should note that the complex conjugate of a number is a mirror image of the complex point across the real-axis. Figure A–3 illustrates an Argand diagram where two ×s are used to show the positions of complex conjugate numbers.

A.3.2 PLOTTING COMPLEX NUMBERS USING POLAR COORDINATES

The various chapters of this book use vectorial representation extensively. For this reason, it is important that you become familiar with this alternate way to represent a complex

FIGURE A–3
Complex conjugates on an Argand diagram.

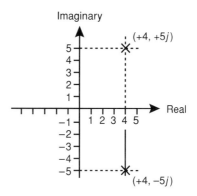

number on an Argand diagram. In this case, we define the complex coordinates using the *Polar coordinate* system. A *vector* of length R that makes an angle ϕ with the real axis is used to represent the complex number:

$$\overrightarrow{z} = R \angle \phi$$

Figure A–4 illustrates a complex vector. We call the length (magnitude) of the complex vector the *modulus,* and we represent it mathematically as

$$R = \text{Modulus of } \overrightarrow{z} = \left| \overrightarrow{z} \right|$$

We can apply the Pythagorean theorem to calculate the modulus of a complex vector:

$$R = \left| \overrightarrow{z} \right| = \sqrt{\text{Real}^2 + \text{Imaginary}^2}$$

For example, we calculate the modulus (or magnitude) of the complex vector of Figure A–4 as

$$R = \sqrt{4^2 + 5^2}$$

$$\cong 6.4$$

FIGURE A–4
Complex number represented as a vector.

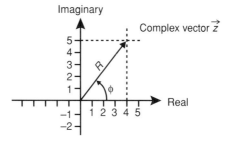

We call the angle that the complex vector makes with the real-axis the *argument* of the vector. An examination of Figure A–5 will help you understand that we can find the value of this angle ϕ by using trigonometry. Notice that the opposite side of angle ϕ in Figure A–5 represents the *imaginary* rectangular coordinate and the adjacent side represents the *real* rectangular coordinate.

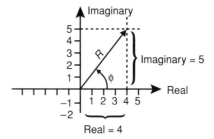

We can calculate the value of the argument by using trigonometry since we know that the tangent of an angle is defined as the ratio of the opposite side over the adjacent side:

$$\tan(\phi) = \frac{\text{Opposite}}{\text{Adjacent}}$$

$$= \frac{\text{Imaginary}}{\text{Real}}$$

We can therefore find the value of ϕ by calculating the inverse of the tangent:

$$\phi = \arctan\left(\frac{\text{Imaginary}}{\text{Real}}\right)$$

For example, we calculate the argument of the complex vector of Figure A–5 as follows:

$$\phi = \arctan\left(\frac{5}{4}\right)$$

$$\approx 51.3° \text{ or } 0.896 \text{ radian}$$

This technique works well for vectors that reside in the first or fourth quadrants.

If the vector resides in the second or in the third quadrant, we need to be careful. Examine the two complex vectors represented in Figure A–6. The vectors illustrated in this figure extend their angle ψ beyond 90 degrees. Because of this, calculating the arctangent of these angles results in finding ϕ. We therefore need to apply a correction of +180 degrees (or π radians) to find the value of ψ:

$$\psi = \arctan\left(\frac{\text{Imaginary}}{\text{Real}}\right) + 180°$$

For example, let's calculate the value of the angle of the vector in the second quadrant:

$$\phi = \arctan\left(\frac{+5}{-4}\right) \qquad \cong -51.3°$$

$$\psi = \arctan\left(\frac{+5}{-4}\right) + 180° \cong 128.7°$$

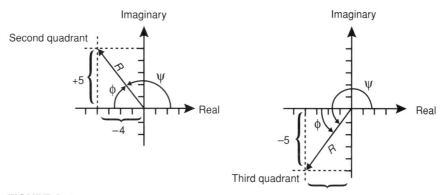

FIGURE A–6
Vectors in the second or third quadrant.

Conjugate complex vectors are drawn in Figure A–7. Note that the complex conjugate vectors illustrated in this figure have angles of opposite signs. Finding the complex conjugate of a vector is therefore a simple matter of inverting the sign of the angle. As illustrated in the figure, converting polar coordinates back to rectangular coordinates is simply a question of applying trigonometry:

$$Polar \Leftrightarrow Rectangular \qquad\qquad Polar \Leftrightarrow Rectangular$$
$$R \angle \phi \Leftrightarrow R\cos(\phi) + jR\sin(\phi) \qquad R \angle -\phi \Leftrightarrow R\cos(\phi) - jR\sin(\phi)$$

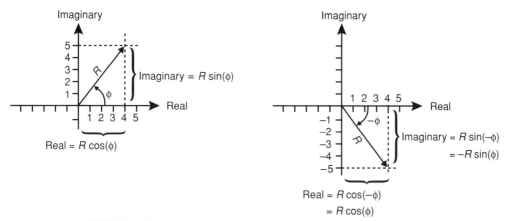

FIGURE A–7
Complex conjugates represented as vectors.

A.4   USING EXPONENTIAL NUMBERS

In the same way that π has a special value 3.14159..., e also has a special value 2.71828..., and we can compute it to as many decimals as we need. The sequence of digits that describes these numbers does not repeat any known patterns, and numbers such as π or e are called *irrational numbers*. Irrational numbers are still constants, and we can raise them to some exponent such as π^2 and $e^{3.5}$. They can also be scaled by any value such as $1.4\pi^2$ and $8.2e^{3.5}$. When the irrational number e is raised to some exponent, we refer to it as an *exponential* number. This book uses complex exponential numbers extensively, and it explores the practical applications that stem from some of the properties of these numbers. For this reason, it is important to be familiar with the manipulation of exponential numbers.

A.4.1   BASIC OPERATIONS USING EXPONENTIAL NUMBERS

We can assemble an *exponential number* by raising the number e to some *argument* x and by scaling the result by some value A:

$$Ae^x$$

Exponential numbers follow all the standard mathematical rules that apply to numbers that are raised to some exponent.

Adding Exponential Numbers
The addition of exponential numbers that have *real exponents* cannot be simplified. However, when the exponent is *complex*, there are important special cases, described in Section A.4.2, in which we can simplify the addition of exponential numbers.

$$Ae^x + Be^y$$

can be simplified only under special conditions.

Multiplying Exponential Numbers
The multiplication of two exponential numbers can be simplified by multiplying the scaling values and by adding the exponent:

$$Ae^x \times Be^y = (A \times B)e^{(x+y)}$$
$$= ABe^{(x+y)}$$

For example

$$5e^3 \times 6e^2 = (5 \times 6)e^{(2+3)}$$

$$= 30e^5$$

Using an Exponent of Zero
Raising the irrational number e to an exponent of zero follows the standard rule for any number being raised to a exponent of zero; as expected, it yields a result of 1:

$$e^0 = 1$$
$$Ae^0 = A \times 1$$
$$= A$$

A.4.2 COMPLEX EXPONENTIAL NUMBERS

In 1748, Leonhard Euler (1707–1783) established a link between trigonometric and exponential functions. Euler found that an exponential with a complex exponent corresponds to the sum of two sinusoidal functions. When an exponential has a complex exponent, it becomes a *complex exponential* number. Euler proved that complex exponentials correspond to the following relationship:

$$Re^{j\phi} = R\cos(\phi) + jR\sin(\phi)$$
$$= \text{Real} + j\text{Imaginary}$$

If you carefully examine the right side of Euler's identity, you will notice that it is the exact formula used to convert a complex vector from polar to rectangular coordinates. This means that the left side of Euler's identity $Re^{j\phi}$ corresponds to a complex vector that has a length of R and an argument of ϕ.

Euler's identity therefore provides an alternate way to express complex numbers. When the exponent becomes negative, we obtain the following:

$$Re^{-j\phi} = R\cos(-\phi) + jR\sin(-\phi)$$

Notice that the argument $(-\phi)$ has a negative value. To simplify this result, carefully examine the cosine and the sine functions that are plotted in Figure A–8. The figure reveals that the cosine function displays even symmetry since the left side is a mirror image of the right side. On the other hand, the sine function displays an odd symmetry since the left side is the negative of the right side. This means that

$$\cos(\phi) = \cos(-\phi) \quad \text{and} \quad \sin(-\phi) = -\sin(\phi)$$

This even and odd symmetry is used extensively in DSP theory. We can use this result to simplify a complex exponential when it has a negative argument:

$$Ae^{-j\phi} = A\cos(-\phi) + jA\sin(-\phi)$$
$$= A\cos(\phi) - jA\sin(\phi)$$

FIGURE A–8
Symmetry of the sine and cosine functions.

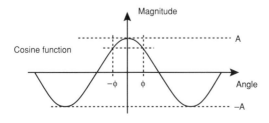
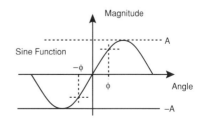

Notice that the sign of only the imaginary part is negative. This means that changing the sign of the argument produces a *complex conjugate*. Since DSP deals with the analysis of sinusoidal waveforms, we need to break cosine and sine functions into components that are more fundamental. Euler's identity provides a way to create pure-real sine and cosine functions from pairs of conjugate complex exponentials.

Let's start by finding how to create a pure-real cosine function. When we add a complex value to its conjugate, we obtain a real value. Adding a complex exponential number to its conjugate produces the following real sum:

$$Ae^{+j\phi} = A\cos(\phi) + jA\sin(\phi)$$
$$\underline{+\quad Ae^{-j\phi} \qquad A\cos(\phi) - jA\sin(\phi)}$$
$$Ae^{+j\phi} + Ae^{-j\phi} = 2A\cos(\phi) - 0j$$

We can also rewrite this result as follows:

$$A\cos(\phi) = \frac{Ae^{+j\phi} + Ae^{-j\phi}}{2}$$

As we can see, a pure-real cosine waveform consists of the sum of two conjugate complex exponentials. We use this result extensively throughout this book.

Let's continue by finding how to create a pure-real sine function. Notice that the expansion of a complex exponential using Euler's identity contains a sine in the imaginary part $j\sin(\phi)$. To turn this imaginary value into a real value, we multiply it by j:

$$Ae^{j\phi} = A\cos(\phi) + j\sin(\phi)$$
$$j \times Ae^{j\phi} = j \times A\cos(\phi) + j \times j\sin(\phi)$$
$$jAe^{j\phi} = jA\cos(\phi) - A\sin(\phi)$$

We then group two conjugate complex exponentials to eliminate the imaginary part:

$$-jAe^{+j\phi} = -jA\cos(\phi) + A\sin(\phi)$$
$$\underline{+\qquad jAe^{-j\phi} = +jA\cos(\phi) + A\sin(\phi)}$$
$$-jAe^{+j\phi} + jAe^{-j\phi} = \qquad 0 \qquad + 2A\sin(\phi)$$

We can rewrite this result as follows:

$$2A\sin(\phi) = \frac{j}{j} \times \left(-jAe^{+j\phi} + jAe^{-j\phi}\right)$$

$$= \frac{-j^2Ae^{+j\phi} + j^2Ae^{-j\phi}}{j}$$

Isolating the sine, we get:

$$A\sin(\phi) = \frac{Ae^{+j\phi} - Ae^{-j\phi}}{2j}$$

A.5 EXERCISES

The following exercises provide practice in the manipulation of complex exponentials.

A.5.1 SHIFTING THE PHASE

As a first exercise, let's examine what happens when we shift the phase of a cosine:

$$\cos(\omega + \phi) = \frac{e^{j(\omega+\phi)} + e^{-j(\omega+\phi)}}{2}$$

We can expand the complex exponentials into two parts:

$$\cos(\omega + \phi) = \frac{e^{+j\phi}e^{j\omega} + e^{-j\phi}e^{-j\omega}}{2}$$

As we can see on the right side of the equation, $e^{+j\phi}$ shifts the phase of the exponential by an amount exactly equal to the cosine phase shift. Since the second part of the complex expression represents the conjugate, the argument is negative.

A.5.2 CHANGING A COSINE INTO A SINE

As a second exercise, let's prove that a sine is a cosine with a phase shift of –90 degrees, or $-\pi/2$ radians:

$$\cos(\omega) = \frac{e^{j\omega} + e^{-j\omega}}{2}$$

$$\sin(\omega) = \cos\left(\omega - \frac{\pi}{2}\right) = \frac{e^{j\left(\omega - \frac{\pi}{2}\right)} + e^{-j\left(\omega - \frac{\pi}{2}\right)}}{2}$$

$$\sin(\omega) = \frac{e^{j\omega}e^{-j\frac{\pi}{2}} + e^{-j\omega}e^{j\frac{\pi}{2}}}{2}$$

Consider the following simplifications:

$$e^{-j\frac{\pi}{2}} = \cos\left(\frac{\pi}{2}\right) - j\sin\left(\frac{\pi}{2}\right) = \qquad e^{+j\frac{\pi}{2}} = \cos\left(\frac{\pi}{2}\right) + j\sin\left(\frac{\pi}{2}\right)$$

$$= 0 - j \qquad\qquad\qquad = 0 + j$$

Using these simplifications, we can rewrite:

$$\sin(\omega) = \frac{e^{j\omega}(-j) + e^{-j\omega}j}{2} \times \frac{j}{j}$$

$$\sin(\omega) = \frac{e^{j\omega} - e^{-j\omega}}{2j}$$

A.5.3 CHANGING A SINE INTO A COSINE

As a third exercise, let's prove that a cosine is a sine with a phase shift of +90 degrees, or $\pi/2$ radians.

$$\sin(\omega) = \frac{e^{j\omega} - e^{-j\omega}}{2j}$$

$$\cos(\omega) = \sin\left(\omega + \frac{\pi}{2}\right) = \frac{e^{j\left(\omega + \frac{\pi}{2}\right)} - e^{-j\left(\omega + \frac{\pi}{2}\right)}}{2j}$$

$$\cos(\omega) = \frac{e^{j\omega}e^{+j\frac{\pi}{2}} - e^{-j\omega}e^{-j\frac{\pi}{2}}}{2j}$$

But we know that

$$e^{+j\frac{\pi}{2}} = +j$$

and that

$$e^{-j\frac{\pi}{2}} = -j$$

therefore

$$\cos(\omega) = \frac{e^{j\omega}(j) - e^{-j\omega}(-j)}{2j}$$

$$= \frac{e^{j\omega}(j) + e^{-j\omega}(j)}{2j} \times \frac{j}{j}$$

$$= \frac{-e^{j\omega} - e^{-j\omega}}{-2}$$

$$\cos(\omega) = \frac{e^{j\omega} + e^{-j\omega}}{2}$$

PRACTICE QUESTIONS

A-1. Consider the following complex numbers:

$$3 + 4j \qquad 5 - 1j \qquad -2 - 3j$$

    (a) Square each of these complex numbers.
    (b) Find the complex conjugate of each of these complex numbers.
    (c) Plot these complex numbers and their complex conjugates on an Argand diagram.

A-2. Consider the following complex numbers:

$$4 + 4j \qquad 4 - 4j \qquad -4 + 4j$$

    (a) Express each of these complex numbers using polar coordinates.
    (b) Give the modulus of each of the complex numbers.
    (c) Give the argument of each of the complex numbers.
    Express the argument in degrees and in radians.

A-3. Examine the following expressions:

$$8e^{4j} \times 2e^{3j} \qquad 8e^{4j} \times 2e^{-3j} \qquad 8e^{-4j} \times 2e^{0j}$$

    (a) Simplify the expressions to a single complex exponential.
    (b) Find the argument of the complex vector that results when each expression is simplified. Express the argument in degrees and then in radians.
    (c) Plot the resulting vectors on an Argand diagram.
    (d) Find the real and imaginary parts of the simplified expressions.

A-4. **(a)** Express $\cos(2)$ as a sum of complex exponentials.
    (b) Plot the two vectors that correspond to $\cos(2)$.

A-5. **(a)** Express $\cos(2 - \pi/2)$ as a sum of complex exponentials.
    (b) Plot the two vectors that correspond to $\cos(2 - \pi/2)$.

A-6. **(a)** Express $\sin(2)$ as a sum of complex exponentials.
    (b) Plot the two vectors that correspond to $\sin(2)$.

Appendix B

PROCESSING BINARY NUMBERS

Digital signal processing (DSP) demands the manipulation of numbers whose value adopts various formats. Although all electrical and computer engineering programs include an exposure to binary systems, these programs do not necessarily cover all the topics that the designer needs to implement DSP systems. This appendix reviews the standard basic topics and elaborates on the more specialized topics that may not have been thoroughly covered in your background of binary logic courses.

We start by introducing the representation of numbers using the binary and hexadecimal notations. This immediately reveals that binary numbers consist of groups of binary digits (bits) that limit the set of possible binary values.

We follow by introducing some of the most popular formats and then examining the properties of these formats in detail. We begin examining the signed and unsigned notations for which we analyze the range and explore how to expand these numbers.

We also examine the integer format and observe that it is inadequate to represent fractional numbers. A large number of DSP applications demand the use of fixed-point processors and, for this reason, we explain the formatting, which supports fractional numbers. This leads us to investigate the precision and tolerance with which we can express numbers. At that point, we acknowledge that the binary format imposes quantization boundaries that create practical limits to our DSP system.

We continue by showing how we can push the range and the precision of the numbers to levels that are acceptable to the functional specifications of our designs. We examine other formats, including the floating-point formats, which give the programmer control over the range and precision parameters.

We continue by examining the properties of two basic mathematical operations whose relevance is essential to the implementation of practical DSP. As we look at the addition and multiplication operations, we analyze bit requirements and overflow limits.

We close by addressing the scaling issues that arise when digital-to-analog converters (DACs) transform formatted numbers into voltages.

B.1 PRESENTING THE BINARY NUMBERING SYSTEM

Digital signals exist as strings of binary numbers that are manipulated using a digital processor. The number of lines available on the processor data bus usually determines the size of the binary numbers that the processor will manipulate with ease. The majority of processors provide data bus sizes that regroup multiples of eight data lines. The processor instructions are designed to optimize the manipulation of binary numbers that match the data bus size. Because of this, we can optimize the processing by using binary number formats that match the size of our processor data bus. Consequently, our binary numbers should consist of multiples of eight binary digits.

B.1.1 EXPRESSING BINARY NUMBERS

By definition, we call the digits contained in a binary number *bits*. These bits can adopt one of the two binary states, 0 or 1. Each of the bits contained in a number carries a different value. This value is evaluated by reading the binary digits from right to left according to the *binary value rule*:

Binary value rule: The value of each bit doubles as we move from the rightmost bit to the leftmost bit.

The rightmost bit is worth: 0 or 1
The next bit is worth: 0 or 2
The following bit is worth: 0 or 4
The one after that: 0 or 8
and so on

We call the bit that is worth the least the *least significant bit* and the bit that is worth the most the *most significant bit*. By numbering the bits, starting at zero for the least significant bit, we can express the binary value rule as

Value of mth bit: 0 or 2^m

For example, we evaluate an 8-bit number using the following calculation:

Binary digits: $b_7 b_6 b_5 b_4 b_3 b_2 b_1 b_0$ where $b_m = 0$ or 1

Value $= b_7 \times 2^7 + b_6 \times 2^6 + b_5 \times 2^5 + b_4 \times 2^4 + b_3 \times 2^3 + b_2 \times 2^2 + b_1 \times 2^1 + b_0 \times 2^0$

We therefore associate a specific *integer* value with the position of each bit in the binary number. For example, consider the following 4-bit number:

$$\textit{Binary} \qquad\qquad \textit{Decimal}$$
$$1101 \quad \Leftrightarrow 1 \times 2^3 + 1 \times 2^2 + 0 \times 2^1 + 1 \times 2^0 = 13$$

The most significant bit b_3 is worth 8, the least significant b_0 is worth 1, and the value of the number totals 13.

You should ask yourself how many *different* values a binary number can adopt. To answer it is important to note that every binary digit can adopt one of two values. The fact that there are two possible values to every binary digit means that every extra binary digit effectively doubles the set of different values that the number can adopt. We can represent this relationship as follows:

Number of different values that a binary number can adopt = $2^{\text{number of bits in the number}}$.

For example, a 4-bit binary number can adopt a set of $2^4 = 16$ different values.

Since most binary numbers are expressed using a large quantity of bits, the *hexadecimal* (hex) notation is used to simplify the expression of large binary numbers. To express a number using the hex notation, the bits are regrouped from right to left in groups of 4 bits and each group is assigned a specific hex digit according to its value as listed in Table B–1. Consider, for example, the conversion of the following 16-bit number:

$$0110 \quad 1110 \quad 1000 \quad 1010 \Leftrightarrow 6E8A$$

and the binary value of this number is expressed in decimals as

$$0 + 16384 + 8192 + 0 + 2048 + 1024 + 512$$
$$+ 0 + 128 + 0 + 0 + 0 + 8 + 0 + 2 + 0 = 28,298$$

TABLE B–1
Binary, decimal, and hexadecimal representation.

Group of 4 Bits		Decimal Value		Hex Representation
0000	$\Leftrightarrow$	0	$\Leftrightarrow$	0
0001	$\Leftrightarrow$	1	$\Leftrightarrow$	1
0010	$\Leftrightarrow$	2	$\Leftrightarrow$	2
0011	$\Leftrightarrow$	3	$\Leftrightarrow$	3
0100	$\Leftrightarrow$	4	$\Leftrightarrow$	4
0101	$\Leftrightarrow$	5	$\Leftrightarrow$	5
0110	$\Leftrightarrow$	6	$\Leftrightarrow$	6
0111	$\Leftrightarrow$	7	$\Leftrightarrow$	7
1000	$\Leftrightarrow$	8	$\Leftrightarrow$	8
1001	$\Leftrightarrow$	9	$\Leftrightarrow$	9
1010	$\Leftrightarrow$	10	$\Leftrightarrow$	A
1011	$\Leftrightarrow$	11	$\Leftrightarrow$	B
1100	$\Leftrightarrow$	12	$\Leftrightarrow$	C
1101	$\Leftrightarrow$	13	$\Leftrightarrow$	D
1110	$\Leftrightarrow$	14	$\Leftrightarrow$	E
1111	$\Leftrightarrow$	15	$\Leftrightarrow$	F

B.2 FORMATTING BINARY NUMBERS

We can regroup the bits that a number contains into fields that carry a special meaning. The position and the meaning of each of the fields define the *format* of the number. Each format has certain advantages and drawbacks, and selecting a particular format depends on the specific requirements of the application.

Binary numbers come in one of *two* families:

1. *Unsigned numbers*: These numbers are always positive. *All* bits in the number belong to a *numerical part* that contains the value.
2. *Signed numbers*: These numbers can adopt either positive or negative values. Some bits are assigned to a *sign field* that represents the polarity of the number; the remaining bits belong to the numerical part that contains the value.

The numerical part of both families includes *three* member types characterized by the following formats:

1. *Integer numbers*: *All* bits in the numerical part of this format are combined to represent a whole number. Some decimal examples of integer numbers are 5, 43, and 349.
2. *Fixed-point numbers*: This format separates the numerical part into an *integer field* and a *fraction field*. Decimal examples of fixed-point numbers are 5.25, 3.9375, and 43.5.
3. *Floating-point numbers*: The bits in the numerical part conform to a particular scientific notation format. This format assigns some bits to an *exponent field* and the remaining bits to a *fraction field*. Decimal examples of floating-point numbers are 0.25×10^3 and 0.8679×10^{12}.

These different formats are illustrated in Figure B–1. As a programmer, you must become familiar with the different formats to be able to plan the DSP steps. The following sections explore the limits of these different numbering systems.

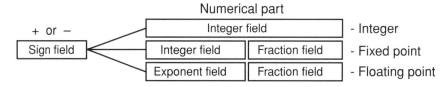

FIGURE B–1
Formats of binary numbers.

B.3 EVALUATING BINARY NUMBERS

Now that we are aware that binary numbers consist of two families (signed and unsigned), each including three member types (integer, fixed point, and floating point), we reason that there are six possible ways to interpret the value of a binary number. Let's examine how to evaluate binary numbers when they adopt different formats.

B.3.1 INTERPRETING THE VALUE OF SIGNED
AND UNSIGNED NUMBERS

One of the best ways to understand the unsigned/signed notation is to visualize in a circle the 16 binary states that are possible using a 4-bit number. These states are illustrated in Figure B–2.

FIGURE B–2
Possible states using a 4-bit number.

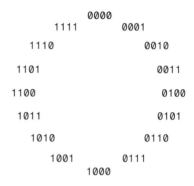

Let's first consider the values that we can assign to a 4-bit *unsigned* number. Referring to Figure B–3, we can start at the binary state of 0000 and move clockwise around the circle through the different values. We reach the maximum at the binary state of 1111, which is worth 15 in decimal. We have to stop there before the 4-bit number overflows.

FIGURE B–3
Values of an unsigned 4-bit number.

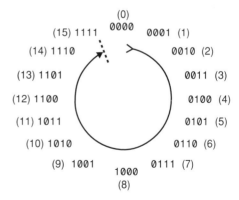

Let's now consider the values that we can assign to a 4-bit *signed* number. In this case, the most significant bit represents the sign of the number. When the sign field bits adopt the binary state of 0 the number is *positive;* conversely, the state of 1 represents a *negative* number.

We can start at the binary state of 0000 and move clockwise around the circle through the different values (see Figure B–4). We reach the maximum positive count at the binary state of 0111, which is worth +7 in decimal. We have to stop there before the count

FIGURE B–4
Values of a signed 4-bit number.

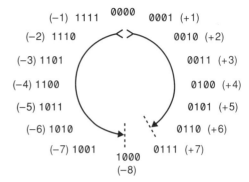

overflows into the negative range. Alternately, we can start moving *down* from the binary state of **0000** through the different negative values. We reach the minimum negative count at the binary state of **1000**, which is worth –8 in decimal. We have to stop there before the count *underflows* into the positive range.

Both the unsigned and the signed notation of a 4-bit number allow for 16 different values. The actual *value* associated with each 4-bit state actually depends on the way that we *interpret* the bits contained in the number.

B.3.2 MANIPULATING SIGNED NUMBERS

Digital signal processors operate primarily using numbers expressed with signed notation. Because of this, it is necessary to learn some of the special manipulations required to handle signed numbers.

Inverting the Sign

Many signal processing algorithms call for inverting the sign of numbers. For example, you may want to convert the number +6 to a value of –6. We call the standard operation that performs a sign inversion the *2's complement*. This operation is achieved by taking the binary complement of the number (inverting the state of all bits) and adding 1 to the result.

As an example, let's use this technique to calculate the negative of the 8-bit representation of +6:

$$
\begin{array}{llr}
8\text{-bit representation of} & +6 = & 00000110 \\
\text{The complement of 6:} & \overline{6} = & 11111001 \\
\text{Adding 1:} & + & \underline{\qquad 1} \\
8\text{-bit representation of} & -6 = & 11111010 \\
\end{array}
$$

Extending the Sign

You may have noticed that the sign bit sometimes repeats itself in the most significant bits. Closely examine the binary representation of +6:

$$+6 \xrightarrow{\quad 8\text{-bit Binary}\quad} 00000110$$

Notice that the five most significant bits of the positive number +6 are all in the zero state. Because of this, the most significant binary digit, which contributes to the value of the number, is the 1 located in the third bit position. The five most significant 0s are considered sign bits. We can consider the sign to be in the fourth bit position (next to the most significant digit) and the following 0s are all *sign extension* bits. Sign extension is necessary here because not all 8 bits are required to express the number +6; all extra bits must extend the sign so that the value of the number is not altered. For example, if we were to extend the expression of +6 into a 16-bit number, we would simply extend the sign over all the extra bits to create a 13-bit positive sign field:

$$+6 \xrightarrow{\quad \text{16-bit Binary} \quad} 0000000000000110$$

Let's examine the sign extension operation for a negative number. As we found previously, the number –6 corresponds to the following 8-bit binary representation:

$$-6 \xrightarrow{\quad \text{8-bit Binary} \quad} 11111010$$

The binary state of 1 corresponds to a negative sign, and we can see that the five most significant digits are all negative signs. Because of this, the most significant binary digit, which contributes to the value of the number, is the 0 located in the third bit position. The negative sign located in the fourth bit position has been extended to create a 5-bit sign field. We could express –6 as a 16-bit number by extending the sign over the extra bits:

$$-6 \xrightarrow{\quad \text{16-bit Binary} \quad} 1111111111111010$$

You should note that extending the sign does not alter the value of the number.

B.3.3 EVALUATING INTEGER NUMBERS

Unsigned integer numbers carry the simplest possible format. In this case, all the bits contribute to the numerical value according to the binary value rule. The bit structure of an unsigned 8-bit integer is therefore as follows:

Structure of an unsigned 8-bit integer: IIIIIIII

The Is that appear in all bit positions indicate that every bit contributes to the *integer* value of the number. The quantity of I-bits determines the collection of different values that this integer number can adopt:

Number of different values that an m-bit integer can adopt = 2^m

For example, an unsigned 8-bit integer can adopt $2^8 = 256$ different integer values that start at the value of 0.

Range of values for an *m*-bit *unsigned* integer: 0 to $2^m - 1$.

For example, let's calculate the range of an 8-bit unsigned integer:

0 to $2^8 - 1 = 0$ to 255

If we now change to the structure to a *signed* binary integer, the most significant bit must correspond to the sign (or the sign extension) of that number. Consider the example of a signed 8-bit integer number that has one sign bit:

Structure of a signed 8-bit number: `SIIIIIII`

The S-bit represents the sign, and the group of I-bits represents the integer value. By convention a sign bit in the state of 0 heads a positive number, and a state of 1 indicates a negative number. Since the integer value depends on the seven I-bits this means that there are $2^7 = 128$ possible positive values; conversely, there are 128 possible negative values.

The positive 8 bits range from `00000000` to `01111111`

$$\dfrac{\text{Decimal}}{} \longrightarrow 0 \text{ to } + 127$$

The negative 8 bits range from `11111111` to `10000000`

$$\dfrac{\text{Decimal}}{} \longrightarrow -1 \text{ to } -128$$

Range of values for an *m*-bit *signed* integer: -2^{m-1} to $+2^{m-1} - 1$.

Notice that positive integers cannot quite reach the value of $+2^m$ because they include the value of 0. We can increase the range of possible signed values by increasing the quantity of I-bits in the number. For example, a signed 16-bit integer contains 1 S-bit and 15 I-bits and produces values that range from $-32,768$ to $+32,767$.

Remember that the first significant binary digit in a *positive* number is the leftmost binary 1. Conversely, when the number is *negative,* the first significant binary digit is the leftmost binary `0`.

Consider the following examples of signed 8-bit numbers:

Case 1

`00010111` is a positive 8-bit integer whose structure includes three sign bits and five integer bits: `SSSIIIII`.

The integer value is

$$10111 \xrightarrow{\quad \text{Decimal} \quad} +(16 + 0 + 4 + 2 + 1) = +23$$

Case 2

`11111001` is a negative 8-bit integer whose structure includes five sign bits and three integer bits: `SSSSSIII`.

The integer value is:

$$-\left(\overline{001} + 1\right) = -(110 + 1) = -(111) \xrightarrow{\quad \text{Decimal} \quad} -(4 + 2 + 1) = -7$$

B.3.4 EVALUATING FIXED-POINT NUMBERS

The format of fixed-point numbers requires that we separate the numerical part into an *integer field* and a *fraction field*. Consider the following unsigned fixed-point number, which has a 12-bit integer field and a 4-bit fraction field:

IIIIIIIIIIIIFFFF

The I-bits belong to the integer field and the F-bits belong to the fraction field.

To evaluate fractional numbers, we must remember that the value of each binary digit *doubles* as we read the number from right to left. It is important to note that we could also say that the value of the binary digits *halves* as we read the number from left to right.

When reading a fixed-point number from left to right, we eventually reach the least significant bit of the integer field, which has a value of 1. Proceeding to the right of the integer field, we must continue halving the value; therefore, the most significant bit of the fraction field is worth $1/2$.

All the bits positioned to the right of the integer field have a value less than 1; therefore, they all contribute to a fractional value.

Using a Binary Point

To illustrate the separation between the integer field and the fraction field, we position a *binary point* between these fields. As an example, let's examine the format of an unsigned 16-bit fixed-point number that has a 12-bit integer field and a 4-bit fraction field:

IIIIIIIIIIII.FFFF

**The binary point is an abstract concept that helps us
to interpret the value of a fixed-point number.
We show it only for convenience; it does not actually use
any of the bits that belong to the number.**

Since the value halves as we evaluate the binary number from left to right, the bits of a fixed-point number are assigned the following values:

Binary position: $\cdots$ I   I   I   I   .   F   F   F   F   F $\cdots$

Value: $\cdots$ 8   4   2   1 $\dfrac{1}{2}$ $\dfrac{1}{4}$ $\dfrac{1}{8}$ $\dfrac{1}{16}$ $\dfrac{1}{32}$ $\cdots$

For example, let's evaluate the following 8-bit unsigned fixed-point number:

$$101.10101 \xrightarrow{\text{Decimal}} 4 + 0 + 1 + \frac{1}{2} + \frac{0}{4} + \frac{1}{8} + \frac{0}{16} + \frac{1}{32}$$

$$= 5\,\frac{21}{32} = 5.65625$$

We can express the individual values of the fraction bits using a common denominator. When we do this, the smallest fraction defines the common denominator.

Binary position: I I I I . F F F F F

Value: 8 4 2 1 $\dfrac{16}{32}$ $\dfrac{8}{32}$ $\dfrac{4}{32}$ $\dfrac{2}{32}$ $\dfrac{1}{32}$

We encode the value of a particular number by turning the bits on and off until we achieve the required value. Since the least significant bit contains the smallest fraction, it represents the smallest value by which the number can change. Consequently, adding more bits to the fraction field lowers the value of the smallest fraction and allows us to encode a value with more precision.

The exact number of bits in the fraction field defines the precision with which we can code the value of the fixed-point number.

$$\text{Precision of a fixed-point number} = \frac{1}{2^{\text{Quantity of bits in the fraction field}}}$$

For example, the precision of a number that uses five F-bits is

$$\frac{1}{2^5} = \frac{1}{32}$$

We could also express this as a tolerance.

The exact number of bits in the fraction field defines the tolerance with which we can code the value of the fixed-point number.

$$\text{Tolerance of a fixed-point number} = \pm \frac{1}{2^{\text{Quantity of bits in the fraction field} + 1}}$$

For example, the tolerance with which we can encode five F-bits is

$$\pm \frac{1}{2^{5+1}} = \pm \frac{1}{64}$$

We can achieve extra *precision* by including more F-bits in the fixed-point number format. When we program our digital processor, the precision issue will determine how many bits we should use to encode our signal. For example, consider Table B–2 in which we list the tolerance for unsigned 16-bit fixed-point numbers that use different fraction field sizes.

Encoding Fixed-Point Numbers

We know that digital processors record information in memory in the form of binary numbers. When a number is stored in memory, it contains no binary point, and consequently its value always corresponds to some particular *integer*. However, we can choose to interpret the number as containing a binary point. This *interpretation* allows us to convert it into a fixed-point number.

TABLE B–2
Tolerance for different fraction field sizes.

$$\text{IIIIIIIIIIIII.FFF} \pm \frac{1}{16} \qquad \text{IIIIIIIIIII.FFFFF} \pm \frac{1}{64}$$

$$\text{IIIIIIIIIIII.FFFF} \pm \frac{1}{32} \qquad \text{IIIIIIIII.FFFFFF} \pm \frac{1}{128}$$

Fixed-point numbers use a binary point to define the boundary that separates the integer and fraction fields. Although the binary point does not use any of the bits, its position tells us how to *interpret* the value of the bits in the fixed-point number.

Consider, for example, that the 8-bit number `00000101` = 5 is stored in a digital processor memory location. Let's arbitrarily interpret this 8-bit number as a fixed-point number that has four I-bits and four F-bits. This number is then worth

$$\texttt{0000.0101} \implies 0 + 0 + 0 + 0 + \frac{0}{2} + \frac{1}{4} + \frac{0}{8} + \frac{1}{16} = \frac{5}{16}$$

The value that we have just obtained depends totally on the format. For example, if we had considered a format of five I-bits and three F-bits, the same 8-bit number would be worth twice as much as before:

$$\texttt{00000.101} \implies 0 + 0 + 0 + 0 + 0 + \frac{1}{2} + \frac{0}{4} + \frac{1}{8} = \frac{5}{8}$$

The rules that we use to interpret the value of a number depend on the size of the integer and fraction fields that we have chosen to use to code the fixed-point number. Consequently, we must choose which bits belong to the integer and fraction fields *before* we store the numbers in memory.

Let's examine how the distribution of integer and fraction bits affects the value of fixed-point numbers. For every fraction bit that we add, there is one less integer, which results in moving the binary point one location to the left. This means that adding a fraction bit effectively halves the value of the individual bits in the fixed-point number. Consider the different interpretations for the following 8-bit number:

$$\texttt{10010110} \xrightarrow{\text{IIIIIIII Decimal}} 128 + 0 + 0 + 16 + 0 + 4$$
$$+ 2 + 0 = 150$$

$$\texttt{10010110} \xrightarrow{\text{IIIIIIIF Decimal}} \frac{128}{2} + \frac{0}{2} + \frac{0}{2} + \frac{16}{2} + \frac{0}{2} + \frac{4}{2}$$
$$+ \frac{2}{2} + \frac{0}{2} = \frac{150}{2}$$

$$10010110 \xrightarrow{\text{IIIIIIFF Decimal}} \frac{128}{4} + \frac{0}{4} + \frac{0}{4} + \frac{16}{4} + \frac{0}{4} + \frac{4}{4}$$

$$+ \frac{2}{4} + \frac{0}{4} = \frac{150}{4}$$

$$10010110 \xrightarrow{\text{IIIIIFFF Decimal}} \frac{128}{8} + \frac{0}{8} + \frac{0}{8} + \frac{16}{8} + \frac{0}{8} + \frac{4}{8}$$

$$+ \frac{2}{8} + \frac{0}{8} + \frac{150}{8}$$

We call the format of a fixed-point number, as determined by the size of the fraction field, the *quantization format* (Q format).

The Q format determines the amount of *scaling* that is applied to the fixed-point number before storing it in memory. The exact amount of scaling is determined according to the following relationship:

Scaling factor of a Q*m* format: 2*m*

The following examples show that interpreting the value of a fixed-point number requires reversing the scaling of its Q format:

Binary number stored in memory: `00001001`

$$\xrightarrow{\text{Decimal}} 9$$

Interpretation as a Q1 number: `0000100.1`

$$\xrightarrow{\text{Decimal}} \frac{9}{2} = 4\frac{1}{2} = 4.5$$

Interpretation as a Q2 number: `000010.01`

$$\xrightarrow{\text{Decimal}} \frac{9}{4} = 2\frac{1}{4} = 2.25$$

Interpretation as a Q3 number: `00001.001`

$$\xrightarrow{\text{Decimal}} \frac{9}{8} = 1\frac{1}{8} = 1.125$$

Interpretation as a Q4 number: `0000.1001`

$$\xrightarrow{\text{Decimal}} \frac{9}{16} = .5625$$

We must apply the scaling factor to a fixed-point number *before* we store it into memory as a binary number. For example, if we choose to store an unsigned fixed-point number that uses a 12-bit fraction field, we must use a Q12 format. In this case, the scaling factor is

$$2^{12} = 4096$$

Table B–3 lists three examples of the scaling calculations that we must perform to store fixed-point numbers as unsigned 16-bit numbers.

TABLE B–3
Scaling fixed-point numbers into binary numbers.

Number	Scaling Factor	Fixed-Point Number Stored Using 16 Bits in Memory
2.390625	Q12: 4096	$2.390625 \times 4096 = 9792 \xrightarrow{\text{Binary}}$ 0010011001000000
6.78125	Q10: 1024	$6.78125 \times 1024 = 6944 \xrightarrow{\text{Binary}}$ 0001011000100000
6.78125	Q8: 256	$6.78125 \times 256 = 1736 \xrightarrow{\text{Binary}}$ 0000011011001000

Note that the Q format corresponds to the position of the binary point. When we read the binary numbers from memory, we need to reverse the scaling to recover the value of the fixed-point number. To do this, we divide the binary value by the scaling factor. Table B–4 shows the unscaling calculations, which we must perform to reverse the scaling that was done in the previous example.

Evaluating the Quantization Error

The numbers in Tables B–3 and B–4 were selected so that the scaling operations would yield exact integer results. In practice, the scaling operations usually produce a result that

TABLE B–4
Unscaling fixed-point numbers from binary numbers.

16-Bit Fixed-Point Number	Unscaling Factor	Value of the Number
$0010011001000000 \xrightarrow{\text{Decimal}} 9792$	Q12: $\dfrac{1}{4096}$	$\dfrac{9792}{4096} = 2.390625$
$0001011000100000 \xrightarrow{\text{Decimal}} 6944$	Q10: $\dfrac{1}{1024}$	$\dfrac{6944}{1024} = 6.78125$
$0000011011001000 \xrightarrow{\text{Decimal}} 1736$	Q8: $\dfrac{1}{256}$	$\dfrac{1736}{256} = 6.78125$

includes an integer and a fractional part. Converting this result to a binary format creates a problem since binary numbers are always pure integers.

For example, let's consider storing the number 1.3 using an unsigned 8-bit Q5 format. To convert this number, we must first scale it by applying the Q5 scaling factor:

$$1.3 \times 2^5 = 1.3 \times 32 = 41.6$$

The second step is to store the scaled result into memory, but this poses a problem since the result is not a whole integer. We must therefore remove the fractional part from the 41.6 result either by truncating it to 41 or by rounding it up to 42. Truncating means that we lose the 0.6 fractional part:

$$41.6 - 0.6 = 41 \xrightarrow{\text{Binary}} \texttt{00101001}$$

Now we can store the binary integer `00101001` into memory. When we read this unsigned 8-bit Q5 number, the unscaling operation will yield an interpreted value of

$$\frac{41}{32} = 1.28125$$

which is a little less than the original value of 1.3.

Clearly, the unscaled value is not quite accurate. The result is slightly smaller than the number that we wanted to store because we lost 0.6 in the truncating process. We can calculate the percentage of accuracy of the stored number as

$$\text{Percentage accuracy} = \frac{\text{Stored value}}{\text{Value we wanted to store}} \times 100$$

For example, in the previous example, 1.3 was stored with a percentage accuracy of

$$\text{Percentage accuracy: } \frac{41}{41.6} \times 100 \approx 98.6\%$$

We call the error that is contained in the stored number the *quantization error*. We can calculate the percentage quantization error as follows:

$$\text{Percentage quantization error} = \frac{\text{Rounding error}}{\text{Value we wanted to store}} \times 100$$

For example, in the previous example, the value of 1.3 was stored with a percentage quantization error of

$$\frac{-0.6}{41.6} \times 100 \approx -1.4\%$$

Note that the error is negative since we *truncated* 0.6 from the scaled value.

If we change our approach by rounding up, we must add 0.4 to bring 41.6 to an integer value:

$$41.6 + 0.4 = 42 \xrightarrow{\text{Binary}} \texttt{00101010}$$

Now we can store the binary integer $\texttt{00101010}$ into memory. When we read this unsigned 8-bit Q5 number, the unscaling operation will yield an interpreted value of

$$\frac{42}{32} = 1.3125$$

which is a little more than the original value of 1.3.

The interpreted result is slightly larger than the number that we wanted to store because we gained 0.4 in the round-up process. This increase corresponds to the following percentage quantization error:

$$\text{Percentage quantization error:} \quad \frac{0.4}{41.6} \times 100 \approx +0.96\%$$

The decision to round up or to truncate is left to the user. Choosing the lower percentage error is *not always* the best choice because it could lead to dangerous situations (see Section 7.6.3) on the Z-plane. The rounding/truncating process results in an error, which is *permanent*; once the number is stored in the binary format, we lose its exact value.

Refer to Section 2.3.3, which discusses the control of harmonic distortion in synthesized signals. There we linked the amount of harmonic distortion to the precision of the table entries. As an example, let's examine the precision that results when we use 8-bit numbers to code a 64-entry cosine table. The 64 entries correspond to cosine values taken at intervals of

$$\frac{360}{64} = 5.625°$$

Since cosine values range between +1 and –1, we use the following fixed-point format:

- One sign bit to accommodate the polarity.
- One integer bit to accommodate the maximum range values of +1 to –1.
- Six fraction bits left for the fractional part.

This corresponds to a signed 8-bit Q6 format: $\texttt{SIFFFFFF}$.

The six F-bits included in the format yield the following precision and tolerance. Precision of a Q6 format:

$$\frac{1}{2^6} = \frac{1}{64}$$

which yields a tolerance of

$$\pm\frac{1}{128} = \pm 0.0078125$$

Table B–5 compares the ideal cosine values and the Q6 quantized values for some of the table entries. Note that the error cannot exceed the tolerance limits of the Q6 format.

TABLE B–5
Entries from a 64-entry cosine table.

Ideal Cosine Value	Stored 8-bit Q6 Values	Stored Value	Error	Percentage Error
$\cos(73.125°) = +0.2903$	00010011	+0.2969	+0.0066	+2.3
$\cos(78.75°) = +0.1951$	00001100	+0.1875	–0.0076	–3.9
$\cos(84.375°) = +0.0980$	00000110	+0.0938	–0.0042	–4.3
$\cos(90°) = +0.0000$	00000000	+0.0000	0.0000	0.00
$\cos(95.625°) = –0.0980$	11111010	–0.0938	+0.0043	+4.3
$\cos(101.25°) = –0.1951$	11110100	–0.1875	+0.0076	+3.9
$\cos(106.875°) = –0.2903$	11101101	–0.2969	–0.0066	–2.3
$\cos(112.5°) = –0.3827$	11101000	–0.3750	+0.0077	+2.0

Determining the Range of Fixed-Point Numbers

According to the section in this appendix on Encoding Fixed-Point Numbers (p. 364), such numbers are scaled binary numbers. The range of values that a particular fixed-point number can adopt is related to the bit-size of the number and its Q format. Once we know these characteristics, we can ascertain the size of the integer and fraction fields from which we will calculate the range of values that the number can adopt.

UNSIGNED FIXED-POINT NUMBERS For example, let's evaluate the range of an unsigned 8-bit fixed-point number that is arbitrarily stored with a Q4 format. Since we know the Q format, we know that this number has four F-bits and since the number is unsigned, the leftover four bits must be integer bits. The format of an unsigned 8-bit Q4 number is IIIIFFFF.

We can now determine the lower and upper limits for this unsigned 8-bit Q4 number. The lower limit is the smallest possible number:

$$00000000 \xrightarrow{\text{Unscaled}} \frac{00000000}{2^4} \xrightarrow{\text{Decimal}} \frac{0}{16} = 0$$

The upper limit is the largest possible number:

$$11111111 \xrightarrow{\text{Unscaled}} \frac{11111111}{2^4} \xrightarrow{\text{Decimal}} \frac{255}{16} = 15\frac{15}{16}$$

Therefore, IIIIFFFF ranges in value from zero to

$$15\frac{15}{16}.$$

We can break this result into separate ranges for the integer and for the fraction fields. The range of the 4-bit integer field is from zero to 15, and the range of the 4-bit fraction field is from 0/16 to 15/16.

SIGNED FIXED-POINT NUMBERS   Now let's modify our example to evaluate the range of a *signed* 8-bit fixed-point number that is stored with a Q4 format. Our number now has a sign bit and, since we know the Q format, we know that this number has four F-bits. This leaves three bits for integers. The format of a signed 8-bit Q4 number is `SIIIFFFF`. We can now determine the lower and upper limits of this number. The lower limit is the most negative number:

$$10000000 = -\left(\overline{10000000} + 1\right) = -(01111111 + 1) = -(10000000)$$

$$\xrightarrow{\text{Unscaled}} -\frac{10000000}{2^4} \xrightarrow{\text{Decimal}} -\frac{128}{16} = -8$$

The upper limit is the most positive number:

$$01111111 \xrightarrow{\text{Unscaled}} +\frac{01111111}{2^4}$$

$$\xrightarrow{\text{Decimal}} +\frac{127}{16} = 7\frac{15}{16}$$

Therefore, `SIIIFFFF` ranges in value from −8 to 7 15/16.

Note that the negative part makes it completely to the −8 value but that the positive part is still a fraction away from reaching +8. We will use a negative superscript to indicate that the range does not quite reach +8:

Range of `SIIIFFFF`: $\pm 8^-$

We can break this result into separate ranges for the integer and for the fraction fields. The range of the signed 3-bit integer field is

$$-2^3 \text{ to } +(2^3 - 1) = -8 \text{ to } +7$$

and the range of the 4-bit fraction field is

$$\frac{0}{2^4} \text{ to } \frac{2^4 - 1}{2^4} = \frac{0}{16} \text{ to } +\frac{15}{16}$$

If we decide to increase the number of F-bits, the *precision* of the fractional part improves, but the range remains approximately the same. For example, adding eight F-bits to the previous example will yield

16-bit Q12 format: `SIII.FFFFFFFFFFFF`

which has a range of

$$-\left[2^3\right] \quad \text{to} \quad +\left[\left(2^3 - 1\right) + \frac{2^{12}-1}{2^{12}}\right] = -8 \quad \text{to} \quad +7\,\frac{4095}{4096} = \pm\,8^{-}$$

Clearly, increasing the number of F-bits brings the positive limit slightly closer to +8. The precision greatly improves, but the range hardly increases.

The precision of a fixed-point number is controlled by the number of fractional bits.

If we decide to interpret the number as having fewer F-bits, there will be more I-bits and the range is extended. Consider the precision and range of a 16-bit Q4 number:

SIIIIIIIIIII.FFFF

which yields a range of

$$-\left[2^{11}\right] \quad \text{to} \quad +\left[\left(2^{11} - 1\right) + \frac{2^4-1}{2^4}\right] = -2048 \quad \text{to} \quad +2047\,\frac{15}{16} = \pm\,2048^{-}$$

The range of a fixed-point number is controlled by the number of integer bits.

When we select the number of bits and the Q format for a fixed-point number, we must consider the range of values that the number can cover. For example, let's assume that we need to select a common fixed-point format to encode the following list of numbers: +5.344, –43.3, –67.567, +0.5678, +128.95.

- We notice that these are signed numbers, which means that we must reserve one bit for the sign.
- To select the required number of integer bits, we must identify the lower and upper limits. The lower limit is –67.567, which requires a minimum of seven I-bits. The upper limit is +128.95 which requires a minimum of eight I-bits.
- The number of F-bits that we need is determined by the tolerance that is acceptable for our application (see the section in this appendix on the Binary Point [p. 363]).

For example, if we arbitrarily select a tolerance of

$$\pm\,\frac{1}{1024}$$

we need nine F-bits. We could code the list of numbers using the following 18-bit Q9 format:

SIIIIIIII.FFFFFFFFF

In practice, the number of bits that we use should be a multiple of 8 bits. In this case, we have to either extend the format to 24 bits (and waste 6 bits) or reconsider the tolerance (remove 2 F-bits) and use 16 bits:

$$\text{SSSSSSSIIIIIIII} \cdot \text{FFFFFFFFF} \pm \frac{1}{1024} \quad \text{or} \quad \text{SIIIIIIII} \cdot \text{FFFFFFF} \pm \frac{1}{256}$$

If we choose the 24-bit format, we should try to use the seven sign bits that are wasted. We could increase the precision and range by respectively increasing the number of F-bits and I-bits. For example, we could settle for a compromise by increasing both the precision and range by making the arbitrary choice of a Q12 format.

$$\text{SIIIIIIIIIII} \cdot \text{FFFFFFFFFFFF} \pm \frac{1}{8192}$$

As we can see, increasing to 12 F-bits increases the precision, and the 11 I-bits give us an increased range. Range of a 24-bit Q12 number

$$-\left[2^{11}\right] \quad \text{to} \quad +\left[\left(2^{11} - 1\right) + \frac{2^{12} - 1}{2^{12}}\right] = -2048 \quad \text{to} \quad +2047\frac{4095}{4096}$$

B.3.5 FLOATING-POINT NUMBERS

When the numbers that we need to process cover a very large range of values, the fixed-point number format becomes inefficient because it requires too many I-bits. In this case, it is better to adopt the floating-point format. Contrary to the fixed-point format, the floating-point format actually uses binary digits to encode the binary point position. The bits in a floating-point number define the sign, the *position of the binary point,* and a pure fractional value. Since this format uses bits to keep track of the position of the binary point, it frees the programmer from having to control the scaling of the numbers.

Formatting Floating-Point Numbers

To allow the exchange of floating-point numbers between applications, the Institute of Electrical and Electronics Engineers (IEEE) and the American National Standards Institute (ANSI) have published the ANSI/IEEE-854 standard, which covers the formats for floating-point arithmetic. This standard covers the single, double, single-extended formats, and double-extended formats. This section covers some aspects of the single-precision format.

The single-precision standard is a 32-bit format that contains three fields. One bit defines the sign, 8 bits define the position of the binary point (the exponent), and 23 bits define the value of a fraction field. Table B–6 illustrates this format.

When compared to a 32-bit fixed-point number, the exponent field provides a large range, but the exponent field bits are taken from the fraction field and therefore the precision is reduced. The value of a single precision floating-point number is found by applying the following relationship:

$$\pm 0.\text{FFFFF}\ldots \times 2^{(\text{Exponent}-127)}$$

The 23 fraction bits therefore provide the following tolerance:

$$\pm \frac{1}{2^{23+1}} = \frac{1}{16,777,216}$$

This corresponds to a percentage accuracy better than 0.000006%, which is plenty for most practical applications.

TABLE B–6
IEEE single-precision format.

Bit 31	Bit 30—Bit 23	Bit 22—Bit 0
Sign Field	Exponent Field	Fraction Field

The exponent field is used to offset the location of the binary point left or right by a certain amount. The exact position of the binary point depends on the *biased* amount, which is determined by taking the value of the exponent and subtracting 127. Since the exponent field contains 8 bits, its value ranges from 0 to 255; consequently, the biased exponent ranges from –127 to +128. Consider the following example:

Sign Field	*Exponent Field*	*Fraction Field*
0	01111100	0.01100000000000000000001

$$+ \qquad 2^{124-127} = 2^{-3} \qquad \frac{01100000000000000000001}{2^{23}}$$

$$= + \frac{01100000000000000000001}{2^{23}} \times 2^{-3} = + \frac{01100000000000000000001}{2^{26}}$$

$$= + \frac{2^{21}+2^{20}+2^0}{2^{26}} = + \frac{3,145,729}{67,108,864} = +0.0468750$$

The range of a single precision floating-point number therefore covers

$$-\frac{1}{2^{23}} \times 2^{-127} \text{ to } + \frac{2^{23}-1}{2^{23}} \times 2^{+128} \approx -7.0 \times 10^{-46} \text{ to } +3.4 \times 10^{38}$$

This represents a very wide range of numerical values, which should be sufficient to accommodate most practical application.

B.4 UNDERTAKING MATHEMATICAL OPERATIONS

The processing of signals requires basic mathematical operations such as addition and multiplication. When these operations are performed on a digital processor, the programmer is faced with the limits of the numbering system being used. The following sections describe these limits and some of the techniques that can be used to optimize the processing.

B.4.1 ADDING BINARY NUMBERS

When we perform signal processing, it is necessary to ensure that the overall calculations do not result in an *overflow*. An overflow occurs when the result of the calculation exceeds the range of values that the number format can accommodate. We must therefore evaluate the worst-case scenario to ensure that our number format can handle the magnitude of the result.

Adding Numbers Using a Fixed-Point Processor

Let's start by analyzing the addition of binary numbers. When we add *two binary numbers*, the worst-case result requires one bit more than that contained in the largest of the two numbers being added. The following examples illustrate the worst-case results of adding two 4-bit numbers (decimal values are in brackets):

Worst - Case Unsigned	Worst - Case Positive	Worst - Case Negative
1111 (15)	0111 (+7)	1000 (−8)
+1111 (15)	+0111 (+7)	+1000 (−8)
11110 (30)	11110 (+14)	10000 (−16)

All three cases require an extra bit (a 5-bit answer) to accommodate the result without producing an overflow. This extra bit is required for all binary additions when we allow the two numbers to adopt any value.

Let's continue by examining the addition of two *fixed-point numbers*. We already know that we must interpret the value of these numbers according to the scaling factor determined by the Q format. To add fixed-point numbers, we must align the binary points of all the numbers. Since the Q format determines the position of the binary point, it follows that we can add only fixed-point numbers that have identical Q formats.

We can modify the Q format of a number by using the *arithmetic* shift operation (not to be confused with the *logical* shift operation).

- The arithmetic shift *left* operation shifts out a sign bit and shifts in a zero in the least significant position. This effectively multiplies the value of the number by 2. Note that an overflow occurs if the sign bit changes during the shifting process. The number of replicated sign bits therefore limits the amount of left shifting.
- The arithmetic shift *right* operation shifts in a replicated sign bit and shifts out the least significant bit. This divides the value of the number by 2.

Figure B–5 illustrates the arithmetic shift operations. For example, consider the operations that are required to add two signed 8-bit fixed-point numbers that have different Q formats:

	Adjust	
1.0100000 Q7 (−0.75)	$\longrightarrow$	111.01000 Q5 (−0.75)
+ 011.10000 Q5 (+3.50)		+011.10000 Q5 (+3.50)
Binary points are misaligned		010.11000 Q5 (+2.75)

As this example illustrates, we used the arithmetic shift-right operation to remove two fractional digits from the Q7 number to translate it into a Q5 number. We lost two fractional bits of precision in the process, and the sign was extended to maintain a total of 8 bits in the number. Note the important fact that the Q format of the result is the same as that of the numbers being added.

**Adding fixed-point numbers produces a result that
has the same Q format as the numbers being added.**

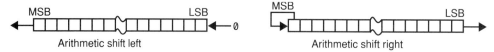

FIGURE B–5
Arithmetic shift operations.

Using Modulo Addition

In the sections in Chapter 9 on Selecting the Q Factors of a Finite Impulse Response System Based on the Value of the System's Coefficients (p. 329) and Selecting the Q Factors of a Finite Impulse Response System Based on the System Response Curve (p. 331), we add a long list of fixed-point numbers when processing FIR filters. Fortunately, in most cases, we know the range of the result *before* we add the numbers. For example, consider an application that synthesizes a signal that sums five cosine harmonics (see Section 2.4 for this type of application). If we know that each harmonic has a maximum digital amplitude of ±1, it is impossible for the synthesized signal to exceed a numerical range of $\pm 1 \times 5 = \pm 5$. Another example is a linear-phase digital filter that we design to have a maximum gain of 5. If the signal input applied to this particular filter has an amplitude of ±1, the worst-case output range cannot exceed $\pm 1 \times 5 = \pm 5$.

When we know the numerical limits of the result, we can choose the Q format to accommodate the magnitude of the digital output. For example, if we know that the result cannot exceed a magnitude of ±5, we know that three I-bits will accommodate the numerical result. If our processor uses signed 8-bit fixed-point numbers, we should use a Q4 format (one sign, three I-bits, and four F-bits).

There is another very important advantage to knowing the numerical range of the result. When we add a long list of numbers, we do not need to worry about intermediate overflows. The following example illustrates this by producing the modulo sum of five signed 8-bit Q4 numbers. In this case, the characteristics of the system make it such that the final result cannot overflow the range of a Q4 format. Note that any value that lies outside the range of –8 to +7.9375 is an overflow for a signed 8-bit Q4 format.

$$
\begin{aligned}
&\texttt{0011.0100}\ (+3.250) \qquad \text{Intermediate results} \\
&\texttt{0111.0110}\ (+7.375) \rightarrow\ (+3.250) + (+7.375)\ =\ +10.625\quad \textit{Overflow} \\
&\texttt{1011.0100}\ (-\ 4.750) \rightarrow\ (+10.625) + (-\ 4.750)\ =\ +5.875 \\
&\texttt{0011.1000}\ (+3.500) \rightarrow\ (+5.875) + (+3.500)\ =\ +9.375\quad \textit{Overflow} \\
&\underline{+\,\texttt{1000.1000}\ (-\ 7.500)} \rightarrow\ (+9.375) + (-\ 7.500)\ =\ +1.875 \\
&\texttt{0001.1110}\ (+1.875) \rightarrow\ \text{Correct final result}
\end{aligned}
$$

Although there are two intermediate overflows in this example, the final result is still correct because we already knew that there would be no overflow.

It is important to note that modulo addition will work only if we ensure that the accumulator saturation mode is disabled.

Adding Numbers with a Floating-Point Processor

A floating-point processor includes special hardware that provides floating-point arithmetic. In this case, when floating-point numbers are used, the processor automatically performs the normalization that is required to properly scale the numbers. These processors are more expensive than their fixed-point counterparts, but they certainly simplify the programming.

When fixed-point processor are used, the programmer can benefit from the large range of floating-point numbers by including a library of software routines that implement the floating-point operations. Unfortunately, the extra computing that is associated with these routines will considerably slow the processing.

B.4.2 MULTIPLYING BINARY NUMBERS

The processing of signals usually requires the multiplication of numbers. Signal processors include a *hardware*-multiplying unit that can perform the multiplication of two numbers within a single machine cycle. We call these two numbers the *multiplier* and the *multiplicand*.

Multiplying Numbers Using a Fixed-Point Processor

Multiplying two binary numbers produces a result that requires a rather large number of bits. The number of bits in the result of a multiplication equals the combined bits of the multiplier and of the multiplicand. For example, multiplying two 16-bit numbers yields a 32-bit result.

Since the result of a multiplication contains more bits than the multiplier or the multiplicand, the format of the result is different from that of the numbers being multiplied. Consider the following 4-bit multiplication operations:

Binary

$$
\begin{array}{r}
\texttt{BBBB} \ \ (4 \text{ bits}) \\
\times \quad \texttt{BBBB} \ \ (4 \text{ bits}) \\
\hline
\texttt{BBBBBBBB} \ \ (8 \text{ bits})
\end{array}
$$

Unsigned Fixed Point

$$
\begin{array}{r}
\texttt{III.F} \ \ (3\text{I and }1\text{ F}) \\
\times \ \texttt{II.FF} \ \ (2\text{I and }2\text{F}) \\
\hline
\texttt{IIIII.FFF} \ \ (5\text{I and }3\text{F})
\end{array}
$$

Signed Fixed Point

$$
\begin{array}{r}
\texttt{SII.F} \ \ (1\text{S}, 2\text{I and }1\text{F}) \\
\times \quad \texttt{SI.FF} \ \ (1\text{S}, 1\text{I and }2\text{F}) \\
\hline
\texttt{SSIII.FFF} \ \ (2\text{S}, 3\text{I and }3\text{F})
\end{array}
$$

Since we are multiplying two 4-bit numbers, the result requires eight bits. Note that the result combines the bit categories of the multiplier and of the multiplicand. For example, when there are two I-bits in the multiplier and three I-bits in the multiplicand, the result contains five I-bits.

Overflow cannot occur when we multiply. It is interesting to note that if one of the two numbers being multiplied is a pure fraction (no integer bits), the result must be smaller than the two numbers. If this is the case, the number of integer bits in the result will not be increasing. When processing DSP algorithms, products must be accumulated. Ensuring that one of the multiplied numbers is a pure fraction simplifies the overflow management of the accumulator. Section 4.1 discusses that ADC input samples are always pure fractions.

The programmer selects a specific format for the numbers that are processed. These choices depend on the range and precision of the numbers being multiplied.

Multiplying Numbers Using a Floating-Point Processor

Since a floating-point processor automatically manages the normalization and scaling of all the numbers, the programmer need be concerned only with overflow. Because of the enormous range covered by floating-point numbers, an overflow is an unlikely event. Overflows happen when unusual operations are performed such as dividing by zero. Most floating-point processors capture these unlikely events with an interrupt trap that triggers the execution of special overflow recovery routines.

B.4.3   OUTPUTTING FLOATING-POINT ANSWERS

When the processor produces a floating-point answer, we must normally convert this result to a digital output format that the data bus will carry to the system DAC. The outputted digital samples must use a *fixed-point* format, which covers only a small part of the range of values that floating-point numbers provide. Because of this, the likelihood of an overflow is high when we convert the floating-point answer to the fixed-point format required by the DAC.

As an example, consider a system that transfers signed 16-bit Q11 output samples to the DAC. This output format corresponds to the following bit assignment for a signed 16-bit Q11 format:

$$\text{SIIII.FFFFFFFFFFF}$$

This output format contains a numerical part consisting of four I-bits and 11 F-bits. We must produce each fixed-point output sample from a single precision floating-point answer, which consists of a signed 23 F-bit pure fractional numerical part that is multiplied by an 8-bit exponent:

$$\text{SFFFFFFFFFFFFFFFFFFFFFFF} \times 2^{\text{8-bit exponent}-127}$$

To convert this floating-point number into the Q11 output format, we first need to adjust the biased exponent so that it generates four I-bits:

$$\text{SFFFFFFFFFFFFFFFFFFFFFFF} \times 2^4 = \text{SIIIIFFFFFFFFFFFFFFFFFFF}$$

When the biased exponent adopts a value of 4, the numerical value contains four I-bits. This means that we must adjust the floating-point answer to ensure that the exponent field contains $4 + 127 = 131$. Using the arithmetic shift operation on the signed fractional field makes this adjustment. Consider the following example in which we adjust a floating-point answer that has an 8-bit exponent value of 129 to a 16-bit Q11 fixed-point format:

$$\text{S.FFFFFF} \cdots \times 2^{129-127}$$

$$\xrightarrow{\text{Adjust}} \text{SSS.FFFF} \cdots \times 2^{129-127+2} \xrightarrow{\text{Output}} \text{SSSII.FF} \cdots$$

The exponent field contains 129 but the Q11 format requires 131. The content of the fractional field is therefore shifted right twice. For every arithmetic shift-right operation, the sign bit is replicated, a fractional bit of precision is shifted out and the exponent increments. The Q11 number consists of the sign bit and the 23 bits in the adjusted fractional field. The resulting 24-bit Q11 number must be truncated to 16 bits. This truncation results in a loss of $24 - 16 = 8$ bits of precision.

$$\text{SSSII.FFFFFFFFFF}\cancel{\text{FFFFFFFF}}$$

In the case where the exponent of the floating-point number exceeds 131, the value of the floating-point number would overflow the 4 I-bits allowed by the Q11 output format of this example.

$$\text{SFFFFFFF} \cdots \times 2^{132-127} = \text{SIIIII.FF} \cdots \text{ overflow—too many integers}$$

SUMMARY

- The binary value rule states that the significance of each bit will double as we move from the least significant to the most significant bit.
- The number of different values that binary number can adopt equals $2^{\text{number of bits in the number}}$.
- Binary numbers can adopt many different formats: signed/unsigned, integer, fixed point, or floating point.
- We find the negative of a number by performing a 2's complement operation.
- Extending the sign will increase the number of bits in a signed number without changing the value of the number.
- Values for an m-bit *unsigned* integer range from zero to $+2^m - 1$.
- Values for an m-bit *signed* integer range from -2^{m-1} to $+2^{m-1} - 1$.
- The *binary point* is an abstract concept that allows us to interpret the value of a fixed-point number. We show it only for convenience since it does not actually use any of the bits that belong to the number.
- The exact number of F-bits defines the precision and the tolerance with which we can code the value of the fixed-point number.
- The Q format determines the amount of scaling applied to a fixed-point number.
- The precision of a fixed-point number depends on the number of F-bits. Converting a decimal number to a binary format can create a small error that we call the *quantization error*.
- The number of I-bits controls the range of fixed-point numbers.

- Floating-point numbers cover a very large range of values.
- When we add *two binary numbers*, the worst-case result requires one more bit than the number in the largest of the two numbers being added.
- We can modify the Q format of a fixed-point number by using the *arithmetic shift* operation.
- We can add only fixed-point numbers that have the same Q format.
- When we add fixed-point numbers, the Q format of the result is the same as that of the numbers being added.
- We do not need to worry about any intermediate overflows when we add a list of fixed-point numbers if we are sure that the final result fits the Q format being used.
- We can multiply fixed-point numbers even if they have different Q formats.
- When we multiply fixed-point numbers, the format of the result will combine the bit categories of both the multiplier and the multiplicand.
- A floating-point number needs to be converted to a fixed-point number to send it to a DAC.
- A DAC always interprets its digital input as being formatted as a pure fractional fixed-point number.

PRACTICE QUESTIONS

B-1. Consider the following binary numbers:

00100100   10100101   11011011   11000010   11101111

(a) Express these numbers using the hexadecimal notation.
(b) What is the decimal value of these numbers if you interpret them as being *unsigned* binary numbers?
(c) What is the decimal value of these numbers if you interpret them as being *signed* binary numbers?

B-2. Express the decimal value −3 as an 8-bit signed integer.

B-3. What is the range of values covered by a 16-bit *unsigned* integer?

B-4. What is the range of values covered by a 16-bit *signed* integer?

B-5. What is the minimum number of bits required to express the following decimal numbers as signed integers?

$$+5 \quad +42{,}000 \quad +128 \quad -1 \quad -32 \quad -513$$

B-6. Consider the following binary numbers:

00100100   10100101   11011011   11000010   11101111

(a) What is the decimal value of these numbers if you interpret them as being *unsigned* Q2 numbers?
(b) What is the decimal value of these numbers if you interpret them as being *signed* Q2 numbers?

B-7. (a) What is the precision of translating a decimal number into a Q2 number?
(b) What is the tolerance of a Q2 number?

B-8. How many F-bits are required to obtain a tolerance better than ±0.0003?

B-9. Consider the following decimal numbers:

+5.3   +42,000.75   +128.21875   −1.01   −32.4213456   −513.3

(a) Determine the *minimum* number of bits required to translate these numbers into *signed* Q4 numbers; then translate the numbers and determine the percentage error of the translation.

(b) Determine the *minimum* number of bits required to translate these numbers into *signed* Q10 numbers; then translate the numbers and determine the percentage error of the translation.

B-10. What range of values can a 24-bit Q15 *unsigned* number adopt?

B-11. What range of values can a 24-bit Q15 *signed* number adopt?

B-12. Compare the *range* of a signed 32-bit fixed-point number that carries a Q25 format to that of an IEEE single precision number.

B-13. Compare the number of significant digits provided by a signed 32-bit fixed-point number that has a Q25 format to that of an IEEE single precision number.

B-14. The following unsigned numbers each carry a different Q format:

00100100 (Q2)   01100101 (Q5)   11011011 (Q6)

In each case, specify whether it is possible to translate the numbers into
(a) Unsigned 8-bit Q4 numbers.
(b) Signed 8-bit Q4 numbers.
When the translation is possible, specify what operation is required to perform the translation.

B-15. (a) What are the binary results of adding the following unsigned fixed-point numbers?

$$\begin{array}{ccc} 00100100 \ (Q4) & 01100101 \ (Q5) & 11011011 \ (Q4) \\ +\ 01001000 \ (Q4) & +\ 10100101 \ (Q5) & +\ 01001001 \ (Q5) \end{array}$$

(b) Which additions in part (a) produce 8-bit overflows?

(c) If the numbers used in part (a) were *signed* numbers, which additions would produce overflows?

B-16. Consider the following signed fixed-point numbers:

A = 00100100 (Q3)   B = 01100101 (Q3)   C = 11011011 (Q6)

(a) How many bits are required to perform the following operation with no overflow?

A + B   A × B   A × C

(b) If the results of the operations in part (a) are expressed as 16-bit numbers, how many signed, integer, and fraction bits are in the results? Specify the Q format of the results.

(c) Which of the result(s) could be stored as 8-bit numbers? Specify the Q format of the stored result(s).

(d) Explain what alignment operation is required to perform the operation A + C.

B-17. Refer to Section 2.1.4. A 16-bit signed DAC can output voltages that cover a range of ±12 v.

(a) If this DAC receives a Q12 digital input of **8AC3**, what voltage will it output?

(b) If this DAC receives a Q10 digital input of **8AC3**, what voltage will it output?

(c) In parts (a) and (b), what is the scaling factor resulting from the DAC translation?

INDEX